ROUTLEDGE LIBRARY EDITIONS:
POLLUTION, CLIMATE AND CHANGE

Volume 4

WHAT BECOMES OF POLLUTION?

WHAT BECOMES OF POLLUTION?

Adversary Science and the Controversy on the Self-Purification of Rivers in Britain, 1850–1900

CHRISTOPHER HAMLIN

LONDON AND NEW YORK

First published in 1987 by Garland Publishing Inc.

This edition first published in 2020
by Routledge
2 Park Square, Milton Park, Abingdon, Oxon OX14 4RN

and by Routledge
52 Vanderbilt Avenue, New York, NY 10017

Routledge is an imprint of the Taylor & Francis Group, an informa business

British Library Cataloguing in Publication Data
A catalogue record for this book is available from the British Library

ISBN: 978-0-367-34494-8 (Set)
ISBN: 978-0-429-34741-2 (Set) (ebk)
ISBN: 978-0-367-36208-9 (Volume 4) (hbk)
ISBN: 978-0-367-36213-3 (Volume 4) (pbk)
ISBN: 978-0-429-34460-2 (Volume 4) (ebk)

Publisher's Note
The publisher has gone to great lengths to ensure the quality of this reprint but points out that some imperfections in the original copies may be apparent.

Disclaimer
The publisher has made every effort to trace copyright holders and would welcome correspondence from those they have been unable to trace.

MODERN EUROPEAN HISTORY

What Becomes of Pollution?

Adversary Science and the Controversy on the Self-Purification of Rivers in Britain, 1850–1900

Christopher Hamlin

Garland Publishing, Inc.
New York and London 1987

Library of Congress Cataloging-in-Publication Data

Hamlin, Christopher, 1951–
What becomes of pollution? : adversary science and the controversy on the self-purification of rivers in Britain, 1850–1900 / Christopher Hamlin.
p. cm.—(Modern European history)
Thesis (Ph.D.)—University of Wisconsin at Madison, 1982.
Bibliography: p.
Includes index.
ISBN 0-8240-7812-8 (alk. paper)
1. Stream self-purification—Great Britain—History—19th century. 2. Water—Purification—Biological treatment—History—19th century. 3. Sewage—Purification—Biological treatment—History—19th century. 4. Water quality management—Great Britain—History—19th century. I. Title. II. Series.
TD764.H36 1988
363.7'394—dc19 87-27711

All volumes in this series are printed on acid-free, 250-year-life paper.

Printed in the United States of America

WHAT BECOMES OF POLLUTION:
ADVERSARY SCIENCE AND THE CONTROVERSY ON THE SELF-PURIFICATION OF RIVERS IN BRITAIN, 1850-1900

by

CHRISTOPHER HAMLIN

ACKNOWLEDGMENTS

People who have been around me while I have been working on this project have given many things. I wish to thank them for inspiration, smiles, books, thoughtfulness, comradeship, patience, typewriters, support, love, meals, toleration, suggestions, listening, proofreading, money, enthusiasm, and so forth. One person, Fern Hamlin, gave all these things and in a very fundamental way made my life such that I could do this work. I am also thankful for the support of our families: Elizabeth, who arrived in time to help with final drafts and typing, and our parents, brother, aunts, uncles, and cousins for their interest and encouragement. I would also like to thank my dissertation committee: Aaron Ihde, Victor Hilts, Bill Coleman, Judy Leavitt, and Fred Finley; Sheldon Hochheiser (who also worked through all the chapters); the denizens -- faculty, staff, and students -- of the University of Wisconsin History of Science and Integrated Liberal Studies Departments and University-Industry Research Program where I learned and worked; several scholars at work in related fields: W.H. Brock, John Eyler, Kathleen Farrar, Bill Luckin, Margaret Pelling, Jeanne Peterson, and Colin Russell, for their generous contributions of wisdom; and finally innumerable librarians and archivists: in Madison at the University of Wisconsin Memorial, Chemistry, Engineering, Medical, and Agriculture Libraries, and the Memorial Library Inter-Library Loan Department; in Dublin at the Trinity College Library Older Printed Books Reading Room, the Science and Medicine Libraries of

University College, Dublin, the National Library of Ireland, and the Library of the Royal College of Physicians in Ireland; and in Britai at the Public Record Office, the Greater London Council Record Office, the City of London Record Office, the House of Lords Record Office, the Institution of Civil Engineers, the Royal Institution, the Imperial College Archives, the Royal Society of London, the British Library, the University of Liverpool Library, the Wellcome Institute for the History of Medicine, the Yorkshire Water Authority, and the Thames Water Authority.

CONTENTS

COMMONLY USED ACRONYMS AND ABBREVIATIONS

Parliamentary Inquiries:

RPPC/1865/1st through 3rd reports	Reports of the [First] Royal Commission on Rivers Pollution, 1865-68.
RPPC/1868/1st through 6th reports	Reports of the [Second] Royal Commission on Rivers Pollution, 1868-74.
RCWS	Royal Commission on Water Supply, 1867-69.
RCMSD	Royal Commission on Metropolitan Sewage Discharge, 1882-84.
RCMWS	Royal Commission on Metropolitan Water Supply, 1892-94.
RCST	Royal Commission on the Sewage of Towns, 1857-64.

Journals:

BMJ	British Medical Journal
CN	Chemical News
JPH & SR	Journal of Public Health & Sanitary Review
JRSA	Journal of the Royal Society of Arts
MPICE	Minutes of Proceedings, Institution of Civil Engineers
SR	The Sanitary Record
TSIGB	Transactions, Sanitary Institute of Great Britain

VNEEM . . . . .Van Nostrand's Eclectic Engineering Magazine

Organizations and Miscellaneous:

FPA . . . . .Fisheries Preservation Association

GBH . . . . .General Board of Health

GLCRO . . . . .Greater London Council Records Office

LCC . . . . .London County Council

LGB . . . . .Local Government Board

MBW . . . . .Metropolitan Board of Works

PRO . . . . .Public Record Office, Kew

PSC . . . . .Previous sewage contamination (an index developed by Edward Frankland in the late 1860s for calculating what proportion of a water had previously been sewage or analogous material from the amount of inorganic nitrogenous decay products of sewage remaining in the water)

INTRODUCTION

A characteristic of the science of the past century, according to environmentalist Garrett Hardin, has been the replacement of a Newtonian with a Darwinian perspective. For Newtonians, the significance of a thing in the workings of the world was a virtue of its mass, while the theme of Darwinian science is that an object's importance is a function of its role in complicated biochemical systems.[1] The recognition of the various beneficent and malign activities of microbes that was achieved during the second half of the nineteenth century is a main component of this new outlook. With regard to the decomposition of organic matter, for example, Pasteur's biological explanation, which emphasized the necessity of a diverse microbial population for the continued cycling of matter, replaced Liebig's physical explanation (actually a product of seventeenth century mechanical philosophy) in which decomposition represented the kinetic energy of vibrating particles overcoming the inertial bonds of organic molecules. Because disease and decomposition were very closely associated by nineteenth century pathologists, a roughly analogous shift took place as the germ theory of disease replaced chemistry in providing an explanation for "the intimate pathology of contagion." This nineteenth century

---

[1] Garrett Hardin, Exploring the New Ethics for Survival, The Voyage of the Spaceship Beagle (Baltimore: Penguin, 1973), pp. 38-48.

recognition of the importance of microbes has also been portrayed as one of the greatest triumphs of experimental science and one of the most far-reaching examples of the benefits of applied science. Better health, better agriculture, better industry have all been results of the germ theory of disease and the microbial explanation of decomposition; Pasteur and Koch are justly among our most exalted scientific heroes.

This study examines a sub-division of the shift from Newton to Darwin, and a sub-division even of the shift from Liebig to Pasteur: the replacement of chemistry with biology in regard to the phenomenon of the self-purification of rivers and the associated issues of water purification and sewage treatment between 1850 and 1900 in Britain. The application of biology to these problems was an important advance both because the impurities to be removed from water are primarily living and dead biological materials -- germs of disease or dead organic rubbish such as sewage -- and because assimilation of these materials by aquatic organisms is one of the main ways in which purification takes place.

During the 1880s the foundations of British water-and-sewage science did change. The broad perspective of bacteriology did replace the broad perspective of chemistry, but to focus only on this change or to look to it as an example of the facility of experimental science for discovering facts and solving technical problems would greatly misrepresent the nature of British water-and-sewage science in the second half of the nineteenth century. Indeed, more striking than the change of ideas and explanations is their failure to change. Chemical and biological concepts of impurities and purification, concepts which if not always mutually exclusive, seem at least to require some reconciliation, coexisted as legitimate scientific concepts from well before Liebig and Pasteur focused their attention on decomposition almost until the end of the century. Pasteur demonstrated the dependence of biochemical cycles on the activity of microbes in the early 1860s, some 25 years before that concept was used to explain self-purification or to provide the basic science for a technology of biological sewage treatment.

Thus, what was conceived as a study of the circumstances of a revolutionary change in scientific ideas turned out to be a study of the circumstances of scientific stagnation. What needs explaining is how the British scientific community could tolerate, and in many cases encourage such ambiguity with regard to an issue of such sanitary importance as the fate of sewage in rivers, and how it could successfully ignore experimental investigations with direct or indirect bearing on this issue.

Although central issues of sanitary science such as the reality of bacterial species were unresolved for much of this period (some British naturalists argued that bacterial characteristics such as pathogenicity were a function of environment as late as the early 1890s), factual uncertainties cannot alone account for the perpetuation of ambiguity. A more important cause of this scientific failure lies with the nature of the self-purification issue and with the nature of the institutions that developed in Britain for applying scientific expertise to sanitary problems. Like many aspects of the microbial enlightenment, river self-purification was an issue with both scientific and social components. Determining the factors that effected the disappearance of pollutants in a river after a certain distance of flow was a scientific question; when polluted water could be safely drunk was an important social question. The scientific question is one of mechanisms, the social question one of conclusions. One reason chemical and biological explanations of self-purification coexisted so comfortably is that the study of mechanisms was subordinate to the obtaining of conclusions.

Alvin Weinberg's concept of trans-science is particularly helpful here. By trans-science, Weinberg means "questions which can be asked of science and yet _which cannot be answered by science_."[2] Weinberg describes several types of trans-scientific questions. In some, scientific solutions, though conceivably obtainable are in practice impossible to obtain (an example is the enormous number of experimental animals that would be required to determine with a high degree of statistical confidence the harmfulness of low level radiation). In others the randomness of nature requires of the scientist probabilistic answers which are not useful to society (as with predictions of the likelihood of a major accident to a nuclear power plant).[3]

---

[2] Alvin Weinberg, "Science and trans-science," _Minerva_ 10(1972): 209.

[3] Ibid., pp. 210-11.

Questions which cannot be answered on a factual basis become essentially ethical or political, and answers a scientist gives to them reflect that scientist's values.

The kinds of issues faced by nineteenth century scientists who considered matters of sewage treatment and water quality fit Weinberg's concept of trans-science. Scientists were frequently asked to say whether a sewage effluent was pure or a water safe to drink. Even where purity was defined -- and it often wasn't -- and even after the pathogenic bateria which made water unsafe had been identified -- which they weren't until the 1880s -- the degree of assurance society was demanding in asking these questions went beyond what facts could supply. Controlled experiments which simulated all factors affecting the survival of typhoid bacteria in a river, for example, were clearly out of the question; similarly, because polluted river water had been used safely for years did not rule out the possibility of a combination of circumstances that would make that water unsafe. Delivering an opinion on such a question required transcending science and making a judgment about what policy was in society's best interest.

Weinberg suggests that political or adversary institutions are the proper forums for resolving trans-scientific disputes because these disputes are basically about social policy.[4] In nineteenth century Britain the issue of self-purification, together with water-and-sewage questions in general, was dealt with by scientists within

---

[4] Ibid., pp. 213-16.

a political and an adversary context, yet discussion of self-purification in such a context worked against an understanding of that phenomenon. Political and legal disputation dominated scientific inquiry, fomenting distrust among scientists and hampering their ability to work cooperatively on sewage-and-water problems. Scientific concepts, such as the theory that diseases were caused by minute living germs highly resistant to most techniques of purification or the notion that aquatic life purified water, were evaluated on the basis of political implications rather than scientific validity. Self-purification, too, fit into this category. It was less an issue of scientific inquiry than an excuse for inadequate sewage treatment or suspect water supply -- a "theoretic conceit, repugnant to common sense and natural instinct"; "a good excuse for non-removal of filth . . . [for] obstinate and recalcitrant dirt-lovers"; a "fallacious doctrine . . . which has probably done more to check and paralyse the prevention of river pollution than anything else."[5]

The first chapter examines the ideals and realities of river use in nineteenth century Britain and the failure of legal and technological remedies for river pollution. In the second chapter, which deals with the involvement of scientists, particularly chemists, in pollution inquiries, the concept of trans-science is discussed further.

---

[5] Arthur Angell, "London water supply," SR 8(1878): 40; "On the removal of nuisances by oxidation," SR 5(1876): 157-58; Percy Frankland, "The upper Thames as a source of water supply," JRSA 32(1883-84): 435.

That chapter considers the effects on the normal workings of the scientific community of scientists' participation in the various adversary forums in which water-and-sewage policy was made. Chapters three and four examine nineteenth century ideas of decomposition, disease causation, and purification, especially biological purification, the concepts scientists employed in debates on trans-scientific water-and-sewage issues. Chapter five, focusing on the period 1850-1860, examines the gap between the abilities of science and the needs of society that developed as the existence of water-borne disease became increasingly clear. It also considers the reasons for the rejection of the idea of biological purification at the end of the 1850s. Chapter six examines the development during the mid-1860s of analytical and purificatory nihilism, the view that the question of whether a water was pure was unanswerable, and the application of that view in water politics. Chapter seven is concerned with the response of water scientists to the cholera epidemic of 1866. Using Edward Frankland as an example, it explores in detail the relations between a scientist's views about social policy and the content of that scientist's work. I argue that Frankland's water analysis procedures were fundamentally trans-scientific. Chapter eight deals with the effect of participation in trans-scientific issues on the mechanisms through which the scientific community evaluates and accepts research. It looks at the response to experimental studies on self-purification by Edward Frankland in 1868-69 and Charles Meymott Tidy in 1880-81. Chapters nine and ten (for which there is a joint introduction) examine what later writers saw as a biological revolution in sewage-and-water matters. Chapter nine deals with the politicization of water bacteriology, chapter ten with the emergence of a technology of biological sewage treatment from a political context.

I. DUTIES, DISPUTES, AND DECEPTIONS: RIVERS AND BRITISH SOCIETY 1850-1900

> There are some things which cannot and will not be shelved, because they are "irresponsible" -- the Negro was one, the pauper is another, and sewage is a third.[1]
> --William Hope

"We believe this to be the foulest river in the known world," boasted Littell's Living Age of the Thames in July 1858, a summer long remembered for the "Great Stink" of the Thames.[2] The Thames, like nearly every other British river, was severely polluted in the second half of the nineteenth century. Indeed, late nineteenth century Britons claimed for their country the dubious honor of having the most polluted rivers in the world, a claim confirmed by appalled French, German, and American visitors.[3] River pollution was an

---

[1] William Hope, "On the use and abuse of town sewage," JRSA 18(1869-70): 304.

[2] "The Thames in its glory," Littell's Living Age 58(1858): 375.

[3] See: M.N. Baker, British Sewage Works (New York: Engineering News, 1904), p. 7; C.-E.A. Winslow and E.B. Phelps, "Investigations on the purification of Boston sewage with a history of the sewage disposal problem," U.S.G.S. Water Supply Paper #185 (Washington, D.C.: G.P.O., 1906), p. 93; W.R. Nichols, "Report to the State Board of Health of Massachusetts," in 4th Annual Report of the State Board of Health of Massachusetts, Jan. 1874 (Boston: Wright and Potter, 1874), pp. 91-92; and C.B. Folsom, "The disposal of sewage," in 7th Annual Report of the State Board of Health of Massachusetts (Boston: Wright and Potter, 1876), p. 307, for American views. For German views see William Dunbar, Principles of Sewage Treatment, translated with the author's sanction by H.T. Calvert (London: Trounce, 1908), p. 1; Martin Strell, Die Abwasserfrage in ihrer geschichtlichen Entwicklung von den ältesten zeiten bis zur Gegenwart (Leipzig: F. Leineweber, 1913), pp. 192-93; and John von Simson, "Die flusseverunsreinigungsfrage im 19. Jahrhundert," Vierteljahrschrift fur Sozial und

historical phenomenon of major importance -- for health, industrial development, and the evolution of municipal administration. It fostered widespread awareness of environmental catastrophe, widespread recognition of the necessity of environmental management. Pollution occasioned a massive, if unsuccessful scientific commitment, our subject here.

Although river pollution itself has not been a focus of historical research, the economic and ideological landscape in which the pollution problem emerged has been thoroughly surveyed: Victorian river pollution resulted from the rapid industrialization and urbanization of the first half of the century. From 1801 to 1831, for example, Leeds' population grew from 53,000 to 123,000, Bradford's from 29,000 to 77,000, Huddersfield's from 15,000 to 34,000. From 1801 to 1841 London grew from 958,000 to 1,948,000 inhabitants.[4] Geography exacerbated urban problems in Britain: British rivers were too small to carry off the industrial and domestic wastes of the cities that developed along them. Engineer James Gordon calculated that dumping into the Rhine the sewage of all the towns between Basle and Köln would give that river a concentration of only one part sewage per 2345 parts water.[5] By contrast, the lower Lea, a tributary

Wirtschaftgeschichte 65(1978): 373-74. For a French view see Theophile Schloesing, "Assainissement de la Seine. Épuration et utilisation des eaux d'egout. Rapport fait au nom d'une commission," Annales d'Hygiène publique et de Médecine Légale, 2nd series, 47(1877): 206-7.

[4] Cited in S.E. Finer, The Life and Times of Sir Edwin Chadwick (London: Methuen, 1952), p. 213.

[5] James Gordon, "The drainage of continental towns," TSIGB 7(1885-

of the Thames, was two-thirds sewage in 1885, and Tottenham, whose ineffective sewage treatment was drawing complaints from surrounding towns, asked permission to discharge its effluent into the Lea on the grounds that it would actually dilute the river.[6] Bradford claimed its raw sewage would dilute Bradford Beck.[7]

Use of river water for canals or for urban water supplies intensified the problem. The lower Lea was so bad because most of the river's water was extracted upstream for the use of Londoners, leaving the lower reaches a relatively stagnant tidal pool, inadequately flushed by upstream runoff. Similarly, the filling of canals with river water converted many a northern river into a series of ponds behind weirs, with very little flow.[8]

Evidence given before inquiries in the 1870s and 1880s suggests pollution became severe between 1830 and 1850. By 1828 fishermen were complaining of the destruction of the lower Thames fisheries by pollution. The last Thames salmon was caught in 1833. Since intensive sewer-building did not begin until after the Public Health Act

---

86): 195. See also "Contributions to our knowledge of natural waters used for drinking and other domestic purposes," CN 20(1869): 41.

[6] "The drainage of the Lea valley," SR n.s. 12(1890-91): 20-21. See also SR 3(1875): 132; and "The river Lea," SR n.s. 7(1885-86): 92, 106.

[7] W.G. McGowan, "The sewage of manufacturing towns," CN 28(1873): 184. A similar claim was made with regard to putting Manchester's sewage in the Rivers Irk and Medlock ("Rev. of G.E. Davis and A.R. Davis, The River Irwell and its Tributaries: A Monograph on River Pollution," CN 62[1890]: 210).

[8] W. Spinks, "The River Mersey," Proc. Assn. of Municipal and Sanitary Engineers 16(1889-90): 222-28; M.M. Paterson, The Pollution of the Aire and Calder: how to deal with it (London: E. & F.N.

of 1848, the pollution problem was probably at first mainly due to industrial dumping.[9]

The building of sewers, enlargement of water supplies, and installation of water closets increased pollution during the second half of the century. A series of public health and local government acts passed between 1848 and 1875 obliged towns and other sanitary authorities to make sanitary improvements and gave them powers to do so. These acts did not permit pollution but since technologies for purifying sewage were ineffective, increased pollution was a consequence of them. Edward Frankland noted in 1893 that the Thames had grown worse (as measured by the organic elements it contained) in the previous 25 years and ascribed this to the increased rapidity with which sewage was getting into the river.[10]

Pollution was a problem throughout Britain. Of 56,000 river basins in England and Wales, 9000 were listed as totally fishless in 1890.[11] There was nevertheless enormous regional variation, both in

---

Spon, 1893).

[9] R.S.R. Fitter, London's Natural History (London: Collins, 1945), p. 94; Royal Commission on Metropolitan Water Supply, Report, P.P., 1828, 9, (267.), pp. 9-10; Royal Commission on Metropolitan Sewage Discharge [RCMSD], Minutes of Evidence, P.P., 1884, 41, [C.-3842-I.], QQ 1258-61; F.T.K. Pentelow, River Purification. A Legal and Scientific Review of the past 100 Years, being the Buckland Lectures for 1952 (London: Edward Arnold, 1953), p. 2; "The Silver Thames," Littell's Living Age 58(1858): 377; [Nathaniel Beardmore], "Water Supply," Westminster Review 54(1851): 190.

[10] Royal Commission on the Metropolitan Water Supply [RCMWS], Appendices, P.P., 1893-94, 40 pt. II, [C.-7172.-II], Appendix C. 13, p. 201.

[11] W. Sergeant, "Pollution of rivers with special reference to the Mersey and the Ribble," SR n.s. 12(1890-91): 166.

regard to the nature of the pollution and the concerns it inspired. Sewage pollution was the greatest concern in the southeast, where the Thames and Lea supplied most of London's water. About a million people lived upstream of London in the Thames basin, and Londoners drank their diluted and marginally purified sewage. Similar problems existed on other rivers. Many towns on the Severn used the river both as sewage dump and water supply. In the late 1850s, 50,000 Birmingham residents were getting their water from the River Tame, a stream half sewage and bearing the wastes of an upstream population of 270,000.[12] In the southwest and in south Wales, mining pollution caused great damage to fisheries and, when rivers overflowed their banks, to agriculture. Most Scottish rivers were less polluted than English rivers simply because the Scottish population was scattered and the rivers were larger. Nonetheless, where population and industry were concentrated pollution was severe: the lower Clyde was described as "a seething puddle, a gigantic cesspool"; flames 30 feet high were reported on the River Almond, heavily polluted by shale oil works.[13]

---

[12] Royal Commission on the Sewage of Towns, Preliminary Report, P.P., 1857-58, 32, [2372.], pp. 9-10; Second Report, P.P., 1861, 33, [2882.], p. 7; Select Committee on the Sewage of Towns (Metropolis), Minutes of Evidence, P.P., 1864, 14, (487.), QQ 4016-17; Rivers Pollution Prevention Commission, Third Report, P.P., 1867, 33, [3850.], p. xlii [To be cited as RPPC/1865/3rd Rept.].

[13] William Wallace, "On the chemistry of sewage," CN 43(1881): 55; Rivers Pollution Prevention Commission, Fourth Report, Minutes of Evidence, P.P., 1872, 34, [C.-951-I], Q 2176. See also John Warren, "Sewage disposal in Scotland," Sanitary Journal [Glasgow], n.s. 2(1895-96): 53-56, 95-100.

By far the worst pollution was in the north of England. Lancashire rivers were described as

> deadly, slimy giant snakes wriggling their slow and tortuous courses through the towns they infest, or besmearing the landscape with their muddy trail, and emitting odours such as man cannot live in.[14]

Early on a July morning in 1869 the Irwell in central Manchester, its water 22 times more concentrated in organic nitrogen than the Thames at London Bridge appeared

> caked over with a thick scum of dirty froth, looking like a solid sooty crusted surface. Through the scum . . . heavy bursts of bubbles were continually breaking, evidently rising from the muddy bottom and wherever a yard or two of the scum was cleared away, the whole surface was seen simmering and sparkling with a continual effervesence of smaller bubbles rising from various depths in the midst of the water, showing the whole river was fermenting and generating gas.[15]

The north was the center of the textile trade and textile works studded the sides of the small streams that rose in the Pennines. Before flowing three miles the Irwell was polluted by 19 cotton factories, two dye works, one print mill, one saw mill, and two flour mills.[16] The Bradford canal periodically caught fire. It was said birds could walk on the Medlock, a river reported to be so thick with organic matter that its water could be sold as fertilizer.[17] Using

---

[14] Joseph Brierley, "The public health and the water supplies," The Builder 26(1868): 584-85. For general descriptions of rivers in the north see RPPC/1865/3rd Rept.; RPPC/1868/1st Rpt., P.P., 1870, 40, [c.-37.]; and RPPC/1868/3rd Rept., P.P., 1871, 25, [c.-347.]; and S.C. Sewage of Towns (Metropolis), Evidence, QQ 3999-4058; and "The session from a public health point of view," SR 5(1876): 133.

[15] RPPC/1868/1st Rept., pp. 16-18.

[16] Ibid., p. 7.

water from the Calder as ink, a Wakefield resident wrote to the Rivers Commission expressing regret that words could not communicate the smell of the river.[18]

Northern rivers were polluted chiefly by industrial wastes, especially textile washing and dyeing water.[19] Aggravating pollution was the common practice of "tipping" cinders into rivers. In some places so much solid debris was put into rivers (or placed on banks to be washed away at high water) that river beds were raised several feet, increasing the likelihood of floods. A large number of animal carcasses also ended up in northern rivers: dead dogs, cats, and pigs were removed from the Aire at the rate of 50 per day.[20] Though not widespread, chemical-works pollution was extreme: the acidity of the St. Helens canal destroyed masonry and iron canal fixtures.[21]

---

[17] A. Sharratt and K.R. Farrar, "Sanitation and public health in nineteenth century Manchester," Proc. Manchester Literary and Philosophical Society 114(1971): 55; Fisheries Preservation Association, On the Pollution of the Rivers of the Kingdom; the enormous magnitude of the evil, and the urgent necessity in the interest of the public health and the fisheries for its suppression by immediate legislative enactment (London: The Association, 1868), pp. 9, 35; RPPC/1865/3rd Rept., Evidence, QQ 10767-69; J.W. Willis-Bund, "The Aire and the Calder," SR n.s. 11(1889-90): 417-18.

[18] RPPC/1868/3rd Rept., opposite p. 12.

[19] Most northern towns were not significantly water closeted until the end of the century. See RPPC/1868/1st Rept., pp. 60-64; Thomas Stevenson, "Sewage disposal," SR n.s. 6(1884-85): 515; Sharratt and Farrar, "Public Health in Manchester," pp. 51-53, 64.

[20] Fisheries Preservation Association, "On the pollution of rivers," p. 35. On tipping see W. Dunbar, Principles of Sewage Treatment, p. 3; S.C. Sewage of Towns (Metropolis), QQ 4003-9; RPPC/1865/3rd Rept., pp. xi-xiii.

[21] RPPC/1868/1st Rept., p. 35; Clement Higgins, A Treatise on the Law Relating to the Pollution and Obstruction of Watercourses; together with a brief summary of the various sources of rivers

Conditions on British rivers did not improve significantly between 1850 and 1900. Successful remedies for pollution were few, local, expensive, and often temporary. Development of biological sewage treatment in the 1890s inspired confidence that a solution was near, but hopes had been raised and dashed many times before. The progress so apparent in retrospect -- a decline in death rates, progress in water-and-sewage science and technology, an improvement in living conditions in general -- was not apparent to many Victorians. Despite "scientific" solutions, rivers got no better and it began to appear that river pollution might be an inevitable accompaniment of civilization. This crisis of faith in the adequacy of science is especially significant here. The relegation of water-purification and sewage-treatment and schemes to the status of perpetual motion machines is a major theme.

1. The Duty of Rivers

To Victorians rivers had a social duty, but what was that duty? Was pollution a problem, or was environmental quality subordinated to prosperity and technology, lost in the preoccupation with getting and spending described and deplored by Wordsworth and Dickens? It is true that one finds frequently such statements as "In a manufacturing district the first purpose of streams is to aid manufacturing industry; and they may reasonably receive all the liquid or soluble

---

pollution (London: Stevens and Haynes, 1877), p. 7.

refuse from works connected therewith," or "rivers upon which towns were built were the natural channels for the disposal of sewage."[22] Yet opposite views are also frequently encountered; despair that "once brawling streams, laughing and glittering over pebbly beds . . . are rapidly disappearing wherever population and factories are on the ascendant."[23]

The assumption underlying both views of river use is as important as the conflicting views themselves. Although they might not agree what it was, Victorians spoke of rivers and other natural resources as having a purpose -- river policy ought to be determined by principle rather than pragmatism. To some -- including both those who accepted pollution and those who opposed it -- this purpose of rivers was preordained; river management was a component of Christian duty. Thus one defender of pollution wrote "amongst the other purposes for which water was given to the world, not the least important . . . is that of a detergent."[24] Similarly, an opponent of pollution complained of the ruination of streams "designed by

---

[22] RPPC/1865/3rd Rept., Q 12318; [Greville Ffennell], "The Lea Conservancy Board," The Field, 20 June 1868, p. 482. See also Henry Robinson, "The past and present condition of the Thames," MPICE 15 (1855): 202; E.D. Potter, The Pollution of Rivers. by a Polluter. A Letter to the Rt. Hon. George Sclater-Booth, president of the Local Government Board (Manchester: Johnson & Rawson, 1875), p. 7; C. Estcourt, "Pollution of Rivers: the Irwell," SR 5(1876): 417.

[23] Beardmore, "Water supply," p. 190; E. Jesse, "The Thames," Once a Week 3(1860): 108-10; Humphrey Sandwith, "Public health," Fortnightly Review 23(1875): 266.

[24] "The pollution of rivers by chemical manufactures," Van Nostrand's Eclectic Engineering Magazine 7(1872): 174.

Providence to minister to the wants and necessities of man, give fertility to the earth, and beauty to the landscape."[25] Complementing theological justifications were utilitarian justifications: rivers should be utilized for maximum societal benefit. Again there was disagreement about what was the best use of rivers. Anglers argued that pollution destroyed a major source of food, advocates of sewage recycling regarded putting sewage into rivers as an "unspeakable waste" of the earth's capital of fixed nitrogen, while Manchester liberals such as Jacob Bright opposed anti-pollution legislation on the grounds that although some rivers were "black as ink. . . . there were interests concerned . . . 10,000 times greater than the interests of the fisheries and the sentimentalists."[26]

Natural theology and utilitarianism were not incompatible so long as one believed, and many did, that God was a utilitarian. In both perspectives, rivers had a purpose, in principle discoverable by rational means. One can envision a compromise; a recognition that some rivers served God and society best as sewers, others as homes for fish or inspirers of poets. Unfortunately, subverting the assumption of purpose was the reality of common law which institutionalized the ancient English tradition of the rights of property, the privilege of the owners of natural resources to use them as they pleased. This tradition of individualism was at odds with utilitarianism. Complain-

---

[25] "Letter . . . relating to the pollution of streams . . . by the Fisheries Preservation Association and the Sanitary Association of Great Britain," P.P., 1864, 50, (224.), p. 327.

[26] Hansard's Parliamentary Debates 177(8 March 1865), p. 1337.

ing of the pollution of the River Wandle by the sewage of Croydon, landowner G.P. Bidder argued that pollution of rivers which were "ornaments" of estates caused "serious detriment to . . . property." Bidder sued Croydon and won. His customary enjoyment of his property was being denied and the social utility of the act of pollution was irrelevant.[27] When sanitary reform interfered with enjoyment of property, erstwhile reformers became ardent defenders of property. C.B. Adderly, famous as a sanitary reformer, was notorious for obtaining and enforcing an injunction forbidding Birmingham to build sewers. An 1873 anti-pollution bill was introduced first in the House of Lords, its sponsors assuming that efforts to protect property rights would get an especially sympathetic hearing there.[28]

A major theme in Victorian debates about river use is therefore the attempt to reconcile utilitarianism, individualism, and natural theology, all concerns often present in the same individual. A common principle was found: recycling. Its founder was the architect of the public health movement, Edwin Chadwick.

As a young man, Chadwick (1800-90) was a protégé of the utilitarian philosopher Jeremy Bentham. In his work during the 1830s

---

[27] Bidder in disc. of F. Braithwaite, "On the rise and fall of the River Wandle; its springs, tributaries, and pollution," MPICE 20 (1860-61): 209-10, 255-58.

[28] J.T. Bunce, History of the Corporation of Birmingham, 2 vols. (Birmingham: Cornish, 1878, 1885), II, pp. 126-29, 135-39; Paul Smith, Disraelian Conservatism and Social Reform (London: Routledge and Kegan Paul, 1967), pp. 224-25; Asa Briggs, Victorian Cities (Harmondsworth, U.K.: Pelican, 1968), p. 165; Hansard's Parliamentary Debates, 3rd series, 224(13 May 1875), pp. 550-53.

as secretary of the Poor Law Commission and in his later experience investigating the health of towns during the 1840s, Chadwick developed the principle that individual self-interest and social good lay in the same direction. He argued, for example, that the small cost of sanitary reform would be offset by increased prosperity, a healthy population being an asset so real that its value could be recorded as a credit in an account book. Likewise, it was in the interest of individuals, manufacturers, and municipalities not to pollute rivers. Here the reason was that what was thrown away was actually valuable matter whose utility had simply not been attended to. Sewage could be used to fertilize agricultural land, manufacturing wastes could be turned into new products.[29]

Recycling also acquired a theological justification. Though arguably present in Chadwick's works, this appears more strikingly in the writings of the cleric, essayist, novelist, historian, and Christian Socialist Charles Kingsley, one of Chadwick's associates. To Kingsley, improvement of health was identical with moral improvement. He spoke of "the dirt which extends itself from the body to the clothes, the house, the language, the thoughts."[30] As he looks into the Thames from a bridge, a character in Kingsley's 1848 novel _Yeast_ muses

---

[29] R.A. Lewis, _Edwin Chadwick and the Public Health Movement_ (London: Longmans, Green, and Co., 1952), p. 115.

[30] [Charles Kingsley], "The water supply of London," _North British Review_ 15(1851): 232. See also [W. O'Brien], "The supply of water to the Metropolis," _Edinburgh Review_ 91(1849-40): 384.

> Only look down . . . at that huge black-mouthed sewer, vomiting its pestilential riches across the mud. There it runs, and will run, hurrying to the sea vast stores of wealth, elaborated by Nature's chemistry into the ready materials of food; which proclaims, too, by their own foul smell, God's will that they should be buried out of sight in the fruitful all-regenerating grave of earth; there it runs, turning them all into the seeds of pestilence, filth, and drunkenness.[31]

In the view of another, had William Paley written his Natural Theology in the 1860s, the recycling of sewage would surely have been a prime example of the design, wisdom, power, and goodness of God.[32]

Despite increasing disillusionment from the 1870s onward due to repeated failures to make recycling work, enchantment with the idea of recycling is prominent in sanitary literature throughout the second half of the century. Recycling was to end poverty, to lead to universal peace. Rome was presumed to have fallen because it had not recycled its sewage. Sewage was recycled in Victorian utopias. Sewage utilization was to make local taxation unnecessary. Manufacturers and miners were promised that their refuse could be recycled, always effectively, often profitably. In future "farmers and landowners [would] pray to the authorities to bring a sewer that way, as they now beg capitalists to bring a railway."[33] After a visit to

---

[31] Charles Kingsley, Yeast -- a Problem (New York: J.F. Taylor, 1903), pp. 279-80.

[32] The Builder, 25 March 1865, p. 201.

[33] J.H. Stallard, On the Sanitary Requirements of Liverpool (Liverpool: Adam Holden, 1871), p. 23; "Utilisation of London sewage," CN 8(1863): 99; S.C. on the Sewage of Towns, 1st Report, P.P., 1862, 14, (160.), p. ii; [E.J. Mills], "The use of refuse," Quarterly Rev. (Am. ed.), 124(1868): 173-85; Fisheries Preservation Association, "On the pollution of rivers," p. iv. For a more complete discussion see Christopher Hamlin, "Recycling as a goal of sewage

Britain, American chemist William Ripley Nichols reported:

> it is no exaggeration to say that this problem of the conversion of the excremental waste of towns and people and the refuse of factories into useful materials, is now engaging as much of the attention of intelligent minds throughout the world as any social question. The English press is burdened with publications on this general subject. Chemists, farmers, political economists, engineers, physicians, and amateur sanitarians are at work upon it. Towns and cities are making costly experiments to test the worth of all the various plans proposed. Stock companies are formed, whose business is first to make money for themselves at any rate, and secondly to benefit the rest of the world by their ventures.[34]

There were good reasons for expecting recycling to work: urban wastes had traditionally been used as fertilizer; there had been recent successes using gas tar as a raw material for the fledgling organic chemicals industry, recovering HCl from alkali manufacture.[35] There were also good reasons for thinking recycling necessary. During the 1860s and 1870s it became clear that plants could only use fixed nitrogen and there seemed no natural or artificial process that could fix nitrogen as rapidly as it was dumped into the sea as sewage. Chemist William Crookes, for whom nitrogen-fixation was the "king-problem" of chemistry, saw every stroke of the engines that pumped London's sewage into the Thames as "destroying potential

---

treatment in nineteenth century Britain," Proceedings of the Annual Conference of the Society for the History of Technology, October 1981 (in press).

[34] W.R. Nichols, "Report . . . to the Board of Health," p. 21.

[35] On traditions of urban wastes recycling see Sidney and Beatrice Webb, English Local Government, v. 4, Statutory Authorities for Special Purposes with a summary of the Development of Local Government Structure (1922; rpt. with a new introduction by B. Keith-Lucas, Hampden, Ct.: Archeon, 1963), pp. 333-36; Mills, "Use of refuse."

life."[36]

To those who undertook to resolve it, the "sewage question" was not simply a problem of sanitary engineering, but a quasi-sacred quest, upon which the future of civilization and even of humanity depended. Their answers had to be righteous as well as right: had to conform to God's plan for the creation, uphold individual rights, and procure the greatest good for the greatest number. William Hope's statement that

> no candid man, who is not committed to some sewage precipitation or earth-closet heresy, can doubt that Adam utilized his sewage and grew his vegetables in the Garden of Eden by the same system of irrigation I practise myself

is, strange as it may seem, a justification of a technology.[37] In Frederick Krepp's view the [sanitary] engineer was a "great trustee of human happiness" and there was a close relationship between the engineer on earth and the "Great Engineer."[38] William Farr, referring to works for obtaining urban water supplies, wrote of the "admirable and wonderful engineering operations which Providence has placed at our disposal."[39] Chemist Edward Frankland referred to the "age of engineering miracles."[40] As Kingsley observed, "science and

---

[36] [William Crookes], "The economy of nitrogen," Quarterly J. of Science n.s. 8(1873): 149-50.

[37] Hope, "Use and abuse of town sewage," p. 298.

[38] Frederick Krepp, The Sewage Question: being a general review of all systems and methods hitherto employed in various cities for draining Cities and utilising Sewage: treated with reference to Public Health, Agriculture, and National Economy generally (London: Longmans, Green, & Co., 1867), p. 207.

[39] William Farr, Report on the Cholera of 1866 in England, P.P.,

usefulness contain a divine element."[41] On the discovery of bacterial sewage treatment in the 1890s, the electrician Sir William Preece wrote:

> Bacteria thus fulfill the highest function of the engineer, and nature asserts her power in fulfilling the clearly-defined will of the great Creator, the biological system has come to stay.[42]

According to Chadwick and his followers, therefore, river pollution was a "public and national nuisance" in no one's best interest:

> it interferes with the convenience and comfort of all classes of the people; it damages various and important interests, as those connected with manufacturing establishments, canals, fisheries, and so on; it deteriorates property to a large extent, and, as interfering with the main source of water supply, it is of serious importance to public health.[43]

In their view rivers ought never to become polluted because "wastes" simply would not exist.

In reality, of course, there was a great variety of opinions about river use and river pollution. As might be expected much of the concern about pollution arose in connection with the public health movement which had a great impact on urban and environmental politics from the 1840s onward. As a group sanitarians were remarkably ambivalent about pollution. Their sewers and water closets

---

1867-68, 37, [4072.], p. xlv.

[40] Edward Frankland, "The water supply of London and the cholera," Quarterly J. of Science 4(1867): 319.

[41] Charles Kingsley, Yeast, p. 94.

[42] William Preece, "Modern methods of sewage disposal," Engineering News 42(1899): 171.

[43] R.C. Sewage of Towns, Second Report, pp. 9-10.

were, after all, partly responsible for the problem. While they deplored its effects on health -- increasingly so as the connection between bad water and enteric disease became clear -- many of them accepted pollution as an inevitable accompaniment of sanitary reform.

For Chadwick, ridding towns of decaying organic refuse was the first priority and he believed the most effective means for doing this was frequent urban flushing, via copious soft water, water closets, and graded pipe sewers. Sewage was to irrigate farmland, and what was to become of it if irrigation didn't work -- it often didn't, as will be considered below -- was a question Chadwick avoided. In fact, solving one public health problem caused another: inadequately treated sewage ended up in rivers, sometimes rivers from which urban water supplies were taken. Some sanitarians believed this was a necessary, if unfortunate consequence of sanitary progress: "the saving of human life ought to outweigh every other consideration, even supposing that it were necessary to foul rivers"; while others saw the solution as worse than the problem: "by adoption of a partial and one-sided measure for the purification of dwellings, one of the results of modern civilisation has been to render the vilest form of this abuse [poisoning another's water supply] an habitual practice in this country."[44]

The issue of whether rivers were to be sewers or water supplies was particularly acute in London and along the upper Thames and Lea,

---

[44] S.C. Sewage of Towns (Metropolis), Q 4185; [B.H. Paul], "Water analysis for sanitary purposes," BMJ i 1869, p. 427.

rivers which supplied most of London's water. Water-closeting began in London about 1810, earlier than in most other cities. Water closets drained into cesspools or ancient "sewers of deposit" (which was much the same thing), causing serious pollution of ground water. During the 1850s successive Metropolitan Boards of Sewers and later the Metropolitan Board of Works (est. 1855) built new graded sewers which intercepted many cesspools and old sewers. While politicians squabbled over the ultimate fate of London's sewage, these newly constructed sewers were emptying into the Thames. During the hot summers of 1858 and 1859 the river reeked with putrefying sewage and it was pointed out (repeatedly) that elimination of private cesspools had merely converted the river into one huge cesspool. The potential for further stinks was eliminated in the mid-1860s with the completion of downstream outfalls for the metropolitan sewage.[45]

Until the mid-1850s, the inner-city Thames, where both new and old sewers had their outlets, was also the source of water supply for several of the seven companies which, with parliamentary sanction, had divided among themselves the London water market. Public outrage at this state of affairs -- the water contained visible animalcules, deposited 8-9 inches of sediment per month in cisterns, and was expensive as well -- led to an investigation and condemnation of the supply by a royal commission in 1828. No changes were made, however, until after the 1848-49 cholera epidemic. In the early 1850s,

---

[45] Royal Commission on Metropolitan Sewage Discharge, First Report, P.P., 1884, 41, [C.-3842.], pp. xi-xxxv.

Chadwick's machinations, combined with public dissatisfaction at the impure, inadequate, and expensive supply, and the belief (it is too early to label it the recognition) that the water contributed to the cholera, led to the Metropolis Water Act of 1852, which required the companies to move their intakes above the limits of the tide.[46]

Moving water intakes upstream and sewage outfalls downstream led to new problems: dreadful conditions on the lower Thames near the outfalls, a constant threat of water-borne disease from the sewage of upstream towns. In an 1851 report on the quality of the London water supply, chemists A.W. Hofmann, Thomas Graham, and William A. Miller noted that although Thames water was presently safe, it was not likely to remain so as upstream towns undertook their own sanitary improvements.[47] Londoners, and particularly the London water companies (ever ready to improve their supply, never willing to admit it needed improvement) were indeed concerned about the sanitary progress of upstream towns. In some instances, the companies helped towns with sewage treatment problems. On the Lea, the sewage works at Hertford was built and operated by the New River Company, which got much of its water from that river.[48] More often concern took the form of legal and parliamentary protection of the water supply. In the

---

[46] Royal Commission on Metropolitan Water Supply, Appendices, App. B-1, pp. 60-63; G. Phillips Bevan, The London Water Supply, its Past, Present, and Future (London: Edward Stanford, 1884); Francis Bolton, London Water Supply, including a history and description of the London Water Works, new edition, entirely revised and enlarged . . . by Philip A. Scratchley (London: Wm. Clowes, 1888), pp. 15-20.

[47] Thomas Graham, W.A. Miller, and A.W. Hofmann, "Chemical report on the supply of water to the Metropolis," J. Chem. Soc. 4(1851): 386.

Thames Navigation and Lea Conservancy Acts of 1866 and 1868 respectively, mechanisms were set up whereby large subsidies from the water companies would pay for enforcement of anti-pollution clauses by elected conservancy boards. Anti-pollution language was strengthened several times in the next decades, and in some cases these boards were so effective (the fine for polluting the upper Thames with sewage was £50/day; some towns accumulated fines of over £100,000) that towns which had invested much money and pride in sewerage systems were forced to revert to cesspools. Citizens of these upstream towns were justifiably bitter -- at their fruitless expenditure, and at having to sacrifice their own health to improve the health of Londoners.[49]

It was thus with considerable trepidation that sanitarians climbed on the anti-pollution bandwagon. A good example is Robert Rawlinson (1810-98), one of Chadwick's cadre of sanitary engineers, later chief engineering inspector of the Local Government Board (eventual successor of Chadwick's General Board of Health) and inspector under the Rivers Pollution Prevention Act of 1876. Rawlinson was also a member of the Royal Commission on Town Sewage of 1857-64 and chairman of the first Rivers Pollution Commission (1865-68).[50]

---

[48] RPPC/1865/Second Report, (The River Lea), P.P., 1867, 33, [3835.], pp. xii, xxv.

[49] RCMWS, Appendices, B-10, pp. 86-100; "The drainage of towns on the Thames," SR 2(1875): 430. On Conservancy powers see "London water supply," JRSA 30(1881-82): 928; RCMWS, Appendices, B-1, pp. 60-63.

[50] "Obituary of Robert Rawlinson," MPICE 134(1897-98): 217-18.

The first reports of the Town Sewage Commission took a strong anti-pollution stance, and during the hearings of the Select Committee on Sewage (Metropolis) of 1864, Rawlinson was questioned on the subject. Given the choice of cesspools or polluted rivers, Rawlinson strongly favored the latter.[51] His vehemence proved embarrassing when his Rivers Pollution Commission held hearings at towns on the upper Thames and Lea in 1865 and 1866. The commissioners repeatedly asked local officials whether they thought it right to poison London's water supply with their sewage. On several occasions the reply was given, sometimes in Rawlinson's own words, that it was better to pollute rivers than to tolerate cesspools; towns must worry about their own health problems before they worried about others'. Rawlinson and fellow commissioners J.T. Way and J. Thornhill Harrison responded to such rejoinders by affirming with even more vigor the viability, profitability, and even necessity of irrigation.[52] Indeed, continued advocacy of the ideal solution of recycling characterized government policy for much of the rest of the century.

Rawlinson's browbeating of officials of upstream towns was satisfactory to the water companies because it was based on the view that the duty of the Thames and Lea must be for the water supply of the Metropolis, an assumption in which the companies had large capital investment. Much less satisfactory to them were the views of another

---

[51] S.C. Sewage of Towns (Metropolis), QQ 3997, 4076.

[52] RPPC/1865/<u>1st Report, Evidence</u>, <u>P.P.</u>, 1866, 33, [C.-3634-I], QQ 2490, 2496, 8857, 8932, 8939.

group of sanitarians who believed that rivers which flowed through populated or agricultural areas were inherently unsuitable for domestic supplies, although deliberate pollution was avoidable and ought to be avoided. The main advocate of this view was chemist Edward Frankland (1825-99), official analyst of the London water supply from 1865 to 1899, chemist for the Royal Commission on Water Supply of 1867-69, most active and controversial member of the second (1868-74) Rivers Pollution Commission, and "the greatest living authority on water supply."[53]

From the late 1860s until the early 1890s, Frankland held the view that urban water supplies should be taken from uninhabited mountainous areas or from deep artesian wells. His arguments were endorsed by others genuinely concerned with the potential of water-borne epidemics, and appealed to those favoring public ownership of the London supply, those outraged by the cost of the companies' water, and those with interests in alternative supplies. Typical of Frankland's stance is his response to an 1867 request from the Thames Conservancy for a set of effluent standards which upstream towns would have to meet to avoid prosecution. Frankland, William Odling, and Henry Letheby defined permissible effluents, but Frankland added a postscript to the report: adoption of these or any other standards would not make river water safe to drink, but would only "preserve the river from being offensive to the inhabitants upon its banks."[54]

---

[53] "Obituary of Edward Frankland," MPICE 139 (1899-1900): 347.

[54] Royal Commission on Water Supply, Evidence, P.P., 1868-69, 33,

This was a particularly irksome response since protecting the water supply was one of the Conservancy's main functions. To Frankland, however, river water was to be assumed bad no matter how good the effluents were that went into it.

The embarrassment and ambivalence of the sanitarians' response to river pollution is in sharp contrast with the outrage of fishermen, landowners and, surprisingly, some manufacturers. To the extent that there existed a nationwide anti-pollution agitation in late nineteenth century Britain, fishermen were responsible for it. Their lobbying was mainly responsible for the passage of the Rivers Pollution Prevention Act of 1876.[55] The Salmon Act of 1865 permitted establishment of Fishery Boards for individual river basins. These boards were drawn from local fishermen (both sportsmen and commercial fishermen), and their tasks were to regulate fishing and to conserve the fishery.[56] In many cases the majority of their efforts were spent battling poachers, but they were also active attempting to enforce the loophole-ridden anti-pollution clauses of the salmon acts. Perhaps more importantly, fishery boards were a ready-made

---

[4169.], after Q 6853.

[55] C.J. Fox, "The principles upon which legislation respecting the pollution of rivers should be based," SR 3(1875): 297-99; A.W. Blyth, "Rivers, streams, pollution of, etc.," Dictionary of Hygiene and Public Health (London: Griffin, 1876), pp. 494-97; J.W. Slater, "Observations on polluted waters," Quarterly J. of Science, 3rd series 5(1883): 389; Charles Warren, Biology and Water Pollution Control (Philadelphia: W.B. Saunders, 1971), pp. 5-7.

[56] Pentelow, River Purification, p. 5. See also Roy Mcloed, "Government and resource conservation: the salmon acts administration, 1860-1886," J. Brit. Studies 7(1968): 116.

local organization for lobbying the government for anti-pollution legislation. Petitions from fishery boards poured in whenever governments considered such legislation.[57]

Acting as a sort of unofficial executive board for the collected fishery boards were, in England, the Fisheries Protection Association [FPA], and in Scotland, the Association for the Preservation of the Rivers and Lochs of Scotland. Both were essentially lobbying organizations, composed of those of high station who presumably could influence those in high places. The FPA was largely responsible for the 1865 Salmon Act and early versions of what eventually emerged as the 1876 anti-pollution law were drafted by J.W. Willis-Bund, a member of the FPA and secretary of the Severn Fishery Board.[58]

Destruction of fish was rarely the main objection fishermen made to pollution. They recognized that the loss of one of the hobbies of the wealthy was not a sufficient justification for meddling with industry. Pollution had to appear as something more than a mere "angler's question."[59] Fisheries advocates posed as

---

[57] The Field, 3 May 1873, p. 419; 22 March 1873, p. 264; 23 Jan. 1875, p. 72; 29 May 1875, p. 536; 24 July 1874, p. 923.

[58] The Field, 22 March 1873, p. 264; 29 June 1867, p. 491; 10 August 1867, p. 120; 21 June 1873, p. 598; 22 Feb. 1873, p. 168; 11 March 1874, p. 186; 23 Jan. 1875, p. 72; 29 May 1875, p. 536 for examples of the activities of the F.P.A. On the Scottish association see SR 8(1878): 41; and "Association for preserving the rivers and lochs of Scotland from pollution," Sanitary J. for Scotland 1 (1876): 17-19.

[59] [Greville Ffennell], "The Lea and the Stort," The Field, 12 Jan 1867, p. 31.

sanitarians and opposed river pollution with far less ambivalence than did many real sanitarians. J.W. Willis-Bund argued:

> I would like to bring out clearly that the drinking water for various towns is taken from rivers, and into the same rivers the pollution and sewage of various towns is allowed to flow. For instance, the Severn; Shrewsbury drains into that river now, and the good citizens of Worcester have to drink its water. Is there no danger of the cholera germ?[60]

In 1864 the FPA and the Sanitary Association of Great Britain issued a "joint letter" condemning pollution.[61] As angling philosopher Greville Ffennell observed in 1867, "the angler is becoming a sanitary inspector." It was the "earnest and continual complaints of the angler" that were becoming "the index and signal for the philosopher and the social reformer."[62] Food supply was another peg on which anglers hung their anti-pollution campaign. The arguments of some anglers against river pollution even verged on socialism: salmon acts inspector Frank Buckland condemned polluters "who for the most part reap no inconsiderable profits from their industrial operations, but while endeavouring to increase their own profits, they treat with indifference the welfare of the public, and an important source of food not only to themselves, but to the public in general."[63]

---

[60] J.W. Willis-Bund, quoted in William Burchell, River Sanitation (London: Krough, [1884]), p. 18. See also Fisheries Preservation Association, "On the pollution of rivers," pp. 12, 28, 38-44; J.W. Willis-Bund, "The Aire and the Calder," SR n.s. 11(1889-90): 417.

[61] See note 25.

[62] "Pollution of the Yorkshire rivers," The Field, 9 Nov. 1867, p. 375.

[63] Frank Buckland, The Pollution of Rivers and its effects upon the Fisheries and Supply of Water to Towns and Villages (London:

Manufacturers' views of the duties of rivers were less uniform than might be expected. Most of them were polluters, yet many recognized that pollution damaged their interests. The problem was one of overcoming short-term concerns and individualism characteristic of dogmatic political economy and working for common long-term benefits. One of the didactic missions of the Rivers Commissions of 1865 and 1868 was to convince manufacturers that it was in their interest that streams be purified.[64] Ironically, it was the northern textile manufacturers, who caused the worst pollution, who were also the most concerned about pollution. The dyeing and printing of cloth required clean water and most works either went to the expense of getting remote water supplies or cleaned the river water they used. In the West Riding hearings of the first commission the point was frequently made that area manufacturers were losing business because river water could not be cleaned enough to allow dyeing of the finest varieties of cloth. The second commission asked Lancashire manufacturers how much money they would save if rivers were clean. Thirty respondents estimated a specific savings and these ranged from £30 to £200 per year.[65] When asked what should be done about river pollution, many manufacturers called for strong anti-pollution laws, on the grounds that necessity would be the mother of invention and

---

C.L. Marsh, 1878), p. 38. See also "Wholesale destruction of trout in the Teviot," The Field, 12 July 1873, p. 32; "Salmon fishery congress," The Field, 8 June 1867, p. 429.

[64] C.B. Folsom, "The disposal of sewage," p. 297.

[65] RPPC/1865/3rd Rept., pp. xvii-xviii, xxi; RPPC/1868/1st Report, pp. 96-97.

lead to discovery of effective and profitable means of dealing with refuse. But while individual manufacturers recognized pollution was to their common detriment, they were wary of the remedies they themselves prescribed. When strong rivers pollution bills came before parliament in 1872 and 1873, deputations of manufacturers succeeded in having them weakened on the grounds that they would give foreign competition unfair advantage. The weak bill that finally passed was with the "bona fide desire in the very centres of manufacturing industry that this general obligation of law should be imposed."[66] It is noteworthy that when county rivers boards were authorized in 1888, the first such boards to be established were in the highly industrialized and heavily polluted areas of Lancashire and the West Riding of Yorkshire. Both boards got from parliament regional anti-pollution statutes stronger than those in place nationally. Apparently "the very centres of manufacturing industry" had found pollution so serious a problem that they had obtained laws to regulate themselves.[67]

2. Failed Remedies: Technology and Law

Recycling was the righteous answer to the sewage problem, but was it the right answer? In fact, there was no technically right answer.

---

[66] Hansard's Parliamentary Debates, 3rd series 230(20 July 1876), pp. 1675-77.

[67] W. Sergeant, "Pollution of rivers," p. 165; W. Dunbar, Principles of Sewage Treatment, p. 9; W. Spinks, "The Mersey," p. 208; Paterson, "Pollution of the Aire and Calder," in passim.

As descriptions of British rivers suggest, nineteenth century pollution abatement technologies were a failure. The failure was not from lack of trying. An immense amount of public and private money was invested to solve the sewage problem during the second half of the nineteenth century. The Rivers Commissions of 1865 and 1868, which represent the largest single application of science to river pollution abatement during the period, spent £39,625 in their eight-year existence.[68] Indeed rarely has so great a gap existed between the expectations and abilities of technology.

To demonstrate the viability of his solution to the sewage problem, sewage irrigation, Edwin Chadwick formed in 1849 a Metropolitan Sewage Manure Company, which was to use some of London's sewage for irrigation. The company folded a few years later, having squandered £50,000 capital, and never paid a dividend.[69] Chadwick's failure did not dim the ardor of irrigation advocates. Enthusiasm peaked in the mid-1860s as sewage farming was endorsed by two Commons select committees and a royal commission. London vestries petitioned parliament to sanction a sewage-farming scheme, claiming it would make local taxes unnecessary.[70] Newspapers published letters on obscure points of irrigation technology, including some from Liebig, a consultant for one of the schemes.[71] By the mid-1870s, following two

---

[68] "Return of all royal commissions from the year 1866 to the year 1874," P.P., 1888, 81, (426.).

[69] C. Greaves in disc. of Braithwaite, "The Wandle," p. 247.

[70] S.C. Sewage of Towns (Metropolis), Appendices, pp. 416-50.

[71] Jacob Volhard, Justus von Liebig (Leipzig: Barth, 1909), II, pp. 57-71.

more favorable government reports, irrigation had become the official mode of sewage treatment in Britain. Until the early twentieth century, the Local Government Board made provision of irrigation land a requirement for towns wishing to borrow money for sewerage.[72] Though reasonably effective as a sewage treatment, sewage irrigation was at best marginally profitable in the wet British climate. Major problems were a lack of skilled sewage farm managers and high land costs. Most experts advised at least one acre per 100 persons; thus for London, upwards of 40,000 acres would be required.[73] Despite claims that "sewage can be pumped to any height, carried any distance," suburban land was a necessity.[74] Since there were often few suitable sites, land prices were usually high. Despite Local Government Board insistence, some towns, particularly those in the uplands of the north, concluded correctly that irrigation was simply not a viable option.[75]

There were alternatives. A large, if diffuse group of sanitarians viewed the water closet as a "national misfortune," "antagonistic

[72] W. Dunbar, Principles of Sewage Treatment, pp. 101, 10-11.

[73] Ibid., p. 110; "Rept. of the B.A.A.S. Sewage Committee," CN 34(1876): 136-37; "London sewage," CN 10(1864): 289-90; RCMSD, Evidence, vol. II, P.P., 1884-85, 31, [C.-4253-I.], Q 18069.

[74] RPPC/1865/1st Rept., p. 13; C.D. Folsom, "Disposal of sewage," p. 342.

[75] W.G. McGowan, "The sewage of manufacturing towns," p. 184; Lewis Angell, "On the treatment of sewage," Assn. of Municipal Engineers and Surveyors 16(1889-90): 23; A.M. Fowler, "The utilisation of sewage water and water of polluted streams," Proc. Assn. of Municipal Engineers and Surveyors 1(1873-74): 193-203.

to a law of nature," and advocated cesspools, dry-earth closets, or other "filth conservancy" methods.[76] More important were several hundred patents for what one irrigationist called "the slovenly and farcical performances which enjoy the title of sewage precipitation"; purification processes in which antiseptic and flocculant chemicals were added to sewage to halt putrefaction (only temporarily, it turned out) and to separate from the liquid a floc containing materials valuable for something, usually as a fertilizer.[77] The more orthodox precipitation processes relied on well-known flocculants such as lime, clay, and alum. More exotic were the Amines process (c. 1885), which utilized herring brine, Webster's (c. 1890), which electrolyzed sewage, and General Scott's (c. 1872), in which hydraulic cement was made from sewage sludge. Even Prince Albert invented an upward filtration, peat-charcoal scheme that was tried briefly at Aylesbury.[78] These too were a failure. At best they removed solids, yet left an effluent sometimes more concentrated in

---

[76] See for example, William Wallace, "On the chemistry of sewage," p. 55; Humphrey Sandwith, "Public health," p. 268; Edward C.C. Stanford, "A chemist's view of the sewage question," CN 19(1869): 254-55; G.V. Poore, "The Shortcomings of modern sanitary methods," TSIGB 9(1887-88): 51.

[77] These methods are detailed in C.M. Tidy, "On the treatment of sewage," JRSA 34(1885-86): 1127-89; J.W. Slater, Sewage Treatment, Purification, and Utilization. A Practical Manual for the Use of Corporations, Local Boards, Medical Officers of Health, Inspectors of Nuisances, Chemists, Manufacturers, Riparian Owners, Engineers, and Ratepayers (London: Whittaker, 1888); and C.D. Folsom, "Disposal of sewage," pp. 329-33.

[78] "Remarks on Prince Albert's proposal for the clarification of sewer water and the preparation of sewage manure," Builder 12(1854): 253-54.

dissolved putrescible matter than raw sewage, and a sludge difficult to dry and nearly worthless as a manure.[79] Since it cost more to make sludge manures than they were worth there was no incentive to treat sewage. Although some processes endured for several decades (most notoriously the ABC process of the Native Guano Company of which William Crookes was a director), most "attracted merely passing notice and then disappeared into that obscurity from which they ought never to have emerged."[80] Some precipitation processes were probably wilful frauds. In any case they almost all failed "leaving behind an abominable smell at the works, and a lawyer's and engineer's bill at the office."[81]

What were towns to do? No process stood out as best; there was a perplexing variety of options, each praised to the skies on general grounds by the "sewage doctor" who had invented it, each rejected on specific grounds by rivals. J.C. Mellis glowingly described his process; the surveyor of Northampton, where the process had been used, countered Mellis' oratory by observing that the process had been given up after the town had been enjoined three times while using it.[82] Sewage-treatment debates were "a maze of conflicting

---

[79] W. Dunbar, Principles of Sewage Treatment, pp. 97-98; C.E.-A. Winslow and E.B. Phelps, "History of sewage disposal," pp. 29-32.

[80] Percy Frankland, "The upper Thames as a source of water supply," JRSA 32(1883-84): 434.

[81] S.S., "Sewer manure," JRSA 2(1854-55): 805-6; J.T. Way, "On town sewage as a manure," J. Royal Agricultural Society (1854): 136.

[82] J.C. Mellis, "The Coventry sewage works," Proc. Municipal Engineers and Surveyors 2(1874-75): 156-66, esp. 163. For other examples of this kind of literature see H.Y.D. Scott, "A new mode of

statements and propositions" which confused the public, according to one observer; sewage-treatment processes "mere costly humbugs; sport, possibly to engineers, but death to the rate-payer," according to another.[83] Surrounded by false prophets and by a central government whose solutions were dogmatic and often unrealistic, towns attempted to establish cooperative institutions to resolve the sewage question. A British Association Committee on the Treatment and Utilization of Sewage, financed chiefly (and with uncharacteristic generosity) by towns, carried out intensive irrigation experiments on a farm at Romford and published eight reports between 1869 and 1876.[84] The Society of Arts annual "Health and Sewage of Towns" conferences, held from 1876 to 1880, were intended to be clearinghouses for sharing towns' sewage-treatment experiences. Deputations of municipal officials were sent to observe how other towns were faring. All this empiricism led only to the correct if unsatisfactory conclusion that no single solution was best in all cases.

Especially pathetic is the fate of the towns on the Thames a

---

dealing with sewage precipitates," JRSA 20(1871-72): 547-58; Frederick Hahn Danchell, Concerning Sewage and its economical disposal (London: Simkin and Marshall, 1872); R.W. Peregrine Birch, The Disposal of Town Sewage (London: Spon, 1870), p. 6; [Henry Letheby], The Sewage Question: comprising a series of reports; being investigations into the conditions of the principal sewage farms and sewage works of the kingdom. From Dr. Letheby's notes and analyses (London: Balliere, Tindall, and Cox, 1872), preface.

[83] C.D. Folsom, "The disposal of sewage," p. 333; "Rev. of Wallace, Air, Water Supply, Sewage Disposal, and Food," CN 40(1879): 179.

[84] "The British Association Committee for the treatment and utilisation of sewage," BMJ i 1870, p. 188.

short distance above London. Here the sewage problem seemed truly insoluble. Pressure not to pollute was exceedingly strong. A large part of the metropolitan water supply was obtained here, fishermen were active, and the river was enjoyed by boaters and landowners who lived along it. The Thames Conservancy fought pollution by imposing huge fines on polluting towns (these probably were never collected). Finding an alternative to pollution was also exceedingly difficult. Great houses, residences of royalty and nobility, dotted the region and their owners were sufficiently influential to ensure that no sewage farms would be built nearby. In 1878 towns in the area formed the Lower Thames Valley Sewerage District, the idea being that opposition could be overcome if the sewage of all was treated at a single works. Eight years and £44,000 later, the District dissolved, having had every one of its proposals blocked by landowners, the water companies, or both.[85]

Towns muddled along, spending large amounts of money, getting negligible results. Many changed processes frequently. Irrigationist William Hope observed that towns often came to irrigation only after trying several precipitation processes and having been served with several injunctions as a result. Leicester, site of one of the first failures of the lime precipitation process, returned to lime after 25 years of unsuccessfully trying other processes. Lime was

---

[85] T.J. Nelson, An Incredible Story, told in a letter, to the Rt. Hon. Earl of Beaconsfield, Prime Minister (London: J. Truscott, 1879); SR n.s. 7(1885-86): 427; n.s. 8(1886-87): 122; SR 6(1876): 236.

no better, but it was cheaper.[86] Tottenham was "notorious as the birthplace -- and in many instances, the grave -- of several sewage processes."[87]

Absence of a technical solution to the sewage question resulted in the seeking of other kinds of solutions. If the problem could not be solved, perhaps it could be made to appear solved, to appear so, sufficiently convincingly that judges would be persuaded the town was not creating a nuisance. In the view of a Luton official, the measure of the success of their lime treatment plant was the thousands of pounds in litigation expenses it had saved.[88] Making sewage treatment appear successful was also important during these years of "civic gospel," as a reassurance to citizens that their investment in the municipality was succeeding: though the sewage works might not work well, it was often, like the city hall, built to advertise the prosperity of the town.[89] Finally, the appearance of success was important to the inventors of sewage treatment processes, whose status and income were a result of reputation.

Much of the contentiousness surrounding sewage treatment was doubtless a reflection of technical failure and a consequent

---

[86] Wallace, "On the chemistry of sewage," p. 57.

[87] Lamorock Flower, "Sewage treatment: more especially as affecting the pollution of the River Lea," Public Health: A Journal of Sanitary Science and Progress 5(1876): 397.

[88] RPPC/1865/2nd Rept., QQ 1606-7.

[89] George Rafter, "Sewage irrigation," U.S.G.S. Water Supply Paper #3 (Washington, D.C.: G.P.O., 1897), p. 10. On the "civic gospel" see Asa Briggs, Victorian Cities, in passim.

reliance on rhetoric and deception. Towns displayed their works when they were running well, tried to hide them when (as was often the case) they weren't. Discharge at the West Ham works was stopped when a lookout spotted an inspector.[90] According to the author of one sewage treatment manual, outfalls should be under water to prevent illicit sampling by those wishing to discredit the works.[91] The attempt to do with words what could not be done with facts is also responsible for the bitterness that emerged when sewage treatment was discussed at scientific meetings and in courtrooms:

> Next to contests about religion there is nothing which waxes so warm as a sewage fight; orators grow apace and become diffuse and excited on this subject, and when much talking is done facts often have a struggle for life, and if the facts do not fit the oratory so much the worse for the facts.[92]

The same blend of contentiousness, moral fervor, and technical incompetence that characterizes the Victorian sewage question is also apparent in two other technologies central in self-purification discussions: water purification and water analysis. "Highest chemical authorities" were "very much at variance as to the . . . processes by which nature removes them [impurities in water]," observed George Stillingfleet Johnson in 1880.[93] Indeed they were. The first sand

---

[90] Greater London Council Record Office [GLCRO], MBW 683, "Minutes, subcommittee on the sewage of the metropolis," 30 Sept. 1885, pp. 704-5.

[91] J.W. Slater, Sewage Treatment, Utilization, and Purification, pp. 168-69.

[92] T.J. Nelson, An Incredible Story, p. 13.

[93] George Stillingfleet Johnson, "The impurities in water, and their influence on domestic utility," JRSA 29(1880-81): 511.

filter went on stream in 1828 in the works of the Chelsea Company, one of seven which sold Thames water to Londoners. Filtration of river water through sand or through some supposedly chemically active substance such as Bischof's "Spongy Iron" or Spencer's "Magnetic Carbide" was common in the last half of the century and was required by law of the London river-water companies after 1852. But even after discovery of pathogenic and saprophytic bacteria in the 1880s the mechanism of filtration and reliability of filters remained a subject of controversy.[94]

Of water analysis it was said "there are few subjects in . . . chemical science which call forth so much difference of opinion, not only with regard to the methods to be employed, but also to the accuracy of the results obtained, and the degree of importance and significance to be attached to them."[95] Chemists of the 1850s and 1860s measured water quality in terms of the inorganic elements of hardness, such as lime and magnesia; organic matter (measured by weight loss from the ignition of an evaporative residue); and/or putrescibility (measured by the oxygen demand from potassium permanganate). In 1869 chemist B.H. Paul wrote that these processes gave results "so broad that they eclipse the truth, . . . so general that they agree equally well with opposite conclusions."[96] In 1867-68

---

[94] RCMWS, Report, pp. 62-63 and evidence of Edward Frankland, Percy Frankland, E. Ray Lankester, German Sims Woodhead, E.E. Klein.

[95] Charles E. Cassall and B.H. Whitelegge, "Remarks on the examination of water for sanitary purposes," SR n.s. 5(1883-84): 427.

[96] [B.H. Paul], "Water analyses for sanitary purposes," BMJ i 1869, p. 544.

more sophisticated processes for the supposedly more hygienically significant entities, organic nitrogen and albuminoid ammonia, were developed, and a period of bitter antagonism between adherents of these rival processes followed which sapped public confidence in water analysis. Widespread scorn toward water analysis was vindicated by studies in the early 1880s which showed that none of the analytical processes could distinguish water polluted by excreta from typhoid victims. Introduction of bacteriological water analysis in 1884 transformed water quality from a matter of milligrams of organic matter to one of bacterial numbers and types, but did not end water-analysis controversy.

Two points must be addressed here that will be considered in detail later. First, and already mentioned, is the despair among scientists and the public that resulted from the inability of science to solve the technical problems connected with river pollution. Second is the inadequacy of science. One of the main reasons for such despair was the vagueness of the task set for technology to perform. There were obvious problems in accepting a technique as a successful means of purification without a consensus of what purification was, or agreeing that an analytical process accurately discriminated between safe and unsafe water when no one was sure what made water unsafe. A glimpse of this technological chaos can be seen in discussions of the purposes of disinfectants. A. Wynter Blyth's 1876 list of the "principles which require disinfection" included

definite chemical products such as sulphuretted

> hydrogen. . . . compound stinking ammonias, . . . nitrogenous bodies, dead but in a state of change, . . . and living, growing cells: some, like the contagion of small-pox, little bits of pus, dry and hard without, soft within, and others . . . soft easily destructible bodies.[97]

William Crookes, appalled by the variety of disinfectants, noted that in some cases disinfectants were used together whose chemical activity totally neutralized one another.[98]

Just as there was no technical solution for river pollution, there were no legal or administrative solutions. Commenting on anti-pollution law, the first Rivers Commission noted that principles were in conflict with realities:

> Theoretically the law recognizes that protection is due to the public and private rights in running water. It prohibits all public nuisances, and imposes on each riparian proprietor the obligation of allowing running water to pass on its course without obstruction or pollution. But a person, judging from the present appearance of the streams in the West Riding, would infer the contrary to be the law, and would conclude that there existed license to commit every kind of river abuse.[99]

Indeed, river pollution was against both statutory and common law. Statutory prohibitions against dumping refuse in streams went back to 1389, during the reign of Richard II.[100] It was also a principle of common law that river water be passed from one riparian proprietor

---

[97] A.W. Blyth, "Disinfection," Dictionary of Hygiene, pp. 187-88.

[98] William Crookes, "On disinfection," 36th Rept. B.A.A.S. (Nottingham, 1866) (London: John Murray, 1867), sections, pp. 34-35.

[99] RPPC/1865/3rd Rept., p. li.

[100] C.E. Saunders, "Legislation for the purification of rivers and its failures," SR n.s. 8(1886-87): 343-47; S.C. Sewage of Towns (Metropolis), Q 3122.

to the next unimpaired either in quantity or quality. Neither legal route was effectual, nor were efforts during the second half of the century to establish workable anti-pollution laws successful. Obstacles to legal remedies were several: a perceived disparity between the requirements of law and social utility, difficulties activating and enforcing laws, and the difficulty of the technical problems to be solved.

Prior to passage of the 1876 Rivers Pollution Prevention Act, most anti-pollution litigation was taken under common law. The plaintiff asserted pollution had interfered with his customary rights. He could ask for an injunction to prevent future pollution and/or damages for past pollution. The most common defense in such cases was that the defendant had acquired a prescriptive right to pollute by means of an easement, an explicit or implied toleration of his pollution by those downstream. Implied easements were most common. The owner of a factory which had been polluting a river since time immemorial -- 20 years for legal purposes -- could claim that pollution was customary, provided no legal complaints had been made in that 20-year period. So long as the pollution did not cause a public nuisance, it could continue. Maintaining prescriptive rights required that pollution be continuous so factories were encouraged to pollute. One textile-mill lessee noted that installation of pollution abatement technology would jeopardize prescriptive rights and thus lower the value of the mill.[101] To anti-pollution activists,

---

[101] RPPC/1865/<u>3rd Rept.</u>, pp. li-liii; Higgins, <u>A Treatise on the</u>

prescriptive rights were simply "privileged abuse," and far from safe-guarding rights they led to anarchy in river management.[102]

Easements were only one of the uncertainties that confronted the individual seeking a legal remedy for pollution. He could not be sure he was suing the right party until the case was over; nor was it clear who had prescriptive right to pollute until that right was tested in court.[103] He could expect no public inspectorate to prosecute even when statutes were broken and even if judgment were favorable he would have to press for enforcement.[104] Litigation was enormously expensive and might take years to complete. In 1841 the Duke of Buccleuch and two others brought proceedings for pollution against a chemical works on the River Almond in Scotland. In 1860, having spent £6,000, they got a favorable verdict.[105] Asked why he had never brought an action for pollution, Knaresboro water works manager Lambert Ellison replied: "I might have to go to Chancery, and I am too old to go to Chancery."[106] Once in court, litigation might fail

---

Law relating to Pollution, pp. 95-96. On prescriptive rights in the U.S. see E.B. Phelps, Stream Sanitation (New York: John Wiley, 1944), pp. 14-15.

[102] RPPC/1865/1st Rept., pp. 11-12; RPPC/1865/3rd Rept., pp. li-liii; "The pollution of rivers," The Field, 12 July 1873, p. 33; The Builder 33(1875): 591-92.

[103] S.C. Sewage of Towns (Metropolis), Q 3290; Jabez Hogg, "River pollution with special reference to impure water supply," JRSA 23 (1874-75): 579.

[104] S.C. Sewage of Towns (Metropolis), QQ 3926-29.

[105] RPPC/1868/1st Rept., pp. 37-38; RPPC/1868/4th Rept., (The Rivers of Scotland), P.P., 1872, 34, [C.-603.], pp. 62-66, 333.

[106] RPPC/1865/3rd Rept., Evidence, p. 180. The labyrinth of Chancery is the subject of Dickens' Bleak House, which would tend to confirm Ellison's fears.

on some obscure point of law. Having for years tried to force Tottenham to treat its sewage effectively, the Lea Conservancy lost its suit because the Tottenham Local Board had changed its name to the Tottenham and Wood Green Local Board.[107] Litigation was indeed "invidious, expensive, and doubtful in its results," and it is remarkable -- and testament to the dreadful conditions that existed -- how much litigation was undertaken.[108]

As early as 1847 there were attempts to reform river law.[109] New laws addressed not simply the problem of catching polluters, but sought to provide definite rules for river use -- to make clear what degree of pollution was permissable, establish mechanisms for river governance and pollution prosecution. In 1876 a comprehensive Rivers Pollution Prevention Act passed, emasculated product of a decade of parliamentary compromise. A Rivers Protection Bill had been introduced in 1865 as a private member's bill and had spurred establishment of the first Rivers Commission which, with its successor, would spend the next eight years investigating river pollution. Recommendations of both commissions were embodied in the 1872 Public Health Bill, in successively weaker bills in 1873 and 1875, and finally in the "weakling of a very pronounced type" that passed in 1876.[110]

---

[107] "Pollution of the River Lea," SR n.s. 12(1890-91): 90-91, 416.

[108] RPPC/1865/3rd Rept., p. li.

[109] Pentelow, River Purification, in passim; Saunders, "Legislation for the purification of rivers and its failures," pp. 343-47.

[110] Bailie Dechan, "River Pollution," SR n.s. 20(1897): 538.

The 1876 Act prohibited dumping into rivers solids which interfered with flow; sewage; poisonous, noxious, or polluting liquids from factories; and liquids and solids of the same description from mines. There were qualifications, however. For existing works, such dumpings were permissable provided "best practicable and available means" were used to purify sewage, or "best practicable and reasonably available means" were used to purify manufacturing or mining wastes. There were inspectors but their job was not to find violators, but to protect towns, factories, and mines from prosecution. Legal action could not be taken under the Act without an inspector's sanction, and when granting this he was to consider "industrial interests involved in the case" and "circumstances and requirements of the locality." Action could be taken against an industry only if "no material injury will be inflicted by such proceedings on the interests of such industry."[111]

The Act failed to provide clear guidelines for polluters, an efficient means of applying the law, or any mechanism for river governance. A chemical definition of pollution drawn up by the second commission was included in 1872 and 1873 versions of the bill, left out thereafter. Absence of a statutory definition of pollution perpetuated the uncertainty in pollution litigation since the definition of pollution (as well as "best means" decisions) were left to county court judges. Applying the law was difficult because proceedings could only be brought by sanitary authorities who, as owners

---

[111] Higgins, A Treatise on the Law relating to Pollution, pp. 25,42.

of sewers, were among the law's principal offenders (19 of the 53 proceedings brought during the first six years of the Act were against sanitary authorities).[112] River governance remained chaotic. Common law rights and abuses, a sacred part of English property rights, took precedence over the Act. Although both rivers commissions had recommended establishment of a conservancy board to govern each river basin, the Act made no provision for these. Several attempts during the 1880s to obtain legislation for such boards were unsuccessful, although the 1888 County Councils Act, which allowed county river boards to enforce the 1876 act, was significant in providing a means whereby regional interests could overcome local squabbles.[113]

The Act's vagueness reflects the impossibility of reconciling differences among the various parties with interests in river use. In essence its message was that except where certain river uses were common law rights and except for the few cases where better technical solutions existed than those in use, rivers would serve the short-term interests of society in accordance with traditions of political economy, and as interpreted by county court judges. If it was weak, the Act was realistic. Parliament had not been deceived by the

---

[112] William Burchell, River Sanitation, pp. 19-22; SR n.s. 11 (1889-90): 418; "The River Thames," Van Nostrand's Eclectic Engineering Magazine [VNEEM] 19(1878): 345; "'Prestonensis' to SR," SR 7 (1877): 289.

[113] Henry Robinson, "River pollution," SR n.s. 6(1884-85): 393; C.E. Saunders, "Legislation for the purification of rivers and its failures," pp. 343-47. See also note 67.

Chadwickian illusion -- perpetuated by both Rivers Commissions and the Sewage of Towns Commission -- that universal recycling would erase conflicts over river use. It recognized that where technology was inadequate, prohibiting pollution was futile, given the momentum of industrial and urban growth. As solicitor Samuel Gael pointed out, the law was powerless to stop London from dumping sewage into the Thames: purification of London's sewage was

> a thing that the law cannot produce; it prohibits the nuisance, but when the arrangements are such that you cannot stop them, it becomes a case beyond the power of the law to check.[114]

Failure of the 1876 Act left resolution of river-use conflicts in much the same state as it had been before the Act. Disputes were resolved not through appeal to national or regional river-use principles, but by whatever backstreets means were available: case-by-case in courts; through threats, deceptions, deals, stalling, dirty politics. Ruinous litigation remained a constant and unpredictable threat. On the Tyne, manufacturers were said to "live in peril" of it.[115] Towns too were constantly at legal hazard for dumping sewage, or as they saw it, "for having effected the purpose which they were appointed to carry out."[116] The Metropolitan Board of Works, dumper of London's sewage, quietly accumulated evidence of dumping by others in the Thames estuary, evidence that would be useful if it were

---

[114] S.C. Sewage of Towns (Metropolis), Q 3287.

[115] "Rev. of Smith's Report to the Local Government Board," CN 45(1882): 220.

[116] RPPC/1865/3rd Rept., Q 10916.

prosecuted.[117] Tottenham managed to have a suit against it transferred to a higher court, where higher legal costs would strap the finances of its nemesis, the Lea Conservancy Board.[118] Upper Thames towns obtained successive waivers from the court-imposed demands of the Thames Conservancy that they treat sewage and avoided paying the huge fines the courts awarded the Conservancy.[119] Staines resisted the Conservancy's efforts by managing not to own the sewers where they flowed into the river.[120]

Such haphazard, unfair, and expensive mechanisms for settling disputes about river use had great effect on scientists employed as witnesses in pollution trials or who certified waters good or bad, technologies effective or ineffective. On the one hand, scientists' discussions assumed the base level at which river use was disputed: the employment of scientists primarily as makers of excuses is a main theme here. Indeed, scientists were in part responsible for the perpetuation of the uncertainty of litigation. Robert Rawlinson noted of pollution litigation:

> An inquiry of that kind is very redundant, costly, and unsatisfactory. With reference to the opinion of medical men, I am sorry to say, you will get the foulest nuisances that can be created in this country supported by professional witnesses who will declare it is no

---

[117] GLCRO/MBW 683, "Minutes, sub-committee on the sewage of the metropolis," 30 Sept. 1885, pp. 704-10.

[118] PRO HLG 50 2079, file 38948 #137.

[119] [Greville Ffennell], "Is sewage detrimental to fish," The Field 39 (1872): 466; "The pollution of rivers," VNEEM 5(1871): 131; [Greville Ffennell], "The Thames," The Field, 30 Jan. 1869, p. 89.

[120] SR n.s. 12(1890-91): 559-60.

> nuisance at all, and when a case has to be supported by such evidence, there is litigation piled upon litigation, and evidence which ought to make the persons blush who give it; and the magistrates cannot convict for the most notorious nuisance.[121]

Moreover, and a main reason for the confusion of British river-pollution science in the late nineteenth century, many of the issues scientists were discussing, both in litigous contexts such as trials or parliamentary select committees, and in the non-litigous arenas of scientific societies, were issues of social utility not of science. By failing to define pollution, the 1876 Act made social utility, not fact, basis for making river policy, just as common law made rights, not facts, basis for judgments. The kinds of questions scientists were asked -- was an effluent acceptable, a water safe, or a purification technique effective -- were in fact trans-scientific questions, matters requiring judgments of social utility which masqueraded as scientifically solvable problems.

The concept of river self-purification that emerged from this social and ideological background was, like the ideal of recycling, a reason why river-use disputes ought not to exist. There was a crucial difference, however. The recyclers claimed that with thorough changes in municipal and regional administration, legal powers and responsibilities; with a re-definition of self-interest; and with opportune technical breakthroughs, pollution and accompanying disputes need not exist. By contrast, the illusion of self-purification

---

[121] S.C. on the Lea Conservancy Bill, Evidence, P.P., 1867-68, 11, (306.), Q 950.

was used to defend the status quo. Since whatever refuse was put in a river quickly disappeared, no problem really existed, and conflicts were illusory. As Edward Frankland recognized, "two classes of persons [polluters and river-water sellers] . . . were chiefly instrumental in the origination and diffusion of this opinion."[122]

---

[122] Edward Frankland, "On the spontaneous oxidation of organic matter in water," J. Chem. Soc. 37(1880): 517.

II. ROUTES TO REPUTATION: SCIENCE IN AN ADVERSARY CONTEXT

> Litigation is, in my experience -- although I may have derived advantage from it -- a more expensive luxury than sewage treatment.[1]
>
> --Charles Meymott Tidy

Reflecting on recent "triumphs of chemistry," R. Green, clerk of the Kensington vestry, remarked in 1858 that it was "only reasonable to infer, that the professors of science, on proper inducements being held out to them, would be able to resolve the great difficulty of the sewage question."[2] Living amid manifest technical prowess, many of Green's contemporaries agreed: scientists could solve sewage and water problems and do so in accordance with the principles of political economy, individual rights, and social morality. The role science took was indeed directed by the inducements held out but these inducements were toward participation in political and legal disputes rather than toward solving scientific problems.

River pollution and the associated water-and-sewage problems were the province of engineers, chemists, and medical men.

---

[1] Charles Meymott Tidy, "On the treatment of sewage," JRSA 34(1885-86): 621.

[2] In Douglas Galton, James Simpson, and Thomas E. Blackwell, Report on the Metropolitan Main Drainage, P.P., 1857-58, sess. 2, 36, (233.), p. 260. Also see Royal Commission on Rivers Pollution, 1865, 3rd Report (The Rivers Aire and Calder), Vol. II, Minutes of Evidence, P.P., 1867, 33, [3850.-I], QQ 3574, 3600. To be cited as RPPC/1865/3rd Rept., Evidence.

Biologists, insofar as such a discipline existed and excluding bacteriologists, were involved in water-and-sewage matters during the 1850s, but not again until the second decade of the twentieth century.[3] Although incidents of proprietary feather-ruffling between professions occurred occasionally between 1850 and 1900, interprofessional conflicts (with a few exceptions that will be considered) did not significantly affect river pollution science. Indeed, professional boundaries were remarkably fluid: the Chemical Society heard papers on bacteriology; the Civil Engineers heard papers on analytical chemistry.[4] The scope of sanitary engineering, itself a product of the public health movement of the 1840s, was particularly broad. Engineers such as Thomas Hawksley and Baldwin Latham spoke with authority on epidemiology or the etiology of water-borne diseases.[5]

Chemists are our main focus here. Before bacteriology, chemistry held a position of authority as regards the sanitary quality of the environment; after the discovery of germs several important British chemists took up bacteriology. Wakefield surgeon James Fowler

---

[3] In Germany, France, and the United States, biologists' participation appears to have been more important and more continuous.

[4] See for example Frank Hatton, "On the action of bacteria on gases," J. Chem. Soc. 39(1881): 247-58; Charles W. Folkard, "The analysis of potable water with special reference to previous sewage contamination," MPICE 68(1881-82): 57-115.

[5] Baldwin Latham, A Chapter in the Local History of Croydon. Presidential Address to the Croydon Natural History Society (London: West, Newman, 1909); Thomas Hawksley in disc. of E. Bryne, "Experiments on the removal of organic and inorganic substances in water," MPICE 27(1867-68): 16-21.

apologized, "I can only give you my opinion as a medical man, not as a chemist."[6] Although there were occasional objections to chemists making essentially medical decisions, chemists were nevertheless besieged by requests to pronounce on whether a water or a sewage effluent was good.[7] The investment of authority in chemistry was despite some blatant failures to provide correct guidance, and despite admissions of some chemists that they could neither make bad water good nor tell when it became so. This clinging to chemistry reflects several factors: an intoxication with precision whatever its insignificance, a preoccupation with ultimate answers, a fascination with the chemical transformations which made matter either agent of disease or agent of fertility. Moreover, anxieties about water-and-sewage matters were such that an authority on these issues was necessary -- had there been no chemistry to provide this authority it would have been necessary to invent one.

Chemists, and experts generally, had two functions with regard to water-and-sewage questions: first, to find answers, and second, to explain, confirm, justify, lend prestige to, or advocate answers already found, and answers which essentially were political rather

---

[6] RPPC/1865/<u>3rd Rept., Evidence</u>, Q 508.

[7] For an example of a controversy on this issue see <u>BMJ</u> 4 April 1868, p. 331; 18 April 1868, p. 391; 25 April 1868, p. 413 (letters of William Odling and responses of the editors). In fact many of the chemists involved in water-and-sewage matters had medical training, viz. William Odling, M.D., 1851, London (W.H. Brock, "William Odling," <u>D.S.B.</u> 10, p. 177), Henry Letheby, M.B., 1842, London (<u>D.N.B.</u> 11, p. 1010), and Charles Meymott Tidy, M.B., 1851, Aberdeen ("Charles Meymott Tidy," <u>J. Chem. Soc.</u> 63(1893): 766).

than scientific. An example of the first function is the service of chemists on royal commissions. J.T. Way and J.B. Lawes served on the Sewage of Towns Commission (1857-64), Way on the first Rivers Commission (1865-68), Edward Frankland on the second Rivers Commission (1868-74), A.W. Williamson on the Metropolitan Sewage Discharge Commission (1882-84), and James Dewar on the Metropolitan Water Supply Commission (1892-94). Such commissions were to make intensive investigations of social problems and to suggest remedies for them. Ideally composed of individuals representing a range of professions with well-established reputations and no political interests in the issues they were to investigate, commissions were to be, in the highest sense, independent and non-partisan. Although many scientists, medical men, and engineers served on commissions, they were not always paid (other than expenses) for their service. The two Rivers Commissions were unusual in being compact, highly professional, salaried bodies: Frankland and Way received £800 per year plus expenses.[8]

It is the second function of chemists that is most important here. Chemists were active as witnesses in adversary proceedings: actual trials as well as quasi-litigious proceedings such as Local Government Board hearings and parliamentary select committees.[9]

---

[8] "The rivers commission," The Field, 30 March 1867, p. 227.

[9] Parliamentary select committees were convened to give intensive scrutiny to a bill. They held hearings at which witnesses were presented by the various interests involved in the bill, and examined and cross-examined by counsel representing those interests and by committee members. Occasionally select committees would be established to investigate a particular issue, sewage recycling, for

Chemists made analyses and interpreted them, testified on the methods by which the analyses were made, and the assumptions made in interpreting them. As one contemporary complained, it was

> not at all infrequent in disputed cases in the law courts for chemists to give evidence with regard to the supposed condition of streams; when one would state the water was dirty, and another would state it was beautiful, and both had made analyses, and both were men of science.[10]

A great deal of money was spent on such testimony. A.W. Hofmann, the Liebig student who presided over the Royal College of Chemistry, estimated an expert witness could make £8,000 to £9,000 per year in Britain.[11] During its eight-year existence, the Lower Thames Valley Sewerage District spent £18,000 on scientists and engineers, much of it for testimony before parliamentary select committees and Local Government Board hearings and for investigations to support that

---

example, rather than a given bill. Private bills, those introduced by individuals, towns, or corporations desiring some special powers -- say the right to use water from a river for town supply -- were expensive, and one of the advantages of sanitary legislation such as the 1848 Public Health Act was that expense could be lessened by having the principal inquiry being done at local hearings by Local Government Board engineering inspectors (or those of antecedant organizations). Again witnesses representing various interests were examined (P. and G. Ford, A Guide to Parliamentary Papers, 3rd ed. [Totowa, N.J.: Rowman and Littlefield, 1972], pp. 5-6; Royston Lambert, "Central and local relations in mid-Victorian England: The Local Government Act Office, 1858-71," Victorian Studies 6(1962): 121-50; Robert M. Gutchen, "Local improvements and centralization in 19th century England," Historical Journal 4(1961): 85-96).

[10] S. Haughton in disc. of William Odling, "On the chemistry of potable water," JRSA 32(1883-84): 979-80.

[11] W.H. Brock, "The spectrum of scientific patronage," in G. L'e Turner, The Patronage of Science in the 19th Century (Leyden: Noordhoff, 1976), p. 186.

testimony.[12] Water bills were called a "milch cow for the parliamentary bar and for the experts."[13]

There is a good deal of evidence indicating how much money chemists actually did make. The cornerstone of a water chemist's practice was water or sewage analysis. Fees for analyzing a single sample ranged from 2-3 shillings for an analysis done at a local chemist's shop to 10 guineas for a complete analysis, including several minerals, done by a top analyst of the stature of J.A. Wanklyn or Edward Frankland.[14] Frankland, who had an extremely active water-analysis practice, charged five guineas for a normal sanitary water analysis (organic carbon and nitrogen, chlorine, nitrogen as nitrites and nitrates, ammonia, total solids, and permanent and temporary hardness), but dropped the fee to two guineas if the work was for science or for a public body such as a sanitary authority.[15]

More common than single analyses, especially in adversary proceedings, were mini-investigations: analyses of a few samples, possibly a personal examination of their source, and a report,

---

[12] "The Lower Thames Valley Sewerage District," SR n.s. 7(1885-86): 427.

[13] Herbert Preston-Thomas, The Work and Play of a Government Inspector (Edinburgh: William Blackwood and Sons, 1909), p. 161.

[14] Brock, "Spectrum of patronage," p. 182; "Water analysis," CN 55 (1887): 154; J.A. Wanklyn, "Chemical analysis for the Local Government Board," CN 35(1877): 105; Edward Frankland to T. Mellard Reade, 19 December 1884, Liverpool University, Reade Papers, #645.

[15] PRO MH 29 19, #47529/94, St. Mary Abbots, Kensington, to LGB, 8 May 1894; PRO MH 29 2, #16558/75, Frankland to LGB, 8 March 1875.

presumably based on the results of the investigation and representing the wisdom and experience of the analyst. Having asked Frankland to run a special check on the water of one of the London companies, the Local Government Board was astonished to receive a bill for 50 guineas for a one-page report of a single sample. Frankland, they learned, had consulted Huxley (without authorization) and investigations of "special and searching character" commanded 25 guineas for each.[16] In an 1893 letter to H.E. Armstrong, who was embarking of a sub-career as a sewage expert, Frankland explained his fees: 100 guineas for an examination and report or 10 guineas per day if less than three days, plus railway and hotel expenses, and five guineas per water or sewage analysis, three guineas per analysis of mud.[17] Little wonder that among the reasons for celebrating the arrival of bacteriological water analysis in the mid-1880s was the belief that it would bring relief from chemists' fees.[18]

Given these fees, and the large number of water-and-sewage-troubled towns, it is likely that analyzing and advising on water-and-sewage matters was a source of substantial income for chemists.

---

[16] PRO MH 29 2, #13119cc/75; Frankland to LGB, 24 Feb. 1875, #16558/75; Frankland to LGB, 8 March 1875, #20986/75; Frankland to LGB, 20 March 1875. See also RPPC/1865/<u>2nd Report, (the River Lea), Evidence</u>, <u>P.P.</u>, 1867, 33, [3835-I], Q 5020.

[17] Royal Society of London, Miscellaneous Mss, MM10 #98, Frankland to H.E. Armstrong, 26 July 1893. The letter appears to read 10 guineas per day for investigations less than three days, though 30 would make more sense.

[18] C.E. Cassall in disc. of Louis Parkes, "Water analysis," <u>TSIGB</u> 9(1888): 392.

In establishments like Frankland's the actual work was done by assistants (often students) who worked long hours for stipends which could be lived on but represented a very small portion of the laboratory's income.[19]

From occasional analyst of waters to highly paid, much-in-demand, sewage-and-water witness was a huge step. Water-and-sewage matters were lucrative for men like Frankland, Charles Meymott Tidy, and J.A. Wanklyn thanks to their reputations -- but for what? and from what work? The kinds of behavior that brought a scientist reputation mirror the views of river use of various segments of society. The last thing wanted from an expert was an ambivalent response. He had to be decisive, to back that decisiveness with facts (not necessarily with all the facts), arguments, and confidence. Moreover, his answers to water-and-sewage policy matters had to have some foundation in moral principle. The most successful experts managed to combine activism with authority. Ironically, a chemist's success in the water business was as much a product of his infamy as of his fame.

Here Edward Frankland (1825-1899), most famous and most infamous of the Victorian water scientists, is the best example. Although Frankland dated his interest in water-and-sewage matters from the late 1840s, the period of his career during which he was influential began only in 1865, when he was appointed analyst of the London water supply

---

[19] The experiences of a staffer of Frankland's water laboratory are chronicled in six letters from J.J. Day to H.E. Armstrong between November 1867 and March 1869 (Imperial College, Armstrong papers, series 2, C241-C246).

by the Registrar General. Like many other positions in public-health officialdom, the analyst's position was ambiguous with regard to the posture of its incumbant toward sanitary reform: ought he to be a reformer, an intervenor only in crises, or a referee in disputes between reformers and companies that supplied water? Despite a long series of investigations of London water, this issue had never been effectively confronted. In 1851 the "government commission," chemists A.W. Hofmann, Thomas Graham, and W.A. Miller, undermined the efforts to reform the London water supply of another government body, Chadwick's General Board of Health, which opposed use of the Thames and favored a publicly owned supply. Both bodies were official, but the Board of Health saw its function as reform, while Hofmann, Graham, and Miller were concerned only with the adequacy of the existing supply. Another "judicial" study (the term was used by defenders of the London water companies to describe studies concerned with adequacy rather than reform; equally studies which found the existing supply adequate) was done in 1856 by Hofmann and Lyndsay Blyth, who found the supply much improved due to the upstream relocation of intakes required under the 1852 Metropolis Water Act. Activism crept back into official reports on the London supply after 1857, when physician-chemist Robert Dundas Thomson, lecturer in chemistry at St. Thomas' Hospital and medical officer of health for Marylebone, began regular monthly analyses of the metropolitan supplies for the Registrar General, the post to which Frankland would eventually be appointed.[20]

---

[20] "Reports on the examination of Thames water," CN 47(1883): 31;

The format Thomson used to report his analyses compared Thames water with distilled water, a not-too-subtle reminder to Londoners that the quality of their water could be improved.[21]

Thus, both in the traditions of the office to which Frankland was appointed in June 1865 and in official sanitary investigations in general, there was toleration, but not an expectation of activism. As will be detailed in later chapters, Frankland was extremely aggressive in using the water analyst's position as a vehicle for promoting water-supply reform.

More important than his highly visible position as the London water analyst in forging Frankland's reputation, however, was his six-year service on the second Rivers Pollution Commission. In 1865, under pressure from the fisheries lobby, the Liberal government had established a three-man commission to tell it how to end river pollution without harming industry. When that commission dissolved itself due to internal bickering in January 1868, the Tory government replaced it with a new commission, chaired by Sir William Denison, a military engineer and colonial administrator, and including John Chalmers Morton, an agriculturalist, and Frankland.[22]

By this time Frankland had become strongly opposed to domestic

---

"Robert Dundas Thomson," D.N.B., 19, p. 748.

[21] George R. Burnell, "On the present condition of the water supply of London," JRSA 9(1860-61): 169-77; RPPC/1868/6th Report (The Domestic Water Supply of Great Britain), P.P., 1874, 33, [C.-1112], p. 250.

[22] On the demise of the first commission and establishment of the second see PRO HO 74 3, pp. 409-77; The Builder 26(1868): 59.

use of river water. He also believed that irrigation and analogous techniques could be used to recycle most municipal and industrial wastes, regarded precipitation sewage treatments as ineffective and frequently fraudulent, and advocated strong and definite anti-pollution laws. The Rivers Commission proved an excellent vehicle for the development and dissemination of these radical views. It gave Frankland £800 per year plus a laboratory and £700 per year to spend on it.[23] Owing to the death of Denison in 1871 and to Morton's ignorance of chemistry and medicine, Frankland's views dominated the commission's six reports: on Lancashire rivers (1870), on the ABC sewage precipitation process (1870), on pollution by the woollen industry (1872), pollution in Scotland (1872), pollution from mining and metallurgical industries (1874), and national water supply (1874).

These reports established Frankland's reputation as a sewage-and-water radical and as the most knowledgable authority on sewage-and-water matters. In the first place, the commission's work was exceedingly well done. Its inquiries were well-focused, its reports concise. Where the reports of the first commission had been bundles of opinions, those of the second were bundles of facts -- reports of several thousand analyses, comprehensive statistics on the sewage-and-water dealings of hundreds of towns and industrial plants, well-designed and relatively thorough experimental and statistical investigations.[24] In part, therefore, Frankland's reputation rested on the

---

[23] "The Rivers Pollution Inquiry," The Field, 11 April 1868, p. 286.
[24] "The pollution of rivers," Nature 9(1874): 197.

quality of the commission's work. It was with this great compendium of evidence that he supported his radical views on water-and-sewage policy.

Aside from the quality of the work, however, the very imprimatur of "Royal Commission" -- with its implications of political disinterestedness -- gave Frankland's views a prestige which an individual chemist could not hope to acquire. Frankland exploited this situation adeptly. He represented the commission's conclusions as confirming his own, to the outrage of opponents who pointed out that, for practical purposes, Frankland and the Rivers Commission were identical.[25] Yet even while he sustained the illusion of independent inquirers (himself and the commission) coming to identical conclusions, Frankland was exploiting his commissioner (or later his ex-commissioner) status. Parts of the commission's reports appeared in Frankland's collected papers.[26] As late as 1891 Frankland was sending

---

[25] _CN_ 64(1891): 222-23; "London water supply," _CN_ 45(1882): 180-81; C.M. Tidy, "The treatment of sewage," _VNEEM_ 35(1886): 3; Tidy in disc. of Percy Frankland, "The upper Thames as a source of water supply," _JRSA_ 32(1883-84): 446. This was also recognized by those who wished to give Frankland credit for his work (W. Dunbar, _Principles of sewage treatment_, translated with author's sanction by H.T. Calvert [London: Griffin, 1908], p. 23; George Rafter, "Sewage irrigation," _U.S.G.S. Water Supply Paper #3_ [Washington: G.P.O., 1897], pp. 29, 45-48; Theophile Schloesing, "Assainissement de la Seine. Epuration et utilisation des eaux d'egout. Rapport fait au nom d'une commission," _Ann. D'Hygiene Publique et de Medicine legale_, 2nd ser. 47(1877): 207).

[26] Edward Frankland, _Experimental Researches in Pure, Applied, and Physical Chemistry_ (London: van Voorst, 1877), pp. 662-74, 685-700, 709-829. See also Frankland, "Annual Report for 1869," in _Reports on the Analysis of Waters supplied by the Metropolitan Water Companies during 1869, 1870, and 1871, by Professor Frankland; copy of his letter to the Registrar General, dated 10th July 1869, and analyses of the Metropolitan water supply for October 1871 and January 1872_, _P.P._,

out the Commission's "Memorandum #3" (in part a polemic defending controversial aspects of Frankland's analytical procedures, in part instructions for interpreting analytical results) with reports of analyses he was doing as a private consultant.[27] As late as 1892 he was using Commission stationery to report analytical results.[28]

Much of Frankland's success as a water analyst was probably due to his connection with the commission. As commissioner he had supervised analyses of the water supplies of most of the towns in England, Wales, and Scotland. He continued to serve as London's official water analyst until his death in 1899 and it is probable that much of the business that came to him in private practice came under the impression that he was particularly safe, his official status reflecting a disinterestedness and a degree of reliability to which other chemists were not subject.[29]

Frankland was also successful as a witness in water-and-sewage matters. Here his value came from his enormous knowledge, membership in the scientific elite, and experience under cross-examination. Perhaps most importantly, Frankland bore the halo of a royal commissioner. This made him the perfect witness: one whose position

---

1872, 49, (99.), p. 6. To be cited as Frankland, Metropolitan Water Supply Papers.

[27] PRO MH 29 15; #49387/91, Frankland to LGB, 30 May 1891. From the printer's mark on this form it appears that 500 copies were made up in 1884, a decade after the commission had ceased to exist.

[28] PRO MH 29 16; #13047/92, Frankland to LGB, 8 Feb. 1892.

[29] Frankland asked to be included in an LGB directory as a water analyst; the circumstances of this are not clear (PRO MH 29 16; #90352/91, Frankland to LGB, 19 Oct. 1891).

represented impartiality, but whose impartiality was well-known to be of a particularly radical variety. In a legal or political situation where the views of one of the parties corresponded with the views of the second Rivers Commission, Frankland's testimony was immensely valuable. It is noteworthy that the heart of Frankland's reputation was not the thoroughness with which he investigated each new case, or the wisdom, experience, and open-mindedness he brought to it, but instead, his dependability: Frankland could present certain points of view on water-and-sewage matters more effectively than anyone else.

The nature of Frankland's reputation is highlighted by contrasting it with the reputation of Charles Meymott Tidy, who made his mark opposing the positions Frankland advocated. After taking an M.B. at Aberdeen in 1866, Tidy (1843-92) became assistant to Henry Letheby, professor of chemistry at the London Hospital. When Letheby died in 1876, Tidy took his place, eventually becoming professor of chemistry, public health, and medical jurisprudence. In the early 1880s Tidy began legal studies and in the middle of the decade was admitted to the bar and appointed reader in medical jurisprudence at the Inns of Court.[30]

When Tidy succeeded Letheby at the London Hospital he also took over a series of analyses of the London water supply Letheby had been carrying out since 1864, ostensibly on behalf of the Association of Medical Officers of Health, but financed by the water companies.[31]

---

[30] "Charles Meymott Tidy," J. Chem. Society 63(1893): 767.

[31] Royal Commission on Water Supply, Evidence, P.P., 1868-69, 33, [4169.], QQ 3861, 3909.

In 1880 the veneer of objectivity was lost when the medical officers ended their sponsorship.[32] In 1880 William Odling, professor of chemistry at Oxford, and William Crookes, editor of Chemical News, joined Tidy in undertaking the analyses, which were now justified with the argument that it was necessary to counter Frankland's extremism.[33] By this time both Frankland, and the Crookes, Odling, Tidy team were using the same analytical processes and getting similar results, the companies' analyses showing the water even slightly more impure than Frankland's did.[34] What differed were the interpretations: Frankland's monthly reports contained frequent reminders that there was no protection from water-borne disease; Crookes, Odling, and Tidy struggled to find new superlatives with which to describe the quality of the supply in each succeeding report.[35]

Tidy's reputation was based on the existence and the trustworthiness of self-purification, that "no matter how foul or dangerous water is originally, it will become innoxious if allowed to stand or to run for a certain distance."[36] (Notably, the time or distance

---

[32] "Association of Medical Officers of Health," SR n.s. 2(1880-81): 224.

[33] PRO MH 29 4; #52429/83, Crookes, Odling, Tidy to LGB, 17 May 1883.

[34] W.C. Young, "A comparison of the organic carbon and nitrogen results obtained by Dr. Frankland and the companies' analysts from the waters supplied by the metropolitan water companies," The Analyst 20 (1895): 159-64.

[35] "Impure London water," BMJ, 12 November 1892, pp. 1069-70.

[36] "Biological examinations of water supplies," SR n.s. 8(1886-87): 120.

required for purification was invariably less than the time the water had stood or the distance it had flowed.) As with Frankland there was a substantive basis for Tidy's reputation. During the late 1870s and early 1880s he published important studies on water analysis and river self-purification.[37] As with Frankland, Tidy found a moral foundation for his position: it was unethical to deceive the public by insinuating that the water was bad when there was no actual evidence showing it so -- in Tidy's view what Frankland was doing.[38]

Thus, where Frankland was the person to get to condemn domestic use of river water and, correspondingly, to affirm the inadequacy of precipitation sewage treatments, Tidy was the person to get to uphold the safety of polluted river water and the acceptability of precipitation effluents. Unlike Frankland, Tidy's success owed much to his glibness, common touch, and willingness to make definite, if outrageous statements. As late as 1886 Tidy scorned the existence of cholera and typhoid germs.[39]

As Tidy's career in particular reveals, reputation was not to be gained from moderation. The characteristics of the successful scientific advocate were different from those of the successful

---

[37] Tidy, "River water," J. Chem. Soc. 37(1880): 267-327; idem., "The processes for determining the organic purity of potable waters," J. Chem. Soc. 35(1879): 46-106.

[38] Tidy, The London Water Supply, being a report submitted to the Society of Medical Officers of Health on the Quality and Quantity of the water supplied to London during the past ten years (London: Churchill, 1878), pp. 3-4.

[39] S.C. Rivers Pollution (River Lea), Evidence, P.P., 1886, sess. 1, 11, (207.), QQ 3796-97.

scientist. Railing against the "professionalism" implicit in the establishment of the Institute of Chemistry, Nature listed the "qualities most valuable in an expert":

> coolness under cross-examination, verbal dexterity, a ready wit, not too much knowledge or conscience, the fidelity of a partisan . . . and a dash of impudence are quite as frequently the passport to the 'professional eminence' of an expert as scientific ability.[40]

The public image of science suffered greatly as a result of scientists' participation in adversary water proceedings. Chemistry could solve water problems, it was felt, but chemists were dealing unconscionably with the public. They made their livings promoting confusion in sanitary litigation while the real sanitary problems remained unsolved. The contempt for experts common in the sanitary literature is evident in Henry Cole's comment on the Metropolitan Board of Works' response to allegations that it was responsible for polluting the lower Thames:

> They set up all kinds of men of science who are, no doubt, very profound; and treat the question so microscopically and scientifically that you get into such a fog that you do not know what to do. They get science to prove that this stuff going into the river does no harm at all. No doubt the next thing that they will do will be to prove that it is the odour of roses, and I dare say that I shall get so confused on the subject that I shall believe it.[41]

---

[40] "The whole duty of a chemist," Nature 33(1885-86): 74. For a similar view of modern expert witnesses see Philip L. Bereano, "Courts as institutions for assessing technology," in William A. Thomas, ed., Scientists in the Legal System -- Tolerated Meddlers or Essential Contributors (Ann Arbor: Ann Arbor Scientific Publications, 1974), p. 88.

[41] In disc. Health and Sewage of Towns Conference, JRSA 26(1877-78): 732.

Similarly, with regard to water supply, The Field observed that Londoners were submitting "to extortion after extortion without resistance and [are] left a witless, helpless prey to every scientific humbug."[42] The Field scorned the altruism which invariably accompanied scientists' statements:

> All, too, is done for virtue's sake. There are no base interested reasons for taking all this trouble; all is philanthropical, paternal, Pecksniffian. Health, cleanliness, purification, moral and physical beauty, the general good, science, art, progress and civilization -- these form our formula when we want to convert our neighbour's goods to our own uses.[43]

"Incontestable evidence" could be procured that sewage was not harmful, noted Field correspondent Greville Ffennell. Upon hearing such claims he suggested readers recall the testimony of one engineer: the "water was as pure as ever -- it was only that which was in it that was objectionable."[44] At its most extreme this attitude was fundamentally anti-science. The Builder complained that Frankland's Rivers Commission reports left water issues too ambiguous. What was needed was an engineering report, stating exactly what needed to be done.[45] Science, observed novelist Thomas Peacock, was "an edged tool, with which men play like children and cut their fingers."[46]

---

[42] "The water-supply of towns," The Field, 3 June 1871, p. 441.
[43] Ibid.
[44] [Greville Ffennell], "The Thames," The Field, 7 August 1869, p. 121.
[45] "The water supply and the Rivers Pollution Commission," The Builder 33(1875): 1095-96.
[46] Thomas Love Peacock, Gryll Grange (New York: AMS Press, 1967), pp. 186-87.

Judges and commission chairmen confirmed the public's impression that employment of scientists in water-and-sewage disputes merely confused matters. Justice Watkin Williams, who presided over the Lea Conservancy's 1884 prosecution of Hertford for polluting the Lea, found Tidy's glowing description of a sewage-polluted ditch "such a flat contradiction of my nose and eyes that I distrust myself altogether."[47] On another occasion Williams remarked, "there are some things which neither engineers nor chemists can drive out of the head of a common person."[48] No one who was not "a chemist and a witness" could think the river anything but polluted, commented a Lancashire judge.[49] G.W. Hastings, M.P., noted that in all the sewage testimony he'd heard in seven years in the House of Commons he had only believed the experts on two occasions and had never felt right about allowing towns to spend money on the basis of experts' claims.[50]

The critics were right. A great deal of the testimony of expert witnesses was devoted to the establishment or destruction of credibility, to the presentation of vast quantities of evidence which might be irrelevant but were certainly impressive, or to discussion of reasons why such and such must be so or could not possibly be so. Perhaps the most blatant examples of the use of science to obscure

---

[47] Herts. Mercury, 23 February 1884.

[48] Ibid.

[49] The Builder 68(1895): 177.

[50] In disc. of R. Godfrey, "The Amines process of sewage treatment," TSIGB 10(1888-89): 217.

matters was the first series of hearings of the Royal Commission on Metropolitan Sewage Discharge, which adjudicated a dispute between the Metropolitan Board of Works, which dumped London's sewage into the lower Thames and claimed there was no pollution, and the Port of London Sanitary Authority which claimed there was pollution. The contending parties spent a total of £40,000 preparing and presenting their cases. Both sides collected huge amounts of data, but as Lord Bramwell, the commission's chairman, pointed out, much of these data seemed to have little connection to the issues. Such questions as the size of the river basin, about which there was virulent disagreement, seemed to Bramwell unconnected with the issue of whether the river stank.[51]

To Bramwell and many others expert witnesses seemed to have renounced the "philosophic spirit" which presumably characterized the way scientists tackled problems. Instead, they had made themselves an appendage of the legal profession. Now abstruse technical points as well as "all the resources of the English language" were used to mangle truth. The ethics of such practice caused considerable concern both among scientists and the public.[52]

Many felt scientists' contradiction reflected more than the

---

[51] Royal Commission on Metropolitan Sewage Discharge, Evidence, Vol. I, P.P., 1884, 41, [c.-3842-I.], QQ 15061-62; the £40,000 figure is from a solicitor's report (Greater London Council Record Office, London County Council, Main Drainage Committee, Presented Papers, Vol. 3, for 23 Jan. 1890, #5, p. 1).

[52] John Chalmers Morton in disc., Royal Society of Arts Rivers Pollution Conference, JRSA 23(1874-75): 84.

unresolved natures of the questions discussed, but was accentuated by some form of unethical behavior by scientists: manipulation of data, disregard for relevant data, employment of "a highly cultivated faculty for evasion."[53] One judge classified witnesses as "liars, damned liars, and experts."[54] Chadwick, outraged by the betrayal of engineer Thomas Hawksley, noted that in different select committee hearings during a single session of parliament, the same expert appeared advocating what were essentially opposite views.[55]

But what were the rules for scientific witnesses? An exchange between Edward Frankland and Board of Works' counsel George Parker Bidder, jr., during the 1883 hearings of the Metropolitan Sewage Discharge Commission is particularly enlightening in this regard:

> [Bidder]: "You have done something even worse or more inaccurate here; you have struck out all the low results, and so got a high average?"
> [Frankland]: "Yes; but you did not complain about me striking out the high ones in a former table."
> [Bidder]: "That is a very smart remark, but I would point out this. I did not ask you to strike out the high ones before; I suppose you had your own reasons for doing it."[56]

Appalled at so brazen a disregard for facts, Lord Bramwell asked Frankland why he had rejected the data. Frankland's reply was weak:

---

[53] "The whole duty of a chemist," p. 74.

[54] Ibid., p. 74.

[55] E. Chadwick in disc. of R. Angus Smith, "Science in our courts of law," JRSA 8(1859-60): 142-43; R.A. Lewis, Edwin Chadwick and the Public Health Movement (London: Longmans, Green, 1952), pp. 88, 120, 132-33, 274. See also Herbert Preston-Thomas, The Work and Play of a Government Inspector, p. 161; and "The strange evidence of Prof. Brande," Lancet, ii, 1851, p. 66.

[56] RCMSD, Evidence, Vol. 1, QQ 11028-30.

the data were clearly not good and even had they been accepted they would not have changed the average significantly. While Bramwell recognized the legitimacy of a scientific witness expressing the emphases of his employer, he drew the line at what seemed a falsification of the results of investigations.[57] For Bidder on the other hand, the expression of righteous outrage was feigned; simply a technique for destroying the credibility of an opposition witness. On another occasion Bidder objected to a Frankland average in which extreme and unrepresentative values were included (in that case to Frankland's advantage).[58] Reconstructing Frankland's perspective toward such practices is more difficult. Selecting data was necessary in pollution investigations because composition of polluted water varied enormously (by day of the week, time of day, recency of rainfall, etc.). Thus what appalled Bramwell and appeared to appall Bidder was a legitimate part of Frankland's job. The basis for selecting data was more problematic, however. It is probable that, as Bidder implied, political utility rather correspondence to nature was the

---

[57] Ibid., Q 11031. Bramwell was sympathetic when Odling was pressed to explain why his views of 1883 contradicted his views of 1858 (Q 16531).

[58] Thames Navigation Act, 1870. In Arbitration. The Thames Conservators and the Metropolitan Board of Works. Minutes of Proceedings, Report, and Determination of the Arbitrators, Evidence, QQ 4661-62. The importance of experienced counsel in hearings on scientific matters cannot be examined fully here. Bidder was probably the best, however: 7th wrangler at Cambridge in 1858, he was said to be able to multiply 15 figures by 15 figures quickly in his head ("The late George Parker Bidder, jr.," Engineering 61(1896): 202).

basis for Frankland's data selection. Here again, however, the mingling of science with issues of social policy confused matters. What natural phenomena were significant and deserving of measurement depended on one's views of river use. Typically the scientific studies of opponents of pollution were addressed to worst cases, while defenders were concerned with the normal state of affairs. During the hearings of the Royal Commissions on Metropolitan Water Supply of 1892-94 and on Water Supply of 1867-69 opponents of river water used science to prove that such water might, under extreme circumstances, cause an epidemic, while defenders used science chiefly to show the general good quality of the water.

A 1904 letter from William Crookes to James Dewar is particularly enlightening with regard to scientists' views on the ethics of their participation in adversary water politics. Crookes was complaining about not having been paid by the Lambeth company, one of the London water companies. He wrote: "but they are a rich company, and owing to our exertions they are about half a million in the pocket."[59] Since 1880, Crookes, first with Tidy and Odling, then with Dewar, had been analyzing daily samples of the waters supplied by the river water companies. This series of analyses had been one of the companies' main strategies for countering the bad publicity that resulted from Frankland's periodic condemnations. As Jabez Hogg observed, the sets of analyses were not merely contradictory, but "appear intended to

---

[59] Quoted in E.A. Fournier D'Albe, The Life of Sir William Crookes (London: MacMillan, 1923), pp. 377-78.

discredit each other." To discredit Frankland, Crookes, Odling, and Tidy employed the same methods Frankland used and drew attention to his extreme interpretations. Despite frequent complaints about the water and the efforts of public and private organizations to obtain public control of the supply, the companies held out until 1901 when, under an act of parliament, they were bought out for £41,000,000, a settlement highly advantageous to them.[60] One reason public takeover took so long to accomplish was the continual victories of the water companies on the scientific front, a result of the neutralization of Frankland's monthly reports with daily reports, and of the quality of the scientific and legal staff and the thoroughness of their preparation for Royal Commissions in 1867-69 and 1892-94 as well as periodic select committee hearings. Crookes' claim of partial responsibility for the advantageous settlement is therefore an admission that selling respectability or credibility as well as information or problem-solving skills were legitimate roles for a scientist.[61]

The main focus here, however, is not whether water-and-sewage scientists recognized a distinction between "truth" and "science for

---

[60] Jabez Hogg, "Chemical analysis of drinking water," JRSA 31(1882-83): 414. The figure of £43,000,000 is from S.E. Finer, The Life and Times of Sir Edwin Chadwick (London: Methuen, 1952), p. 408.

[61] Much the same comments might apply to Percy Frankland's water bacteriology (see Chapter 9). His biographer notes that "these results turned out to be of considerable value to the water companies of this country" (W.H. Garner, "Percy Faraday Frankland," J. Chem. Soc., 1948, pt. iii, p. 1997).

sale." That quest requires a consideration of whether scientists have bins in their brains marked with these labels, and it also requires access to a kind of off-stage evidence of a scientist's most private views of himself, his career, and his profession which is rarely available. More important here is the effect of adversary science on the functioning of the scientific community -- on scientists' morale and on the development of scientific knowledge.

It was clear to many scientists that the adversary process damaged the reputation of science.[62] It was also clear that scientific expertise was valuable in the resolution of many issues settled through adversary proceedings -- not just in water-and-sewage matters, but in regard to railways, patents, insurance, forensic medicine, etc.[63] The problem of how best to apply scientific expertise in adversary proceedings was intensively discussed in 1860, when Robert Angus Smith read a paper on "Science in our courts of law" to the Society of Arts, and again in 1866, when a British Association Committee on "Scientific Evidence in Courts of Law" reported. The issue again became controversial in the aftermath of William Odling's 1885 presidential lecture to the Institute of Chemistry on "The Whole Duty of a Chemist."[64]

---

[62] "Science in the law-courts," CN 52(1885): 299; F.S.B.F. de Chaumont, "On certain points with reference to drinking water," SR n.s. 1(1879-80): 165; Hogg, "Chemical analysis of drinking water," p. 414.

[63] See June Z. Fullmer, "Technology, chemistry, and the law in early 19th century England," Technology and Culture 21(1980): 1-28.

[64] "William Odling," Proc. Royal Society of London, series A, 100 (1922): ii.

Robert Angus Smith, who was uncomfortable as a witness and one of the few sanitary scientists who did little witnessing, saw contradictions between science and advocacy. In his view the value of scientific testimony lay in its precision, and precision was obliterated during cross-examination. Smith recognized several alternative modes for utilizing science in adversary proceedings, including a sort of science court.[65] He concluded that if scientists were to be witnesses, they should stick only to facts. Existing use of scientific witnesses was antithetical to the principles of science:

> we stand aside from a man who twists the expression of natural law for his own interests, as from one, who before His eyes, has neither the fear of God, nor the love and admiration of nature.[66]

The British Association Committee was less alarmed. It recognized that witnesses presented only one side of a case, and that they looked only for facts and arguments which upheld that side. It was bothered that because some scientists found adversary proceedings unethical, the most qualified experts were not necessarily participating in the solution of the country's socio-technical problems. Perhaps concerned most of all with the financial support of science, the committee recommended that courts appoint scientists as referrees to help them reconcile the contradictory testimony of other scientists

---

[65] R.A. Smith, "Science in our courts of law," pp. 138-41. See also for suggestions of science courts "Science in the law courts," CN 53 (1883): 1-2, 39; CN 52(1882): 299.

[66] R.A. Smith, "Science in our courts of law," p. 139. For a similar statement regarding sewage testimony in particular, see "On Whitthread's Sewage Process," VNEEM 9(1873): 316.

who appeared as witnesses for rival litigants.[67]

One of the things that bothered scientists most was that advocacy could not be contained within the court room: in 1885 Percy Frankland described the literature on water quality as consisting in general of "skillfully manipulated facts and theories intended to present some particular point of view in its most plausible form."[68] (As we shall see, far from being an exception, Percy Frankland's own writings exemplify this.) Percy Frankland characterized Charles Meymott Tidy's "scientific" papers -- i.e., those published in scientific journals -- as "strained and partisan evidence, professing, indeed to rest upon impartial and scientific foundation" used to "prop up inferior sources of water supply."[69] Tidy replied in kind: "Dr. Frankland absolutely ignored all arguments used, and all facts proved upon the other side. . . . Dr. Frankland did not like statistics when they did not prove his case though he did not mind them when they did."[70]

Both men were accurate in their descriptions of the other's motives. Each put forward a contrary view of what happened to sewage in rivers, and presented only the facts and arguments that justified that view. In a court room such presentations would have been expected. The problem of ascertaining truth would have been left to the

---

[67] "Report of a committee on scientific evidence in courts of law," 36th Meeting of the B.A.A.S. (Nottingham, 1866), (London: John Murray, 1867), pp. 456-57.

[68] Percy Frankland, "The selection of domestic water supplies," SR n.s. 6(1884-85): 547.

[69] Ibid., p. 551.

[70] C.M. Tidy in disc. of Percy Frankland, "The upper Thames as a source of water supply," p. 446.

judge, for whom scientific testimony would probably be incomprehensible as well as irrelevant both to the facts and the legal points at issue.

But how were the colleagues of Frankland and Tidy to evaluate such an exchange? For scientists questions of social utility or legal rights were peripheral. Scientific discourse was to uncover truth, not to disguise it. Then, as now, the scientific community functioned on a delicate balance between advocacy and honesty. If shared results were to contribute to the progress of science, it was necessary that the reports scientists brought before their colleagues be assumed to be honest, if perhaps honestly wrong. In the water-and-sewage science of the latter nineteenth century no such assumption was justified. Such institutions as the Chemical Society and the Institution of Civil Engineers, which ideally were institutions in which the sharing of research results was to advance knowledge, became forums for the kinds of polemics typically delivered by expert witnesses in courts of law or for formal presentation of research which would provide a foundation for testimony in adversary proceedings. Tidy's 1880 "River Water," given to the Chemical Society, was essentially a version of testimony he gave on self-purification in cases involving use of polluted river water. So pervasive was this infection of scientific institutions that there was no longer any social or institutional badge to distinguish natural philosopher from advocate. Research reports were assumed to represent some interest and were evaluated not on their own merits but on the merits of that interest. William Dibdin's 1887 exposition of the principles of biological sewage treatment was ridiculed not because it was a ridiculous idea, but because it seemed so blatantly to serve the political and economic ends of Dibdin's employer, the Metropolitan Board of Works.[71] The problem was that of the boy who cried wolf. Deceptions were so regularly paraded as truth that there were no longer easy means for recognizing truth when it came along.

Since objectivity did not automatically go along with the territory of a water scientist, it was necessary to lay special claim for the objectivity of one's work: Percy Frankland claimed his research, unlike that of others, was made "all for truly scientific spirit, for no other purpose than service to the truth."[72] Tidy prefaced a review of sewage-treatment alternatives with the qualification that he was here appearing as judge rather than in his usual role of advocate: "I have been retained by a good many people on both sides in my time . . . I am [here] in a perfectly independent position on this matter."[73]

Many scientists recognized the problem and understood that its solution must come from within the scientific community. In Alfred Smee's view the confusion regarding water-quality matters would disappear "if nature is interrogated with sincerity and truth."[74]

---

[71] August Dupre in disc. of William Dibdin, "Sewage sludge and its disposal," MPICE 88(1886-87): 215. See Chapter 10, below.

[72] Percy Frankland, "The upper Thames as a source of water supply," p. 565.

[73] C.M. Tidy, "On the treatment of sewage," JRSA 34(1885-86): 612.

[74] Alfred Smee, in Elizabeth M. Odling, Memoir of the late Alfred

August Dupre's solution to the "maze of conflicting statements and propositions" comprising the sewage question was simple: "Let everyone fairly cooperate in the question, and above all let everyone give to those who might be opposed to him credit for the same honourable motives as he was himself."[75] In retrospect, we can see that Dupre had put his finger on the essence of the problem. Scientists were crediting their colleagues with their own motives. Involvement in trans-scientific questions of water policy was so fundamental a part of water-and-sewage science that scientists came to expect their colleagues to be representing some particular social-policy alternative. Moreover, neither Dupre nor Smee were exemplars in taking the medicine they prescribed. Both were thoroughly mixed up in partisan water-and-sewage matters. In essence, the mark of "sincerity and truth" for Smee was correspondence with his own opinions.

In the view of chemist Frederick Abel, the problem was that scientists were not sticking to facts.[76] In fact, one of the main reasons that application of science to water-and-sewage questions was so ineffective was that the available facts were little help in resolving matters. Where the issue was the salubrity of a particular water, and the entity that made water unsafe was unknown or unidentifiable, facts, such as the amount of mineral matter the water contained, were of little use. In Edward Frankland's complicated system

---

Smee, F.R.S. By his daughter (London: Bell, 1878), p. 393.

[75] August Dupre in C.M. Tidy, "On the treatment of sewage," p. 670.

[76] F.A. Abel in disc. of Percy Frankland, "The upper Thames as a source of water supply," p. 566.

of water analysis (to be described in chapter seven), the facts of analysis were not ultimately the basis for the decision of whether or not a water was good. Instead, that decision was based on Frankland's conviction that society was better served by playing safe and using only water which had never been polluted and whose disease-transmitting potential was therefore nil. Again the concept of trans-science is valuable. At root, the paralysis of the scientific community was not caused by the duplicity of its members but by the nature of the issues it faced, a result of the attempt to use idioms and methods of science to resolve ethical, legal, political, and social problems.

In Weinberg's view, adversary proceedings are the proper forum for trans-scientific disputes, in part because contradiction clarifies, both for scientists and the public, the boundaries between science and trans-science. Also implicit in Weinberg's analysis is the notion of trans-science as a form of applied science. From some base of agreed-upon facts, assumptions, goals, and methods, scientists sally forth to apply their expertise to trans-scientific problems; discover (through adversary or political proceedings) that they are on alien ground; and then return to science to lick wounds, repair strained friendships, and work together in laboratories and scientific societies to expand the boundaries of science to include the trans-scientific. The presumption is that such people will be scientists first, that their loyalty to the "republic of science" will take precedence over their participation in trans-scientific issues. In such a system there will always be scientific institutions that remain aloof from trans-science. Research presented to such bodies will be impartial with respect to trans-scientific issues; the connection of such an institution with that research will be a main factor in giving credibility to the research.[77]

The interaction between science and trans-science in Victorian sewage-and-water matters was quite different. No warm hearth where colleagues agreed on basic principles and acknowledged one another's credibility existed. What was said at scientific meetings resembled -- a few significant instances to the contrary -- what was said before parliamentary select committees or judges. What scientific progress was made came from trans-science, a process haphazard and inefficient because trans-science contained no mechanisms for recognizing scientific advance -- again the most glaring example is the reception of the concept of biological sewage purification. The evolution of an understanding of water pollution and water purification in Victorian Britain is therefore not a case of some existing body of science being harnessed, however unsuccessfully, to social problems, but of a development of a science through the investigation of those problems. It was a very odd sort of development, distorted by a force field of politics, law, and religion, which blocked the normal functioning of the scientific community by dissolving scientists' credibility in

---

[77] Alvin Weinberg, "Science and trans-science," Minerva 10(1972): 216, 220-22. See also Michael Polanyi, "The republic of science, its political and economic theory," Minerva 1(1962): 54-73.

one another's eyes.

Finally, it is noteworthy that adversary science had its defenders. Responding to Angus Smith's argument that advocacy and science were antagonistic, William Odling maintained that the adversary process was healthy for science. Impartial investigations had no incentive to discover and tended to rely on stale authority. Requiring each side to prove its case, on the other hand, led to advance. It was therefore desirable to have a subject "investigated by men desirous of establishing different conclusions."[78] In Odling's view the conflict was not between truths and falsehoods but between different perspectives.

As we shall see, there are cases where Odling's view is justified. Adversary proceedings certainly provided incentive for rigorous examination of opposing concepts. In some cases better understanding of the limits within which certain concepts were accurate emerged from episodes of contradiction. The 1893 hearings of the Metropolitan Water Commission, for example, explored the issue of under what circumstances it was true that water filters removed bacteria. A view of what more commonly happened when scientists clashed was offered by journalist Francis R. Conder:

> Truth, no doubt, may be threshed out by conflict of opinion, but there is no certitude that the whole truth will so come out. As a rule, only those statements will be cleared up which it is in the interest of some one litigant to state or to oppose. Points

---

[78] William Odling, "Science in our courts of law," JRSA 8(1859-60): 167-68.

> of controlling interest may thus be overlooked, and so much attention concentrated on certain details, as to leave little or no time for the elucidation of others of equal or even greater importance.[79]

In Conder's view, whatever clarification emerged from contradiction was accidental.

For this mode of dysfunction of the scientific community, I suggest the rather dramatic designation, "the context of distrust." Distrust of scientists for one another dominated water-and-sewage science during the second half of the nineteenth century. There are indications that in the early years of the twentieth century this domination was lessening, owing to the new certainty provided by bacteriology, to the success of biological sewage treatment, and to thorough and non-partisan investigations such as those carried out by the 1898-1911 Royal Commission on Sewage Disposal.[80] Indeed, lack of certainty about fundamentals of water-and-sewage purification was, at root, responsible for distrust. The next chapter examines the confusion about pollution and purification and the raw material of ideas which scientists applied in adversary proceedings.

---

[79] F.R. Conder, "The water supply of London," Fraser's Mag., n.s. 22(1880): 185.

[80] G.B. Kershaw, Modern Methods of Sewage Purification (London: Griffin, 1911), pp. 33-45 details the work of the Royal Commission on Sewage Disposal.

III. THE ANTITHESIS OF LIFE: DECOMPOSITION IN VICTORIAN SCIENCE AND MEDICINE

The fatal fascination of corruption.
--Henry Letheby*

No reader of nineteenth-century British sanitary literature gets far without encountering a preoccupation with decomposition: the processes of fermentation, putrefaction, and slow combustion which transform organic materials into inorganic matter. To Victorian sanitarians these processes were not parts of the normal workings of the world but instead represented the essence of morbidity. All decay was dangerous. Filth -- for the Victorians the word meant the unpleasant phenomena of decomposition -- corrupted whatever came in contact with it. Vegetation grown on sewage-fertilized farms was believed to be "rank," fish that fed near sewer outfalls considered unclean, and the animalculae -- "Creatures bred of rottenness" -- beneath contempt.[1] One writer warned that the autumnal fall of leaves constituted a threat to health:

> The ground that is strewn with leaves becomes a nursery of morbid influences. The delightful odour that fallen leaves diffuse in the wood suggests their harmlessness; but on the roads and walks, where the leaves are hourly crushed and the dropping rain helps to make a paste of them, they are, without doubt, pestiferous nuisances, which should be removed as quickly as possible by parochial authorities

---

* Henry Letheby, "The city pumps," CN 4(1861): 261.

1 Greville Ffennel, "Thames pollution," The Field (1867) ii, p. 193.

and private proprietors.[2]

Such views were not medically unorthodox. The paludal diseases, like ague and malaria, had traditionally been associated with the decomposing vegetation of marshy areas, and in the 1850s Pettenkofer had stressed the importance of a suitable soil for the nurturing of the decomposing matter that was the specific poison of cholera. Indeed, the association of filth with disease was not new, nor was it devoid of sound hygienic basis.[3] Old Testament hygienic laws, which frequently reflect this association, were often invoked by Victorian sanitarians as representing revealed rules for sanitary practice.

Yet decomposition was much more than a medical concept. The phenomena of decomposition had significant implications in issues of natural theology, particularly in regard to the goodness of the Creation. It provided a metaphor for changes that went on in societies and within individuals. The individual's intuitive judgment that filth was disgusting was shown to conform exactly with the teaching of medical science that filth was preeminently dangerous to health. The sanitary movement was in part concerned with moral reform, and

---

[2] "Decaying leaves," SR 1(1874): 216. For a similar statement, attributed to Pettenkofer, see "An inquiry into the causes of typhoid fever in Massachusetts," in Massachusetts State Board of Health, Second Annual Report (January 1871), (Boston: Wright and Potter, 1871), pp. 171-72.

[3] Owsei Temkin, "An historical analysis of the concept of infection," in Johns Hopkins University History of Ideas Club, Studies in Intellectual History (Baltimore: Johns Hopkins University Press, 1953), pp. 123-47.

many reformers believed immorality was bred of the filth and squalor in which unfortunate persons so often lived.[4]

Victorians faced an important scientific dilemma with regard to the propriety of decomposition. Justus Liebig, who assumed for mid-century Britons the role of law-giver, prophet, and systematizer in agriculture and sanitary science (a status analogous to that assumed by Ruskin, Newman, and Mill in other areas), was much concerned with decomposition in his works on Chemistry in its Application to Agriculture and Physiology (1840), Animal Chemistry, or Organic Chemistry in its Application to Physiology and Pathology (1842), and the successive editions of the popular Familiar Letters.[5] A "decomposition dilemma" was implicit in Liebig's science. On the one hand he told the British public that decomposition was a malignant process. His contact theory of decomposition explicitly described the corruption of healthy tissue by contact with decomposing matter. Therefore, in treating sewage or other organic refuse, British sanitarians often

[4] Charles Kingsley's novel Yeast is an example, as the title suggests. Here the hero attempts to stem the spreading corruption of Catholicism; among the remedies -- the one which the heroine requests on her death bed -- is pure water. Likewise, John Morley spoke of Swinburne's "putrescent imagination," implying that this corruption would be transferred by Swinburne's sensual poetry to the susceptible imaginations of readers (quoted in Walter Houghton, The Victorian Frame of Mind [New Haven: Yale University Press, 1957], p. 368). B.W. Richardson accepted the existence of moral contagia which operated by a contact process of infection just as did physical contagia (Richardson, The Field of Disease. A Book of Preventive Medicine [London: MacMillan, 1883], pp. 856, 844, 850).

[5] Walter Houghton has described Victorian dogmatism in a manner very much applicable to Liebig's popular works on organic chemistry and to his status in Britain (The Victorian Frame of Mind, pp. 137-60).

tried to prevent decomposition. At the same time, Liebig asserted that decomposition was a necessary natural process. According to his mineral theory of manures, plants lived by taking their nutrition in an inorganic form from atmospheric gases and soil minerals. Supplies of soil minerals were finite; the survival of humanity depended on the conservation of fertilizer. Victorian civilization therefore depended on decomposition, the very process that was the essence of epidemic disease, to transform its wastes into the inorganic food of plants.[6]

Theories of decomposition underlay the views sanitarians took of purification and impurity, of the activity of aquatic organisms, and of the phenomena of self-purification in general. This chapter examines theories of decomposition held by Liebig and by mid-century British chemists, and considers the "decomposition dilemma;" in particular the fermentative or "zymotic" pathology which was central in the sanitarians' medical theory. The next chapter considers the resolutions of the decomposition dilemma proposed by Victorian sanitarians and scientists.

I. Liebig's Theory of Decomposition

In the view of Lavoisier, the cause and nature of the breakdown of organic molecules into simpler substances was one of the key

[6] The existence of this dilemma is suggested by A. Gibson and W.V. Farrar ("Robert Angus Smith, F.R.S., and sanitary science," Notes and Records of the Royal Society 28(1974): 249).

unsolved problems of chemistry. In the late 1830s Liebig developed a comprehensive explanation for an enormous variety of decompositive phenomena ranging from the mouldering of wood to the putrefaction of living flesh, and including digestion and rusting as well. According to Liebig, all these processes of decomposition were members of a class of reactions which were initiated by substances that were only slightly, if at all, altered during the reaction. In 1836 Berzelius had designated these "catalytic reactions," and suggested that they resulted from some force which did not follow usual rules of electro-chemical affinity. As illustrations of catalysis, Berzelius included inorganic reactions such as the breakdown of hydrogen peroxide in the presence of platinum, as well as fermentation. He also hinted that catalysis had implications for physiological phenomena.[7] Liebig adopted Berzelius' key achievement, the idea of a class of "catalytic" reactions and agreed that the initiation of a reaction in susceptible matter by a non-participating substance characterized a wide range of organic processes and inorganic reactions. He even used Berzelius' examples to illustrate this class of chemical events. But while adopting Berzelius' classification, Liebig made a public and noisy business of rejecting "catalytic force," arguing that it was only an

---

[7] J.J. Berzelius, "Some ideas on a new force which acts in organic compounds," in H.M. Leicester and H.S. Klickstein, A Source Book in Chemistry 1450-1900 (Cambridge: Harvard University Press, 1952), p. 267; Joseph S. Fruton, Molecules and Life -- Historical Essays on the Interplay of Chemistry and Biology (New York: Wiley Interscience, 1972), pp. 47-48; J.R. Partington, A History of Chemistry (London: MacMillan, 1964), Vol. 4, p. 303; W.A. Shenstone, Justus von Liebig, His Life and Work (New York: MacMillan, 1895), p. 65.

occult name whose use would hinder an understanding of how these processes really worked.[8]

For "catalytic force" Liebig substituted a more concrete, if equally unfounded explanation. All catalytic reactions, "phenomena of fermentation [and by implication all reactions classed catalytic] . . . establish the correctness of the principle long since recognized by LaPlace and Berthollet, namely, that an atom or molecule put into motion by any power whatever may communicate its own motion to another atom in contact with it."[9] All processes of decomposition, therefore, which Liebig with some difficulty divided into the classes putrefaction, fermentation, and eremacausis (slow oxidation), were processes in which essentially passive organic molecules, only weakly bound by chemical affinity, fell or shook themselves apart upon receiving some impact, usually a bump from oxygen or from another molecule undergoing a similar shaking.

Opposing decomposition was a vital force, responsible both for the organization of substances into tissues and for the resistance of these tissues to the depredations of the atmosphere during life.[10]

---

[8] J. Volhard, Justus von Liebig (Liepzig: Barth, 1909), Vol. 2, p. 84; A.K. Balls, "Liebig and the chemistry of enzymes and fermentation," in F. Moulton, ed., Liebig and After Liebig -- A Century of Progress in Agricultural Chemistry (Washington, D.C.: A.A.A.S., 1942), p. 32.

[9] Justus Liebig, Familiar Letters on Chemistry in its Relations to Physiology, Dietetics, Agriculture, Commerce, and Political Economy, 3rd ed., revised and much enlarged (London: Taylor, Walton, and Maberly, 1851), #16, p. 209; idem., Chemistry in its Application to Agriculture and Physiology, edited from the manuscript of the author by Lyon Playfair, 2nd ed., (London: Taylor and Walton, 1842), p. 365; and Partington, Hist. of Chem., Vol. 4, p. 302.

> The constituents of vegetable and animal tissues have been formed under the control of a cause of change in form and composition, operating in the organism. This is the vital force, which has determined the direction of attraction, and which opposes cohesion, heat, electricity, in short, all the causes which, out of the body, prevented the union of atoms to form compounds of the highest or most complex order, and annihilates their disturbing influence.[11]

Liebig portrayed vitality as continually besieged by the forces of inanimate nature. Temporarily these forces could be placated with food which was sacrificed as a substitute for the body's own tissues.[12] Sooner or later, however, vitality succumbed, as the body was either depleted of food or unable to deliver food effectively to the places where it was burnt, and life and the organic substances whose existence it sustained were destroyed by the contact process of decomposition.[13] This notion, in which the forces of decomposition and the forces of life opposed one another, became prominent in British sanitarian thought, and has much to do with the conception of decomposition as a fundamentally malignant process.

The succeeding processes of decomposition were not nearly so

---

[10] Justus Liebig, Animal Chemistry, or Chemistry in its Application to Physiology and Pathology, edited from the author's manuscript by William Gregory. From the 3rd London edition, revised and greatly enlarged (New York: John Wiley, 1852), pp. 1, 19, 143.

[11] Ibid., p. 143.

[12] Ibid., pp. 7, 20. But see idem., Agricultural Chemistry, 1842, p. 377.

[13] Justus Liebig, Animal Chemistry or Organic Chemistry in its Application to Physiology and Pathology, edited from the author's manuscript by William Gregory with notes, additions, and corrections by Dr. Gregory and John W. Webster, M.D. (1842; rpt. with a new introduction by Frederic L. Holmes, New York: Johnson Reprint, 1964), p. 242; and idem., Animal Chemistry, 1852, pp. 19-20.

dramatic as the struggle of life against the elemental processes that opposed it. Liebig rejected the view held by most British chemists that the processes of decomposition were entirely spontaneous transformations of dead organic matter. Instead, oxygen was the inciter of decomposition, being "the proximate cause of the changes which occur in organised bodies after death."[14] Oxygen affected "putrescible" constituents, such as albumen, fibrin, caseine, and gelatine. These molecules were highly nitrogenous (nitrogen had little affinity for chemical combination, Liebig noted), and they were weakened by the large number of directions in which the cohesive force had to act to hold them together. Oxygen provided the mechanical disturbance necessary to overcome the weak internal equilibrium which held these big molecules together and they spontaneously decayed into smaller organic molecules.[15] Just as oxygen had been necessary to incite their putrefaction, the presence of these materials in a condition of putrefaction was required to initiate further changes in non-nitrogenous organic materials (sugar) or in non-putrescible nitrogenous organic compounds like urea.[16]

Liebig did not maintain a consistent distinction between putrefaction and fermentation. In 1842 the distinction was trivial. Putrefaction occurred in nitrogenous matter and its gaseous products

---

[14] Liebig, Agricultural Chemistry, 1842, pp. 397, 301; idem., Letters, 1851, #15, pp. 180-81.

[15] Liebig, Letters, #15, p. 184; Agricultural Chemistry, 1842, pp. 302, 272.

[16] Ibid., pp. 272, 268; idem., Letters, 1851, #14, pp. 173, 184.

did not stink.[17] By 1851 the processes seemed more distinct to Liebig. Putrefying matters were ferments and caused fermentation: "All non-putrescible bodies are called fermentiscible, when they possess the property of being decomposed by contact with putrescent matters. The process of their decomposition is called fermentation. The putrescent body, by which this change is caused, is now named the Ferment."[18] Yeast was such a ferment. The putrefaction of nitrogenous matter in dead yeast plants spread to the sugar solution, causing the sugar to ferment. The fermentation resulted neither from the yeast's organized structure nor from any special substance it contained, but was simply an effect of its state of motion.[19] In any case, like putrefaction, fermentation was a stage en route to the eventual inorganic status of matter which had once lived. In both processes, but particularly in fermentation, organic matters were rearranged by the introduction of the constituents of water, but their end products were usually still organic.[20]

---

[17] Liebig, Agricultural Chemistry, 1842, pp. 265-66, 291.

[18] Liebig, Letters, 1851, #15, p. 185. The distinction remained among the more obscure parts of Liebig's chemistry. British sanitarians often amalgamated the processes as when "L.G." spoke of the "putrefactive fermentation" of the Thames (Times, 24 June 1858); or the Royal Commission on the Sewage of Towns spoke of "putrid fermentation" (Third Report and Appendices of the Commission appointed to inquire into the best mode of distributing the sewage of towns and applying it to beneficial and profitable uses, P.P., 1865, 27, [3472.], Appendix 4, p. 204).

[19] The other alternatives represent the views of Pasteur on the one hand, and of Traube and Berthelot on the other. See Fruton, Molecules and Life, pp. 60, 65; and Liebig, Letters, 1851, #19, pp. 233-35; #15, pp. 186-87.

[20] Liebig, Letters, 1851, #15, p. 186; idem., Agricultural

After putrefying and/or fermenting through several stages, the decomposing body reached a temporary equilibrium.[21] The final stage of decomposition, in which still-organic compounds were transformed into simple inorganic compounds such as CO_2 and water, was a slow combustion for which Liebig invented the term "eremacausis."[22] Like putrefaction and fermentation, eremacausis was a contact process, transferred to bodies susceptible to it from bodies in which it was occurring. Although such oxidation might proceed spontaneously to a slight extent, "contact with a substance, itself undergoing the process of decay, is the chief condition of decay for all organic substances which do not possess the power of combining with oxygen at common temperatures."[23] Thus: "It is decaying wood which causes fresh wood around it to assume the same condition."[24]

In this manner Liebig, in far more detail than other contemporary agricultural chemists, explained the transformation of dead organic matter into the inorganic food of plants:

> Fermentation or putrefaction represents the first stage of the resolution of complex atoms into more simple combinations; the process of decay completes the circulation of the elements by transposing the products of fermentation and putrefaction into gaseous compounds. Thus the elements constituting all organised

---

Chemistry, 1842, pp. 277, 291.

[21] Liebig, Letters, 1851, #15, p. 187.

[22] Liebig derived eremacausis from the Greek ērema=gently, slowly; and kausis (kaiein)=to burn. The term is still found in larger dictionaries, such as Webster's Third (unabridged) New International (1967).

[23] Liebig, Letters, #17, pp. 214-15.

[24] Liebig, Agricultural Chemistry, 1842, pp. 303-4.

> beings, which previously to participating in the vital processes were in the forms of oxygen compounds, -- their carbon and hydrogen, reassume the form of oxygen compounds.[25]

This contact model of chemical events was a central part of the chemistry of life Liebig presented to the British public in the early 1840s. Although an unexplained vital force played an important role in his chemical system, Liebig nevertheless offered simple and vivid chemical and physical explanations for an enormous range of vital processes, and the contact theory of fermentation served him well. Unlike the later attempts of Pasteur to restrict tightly the definition of fermentation (in effect to limit it to those processes that could be shown to be caused by microbes), Liebig included under the title fermentation many of the basic processes of digestion in which unorganized ferments, like rennet and pepsin, were active. The contact theory, therefore, explained the decompositive processes of physiology as well as those decompositions which resolved the refuse of cities into the food of plants. Liebig sought to explain life, not just rotting.[26]

II. Decomposition Theory in Britain

Although some British chemists and physiologists quibbled with certain details of Liebig's description of decomposition, most shared with him the conception of decomposition as a condition of

---

[25] Liebig, Letters, 1851, #17, p. 210.

[26] Fruton, Molecules and Life, p. 56.

de-vitalized organic matter which was fundamentally opposed to vital activity. Their view was far closer to Liebig's than to that taken by Pasteur in the early 1860s. Mid-nineteenth century British chemistry and physiology texts rarely explained decomposition, presenting it merely as the name for the transitory condition of once-living matter. The attention of British chemists and physiologists was occupied with the force which organized matter into life and kept it organized during life; decomposition was merely the proof that this force was absent. Thus, for the physiologist W.B. Carpenter, putrefaction was significant as the proof of death.[27]

The tone taken by J.L. South, writing on "Zoology" for the Encyclopedia Metropolitania of 1845, is typical:

> immediately on the occurrence of which [death] the elemental substances composing the previously living being are set free from the control of vital power, and are again subject to the laws governing inorganic matter, and, resuming their natural combinations, the organized mass is soon destroyed by decomposition.[28]

---

[27] W.B. Carpenter, Principles of Human Physiology, 4th ed., (London: J. Churchill, 1853), p. 1104. See also J. Goodfield-Toulmin, "Some aspects of English physiology 1780-1840," J. History of Biology 2(1969): 287-94; Margaret Pelling, Cholera, Fever, and English Medicine, 1825-1865 (Oxford: Oxford University Press, 1978), p. 137; and [F.O. Ward], "The Metropolitan Water Supply," Quarterly Review 87(1850): 477.

[28] J.L. South, "Zoology," in Encyclopedia Metropolitania, or Universal Dictionary of Knowledge (London: Fellowes . . . , 1845), Vol. 7, p. 171. See also S. Brown, "Review of The Human Body and its Connexion with Man, illustrated by its Principal Organs, by James John Garth Williamson," North British Review 17(1852): 139; and A.H. Hassall, writing as Lancet Analytical Sanitary Commission, "Records of the results of microscopical and chemical analyses of the solids and fluids consumed by all classes of the public," Lancet (1851), i, p. 187.

Gideon Mantell, paleontologist and comrade of Lyell, offered a similar view in a teleological treatise on animalcules: "the moment life departs, the materials of which these organisms are composed, released from the dominion of the laws of vitality, are resolved into fluid molecules, and rapidly pass into their original elementary condition."[29] The view is also prominent in chemistry texts. According to W.A. Miller:

> In the living body, ordinary chemical affinities are suspended, and compounds are produced, which, when released from the influences under which they originated, quickly undergo fresh changes, which are manifested by the occurrence of putrefaction and decay, to which both animal and vegetable bodies are liable when they no longer form part of the living frame.[30]

Even G.J. Mulder, with whom Liebig disagreed on many details of agricultural chemistry, accepted both the view of decomposition as the work of inorganic forces on the constructions of the vital force and the disturbance-transferring mechanism Liebig had proposed. Mulder even strengthened the plausibility of the mechanism. One did not have to rely on Liebig's hypothetical vibrating atoms. Instead, the putrefaction-initiating oxygen might be transferred from molecule to molecule, decomposing each as it went on, in a sort

---

[29] Gideon Mantell, Thoughts on Animalcules; or a Glimpse of the Invisible World Revealed by the Microscope (London: John Murray, 1846), p. 89.

[30] W.A. Miller, Elements of Chemistry, Theoretical and Practical, 2nd ed., 3 vols. (London: Parker, 1860-62), Vol. 3, p. 2. See also 4th ed., 3 vols. (London: Longmans, Green, Reader, and Dyer, 1868-69), Vol. 3, p. 2; William Gregory, Elementary Treatise on Chemistry (Edinburgh: A & C Black, 1855), pp. 256-57; Wm. Prout, Chemistry Meteorology and the Function of Digestion considered with reference to Natural Theology (London: Wm. Pickering, 1834), p. 533; and W.T. Brande and A.S. Taylor, Chemistry (London: John Davies, 1863), p. 807.

of de-polymerization process, and in much the same manner as modern chemists believe catalysts to act.[31]

Liebig's analogy between in vivo physiological decompositions such as digestive processes, and the decomposition of dead organic matter in macrocosmic nature is absent from the writings of British chemists and physiologists and this may explain their lack of interest in processes of decomposition. Those who felt need of a process found the ordinary forces of inanimate chemistry, which ever opposed the vital force, sufficient. These insidious chemical forces assailed the living body even during life, but it was only at death, when vitality was extinguished, that they triumphed: "The plant and animal are able, so long as life remains in them, to battle the storm and the breeze, to withstand the corroding effects which air endeavours to exert on both" wrote the Edinburgh chemist Stevenson Macadam.[32] Carpenter agreed:

> In warm-blooded animals the cessation of those changes in which life consists, immediately leaves the way clear for that disintegration which is effected by chemical forces. . . . The differing behavior . . . of the living and dead organism under the same conditions, is commonly accounted for by the supposition, that the vital force, so long as it endures, antagonizes the operation of physical and chemical agencies, which are constantly tending to the destruction of the living

---

[31] G.J. Mulder, The Chemistry of Vegetable and Animal Physiology, trans. P.F.H. Fromberg; introduction and notes by J.F.W. Johnston (Edinburgh: Wm. Blackwood, 1849), pp. 38-55. Mulder was Dutch but his work was made known in Britain by Johnston.

[32] Stevenson Macadam, The Chemistry of Common Things (London: T. Nelson, 1866), p. 186; and Prout, Chemistry, p. 534.

structure.[33]

This assertion that decomposition, and putrefaction in particular, were processes essentially antithetical to life was given great prominence by Liebig in his denial of the "vital" theory of fermentation.[34] He noted that animalculae and fungi, which some supposed were the causes of putrefaction, themselves suffered putrefaction on death, and therefore could not cause that process.[35] Rejection of organisms as the causes of decomposition did not mean that the scavenging organisms at whose activity Pasteur marvelled were insignificant or ignored. Instead, as will be made clear in the next chapter, their function was recognized in a wholly different sense. Microorganisms and plants did not decompose dead organic matter, they utilized it. The predominance of this view in British chemistry and

---

[33] Carpenter, Human Physiology, p. 114. See also Henry Letheby, "Sewage and sewer gases: report to the City of London Commissioners of Sewers, 14 Sept. 1858," J. Public Health and Sanitary Review 4 (1858): 278-79; R.C. Sewage of Towns, Second Report, Appendices, pp. 64-65; Greville F[fennell], "Is sewage detrimental to fish," The Field 39(1872): 466. According to Prout, "organized beings and their laws are in continual opposition to the general laws by which inorganic matter is governed" (Prout, Chemistry, p. 534).

[34] Liebig, Animal Chemistry, 1852, pp. 143-44.

[35] Ibid., p. 144; idem., Letters, 1851, #19, pp. 237-38. This was an important theme in Liebig's satire of the biological theory of fermentation (Volhard, Liebig, Vol. II, p. 86). Here he came up with the right answer biologically -- that other microbes decomposed them -- but in the context of the satire it was presented as a ridiculous answer, and for Liebig it was an unsatisfactory solution philosophically. That there was a paradox here seems to have caused no great concern for those who advocated a biological explanation of decomposition. Thus John Burdon Sanderson, who accepted that the septic ferment was an organism, pointed to its resistance to decomposition as proof of this (Burdon Sanderson, "Introductory report on the intimate pathology of contagion," in 12th Annual Report of the Medical Officer of the Privy Council for 1869, P.P., 1870, 38, [C.-208.], pp. 255,

medicine is a key factor in the British response to Pasteur's conception of the nature of fermentation and putrefaction and the importance of these processes in the economy of nature: British scientists did not reject Pasteur so much as they found him irrelevant.

III. Liebig's Fermentative ("Zymotic") Pathology

The phenomena of pathology were also included within the vast sweep of Liebig's comprehensive system of organic chemistry. In the Animal Chemistry of 1842 Liebig had conceived of diseases in terms of disturbances of the metabolic equilibrium. Oxygen, inasmuch as it held an ambivalent status both as preserver and destroyer of life, was a particularly important factor. Various states of disease were manifestations either of too rapid or too feeble oxidation.[36] In the Agricultural Chemistry of the same year Liebig presented a distinct "zymotic" theory of disease in which he explained the traditional association of disease with decomposing matter in terms of his contact theory of decomposition. By 1852 the vague oxygen pathology had been removed from the new edition of the Animal Chemistry, apparently replaced by the zymotic pathology of the Agricultural Chemistry.[37]

Briefly, Liebig's zymotic theory of disease was that "an animal

---

244-45.

[36] Liebig, Animal Chemistry, 1842, pp. 244-45.

[37] Idem., Agricultural Chemistry, 1842, pp. 364-70; Liebig, Animal Chemistry, 1852, pp. 137-40.

substance in a state of decomposition, can excite a diseased action in the bodies of healthy persons; that their state is communicable to all parts or constituents of the living body. . . . as long as this state continues, as long as the decomposition has not completed itself, the disease will be capable of being transferred to a second or third individual."[38] This communication was accomplished by the blood which Liebig regarded as having the quintessential characteristics of a substance which could suffer transformations and transfer them to other parts of the body.[39] In support of this pathological theory Liebig cited Magendie's observations of the virulent consequences of allowing putrefying substances to come in contact with open wounds. The phenomena of dissecting-room septicemia, of smallpox and syphilis, and even Semmelweis' studies of puerperal fever were all examples of the dire consequences of transferring decomposing organic matter.[40] Liebig made it emphatically clear that it was not the substances which were in each case the morbid poison, but the processes they were experiencing. They affected "the animal economy as deadly poisons, not on account of their power of entering into combination with it, or by reason of their containing a poisonous material, but solely by virtue of their particular condition."[41] The identification of

---

[38] Idem., Animal Chemistry, 1852, p. 137.

[39] Idem., Agricultural Chemistry, 1842, p. 367.

[40] Ibid., p. 370; Liebig, Animal Chemistry, 1852, p. 137; idem., Letters, 1851, #18, pp. 228-31; Appendix 3, pp. 530-31. These and similar experiments are detailed in Charles Murchison, A Treatise on the Continued Fevers of Great Britain (London: Parker, Son, & Bourn, 1862), pp. 447-48.

[41] Idem., Agricultural Chemistry, 1842, pp. 364-65.

the processes of pathology as the processes of decay also explained why those substances which disinfected virulent materials also stopped the decomposition of organic materials.[42] Thus for Liebig, and for many Victorian sanitarians influenced by him, the process of disease and the process of decay were essentially the same.

IV. Decomposition and Disease in Mid-Victorian Britain

During the period 1850-1880 decomposition was the dominant motif of British sanitary pathology. It was the common element that linked distinct views of the nature of morbid poisons, the substances that were the causes of specific diseases.[43] These were variously conceived as products of decomposition -- either as the noxious gases of putrefaction or the stinking organic intermediates that preceded these; or, following Liebig, as organic materials undergoing a specific process of decay. Even the "germs of disease" sometimes mentioned in the sanitary literature were believed to have an intimate connection with decomposing materials. Not only did many regard decomposing organic matter as the source of the morbid poisons and the means by which they were transmitted, they also viewed the pathological changes in a considerable variety of diseases as putrefactions of living tissue. Decomposition transcended the debate between contagionists (who emphasized the direct transmission of disease from the

[42] Idem., Animal Chemistry, 1852, pp. 136-37.
[43] Pelling, Cholera, pp. 115-19.

ill to the healthy and favored quarantine) and anti-contagionists (who believed in general environmental causes of disease and opposed quarantine), and obscured the issues that separated these parties.[44]

Yet while decomposition was the dominant theme of sanitary science, sanitarians' ideas about the exact relationship between decomposition and disease reflect an impressive lack of unanimity, nor, in general, did they recognize the necessity, have the skill, or even feel the inclination to force resolution of this issue.[45] The sanitary movement that evolved from Edwin Chadwick's 1840s Health of Towns investigations reflected concerns about the morals of the working class, the political economy of cities, the administration of social services, and the godliness and naturalness of modern civilization. It was reinforced by the dissatisfaction of the newly-enfranchised urban middle class with the conditions of urban life and tempered by the exigencies of municipal and parliamentary politics.[46] Often sanitary reformers were clergymen, politicians, or general do-gooders, rather than scientists or medical men. Edwin Chadwick, who

---

[44] Pelling, Cholera, in passim., but especially pp. 303-4; and John Eyler, Victorian Social Medicine, the Ideas and Methods of William Farr (Baltimore: Johns Hopkins University Press, 1979), p. 102.

[45] Lloyd G. Stevenson, "Science down the drain -- on the hostility of certain sanitarians to animal experimentation, bacteriology, and immunology," BHM 29(1955): 2, 25-26.

[46] On municipal public health politics in Britain see Derek Fraser's chapter on "The politics of improvement" in his Urban Politics in Victorian England (1976; rpt. London: MacMillan, 1979), pp. 154-75; and Asa Briggs, Victorian Cities (Harmondsworth, U.K.: Pelican, 1968), in passim., and especially pp. 20-21. A fine study of parliamentary public health politics is Paul Smith, Disraelian Conservatism and Social Reform (London: Routledge and Kegan Paul, 1967).

was a barrister, was concerned with the economic viability of England's industrial cities. For Chadwick disease among the poor represented a preventable cost to society because it unnecessarily generated widows and orphans whose welfare became the burden of the state.[47] Others, like Lord Shaftesbury, William Farr, Southwood Smith, and Charles Dickens, possessed less systematic reforming zeal; they recognized the wretched moral and physical condition of the poor and sought to help where they could.[48] Still others regarded an improvement in the living conditions of the poor as an effective deterrent of revolution.

Environmental theories of disease were useful to reformers because they emphasized the seriousness of the problems with which the reformers wished to deal.[49] Chadwick's explicitly aspecific association of disease with filth, for example, justified his meddling not only with water supply and sewerage, but with burial practices, "offensive trades," such as tanning and slaughtering, and housing. Proponents of rival public health reforms needed to court public

---

[47] R.A. Lewis, Edwin Chadwick and the Public Health Movement (London: Longmans, Green and Co., 1952), pp. 43-46.

[48] For Shaftesbury, see ibid., pp. 62, 110-11; and P. Smith, Conservatism, pp. 10, 31, 67. For S. Smith, see Pelling, Cholera, pp. 7-8. For Farr, see Eyler, Farr, pp. 23-27. For Dickens, see Houghton, Victorian Frame of Mind, pp. 274-75. Charles Kingsley also belongs in this group. At an 1870 meeting on the pollution of the Welsh Dee he argued: "Let them make every father and mother who had a family understand what he [Kingsley] knew to be true, that in fighting for the purity of water they were fighting for the moral and physical purity of everyone of their children" ("The Pollution of the River Dee," The Field 36(1870): 103).

[49] Pelling, Cholera, pp. 59-60; Eyler, Farr, p. 100.

opinion and consider political expediency; scientific or medical concepts in their pure form, and Liebig's zymotic analogy is an exemplary case, were sifted for whatever sanitary "capital" they could yield.[50] The reformers' adaptation of scientific concepts to justify their policies, together with the uncertainty of the medical community about such fundamental issues as the origins, modes of transfer, and specificity of diseases fostered the acceptibility of a multi-factoral view of disease, which lowered the expectation that outbreaks of a disease ought to have a single cause or means of transfer.[51] Despite communication of discoveries from the continent or new concepts such as the germ theory, proximate causes -- the equation of disease with filth -- are nearly as prominent in the sanitary literature in 1880 as in 1850.[52] Still, despite the confusion of multi-factoralism, three

---

[50] Pelling, Cholera, p. 6.

[51] Ibid., p. 63.

[52] The consideration of sanitary "rhetoric" is particularly important here, for it is clear that the progressive development usually attributed to English medicine during this period, and especially the successive refinements of the concept of morbid poisons by Snow, Budd, and Lister, is not reflected in the sanitary literature considered en masse. Sanitary rhetoric has a multitude of purposes and exists on a variety of levels of discourse and these are often responsible for apparently unprogressive statements by those who habitually kept abreast of medical and scientific progress and from whom one expects particularly insightful comments. For example, the association of decomposition with disease is discussed in at least four non-exclusive modes of discourse. These might be designated: 1) the epidemiological, as when Chadwick, Farr, and Simon correlate mortality with bad sanitary conditions; 2) the etiological, in which decomposing filth is regarded as the morbid poison, as with Liebig and Farr; 3) the pathological, in which the process of disease is regarded as a decomposition; and 4) the prophylactic, in which the proximate filth-disease relationship is used to justify sanitary improvements. One might add a fifth, the theological, in which the association is pointed to as an illustration of the sanitary rules of the Creation. The fact that

relatively discrete variants of the filth-disease association appear in the sanitary literature: emphasis on the products of decomposition (Chadwick), on the process of decomposition (William Farr and John Simon), and on the germs which breed in filth and may cause its decomposition (Robert Angus Smith, John Tyndall). These variants are important for the distinct concepts of purity and purification implicit in each.

For Chadwick and Southwood Smith -- and probably for most sanitarians in the 1840s and through much of the 1850s -- the harmfulness of putrefying matter was a consequence of its products, some of which might be the morbid poisons of specific diseases. There was some experimental support for this view. The sanitarian and physiologist B.W. Richardson, for example, produced typhoid symptoms (including intestinal ulceration) in a dog exposed to an ammonium sulfide atmosphere.[53] In 1858 T.H. Barker exposed animals to cesspool gases and produced symptoms of "vomiting and purging," as well as "symptoms not unlike those of enteric fever [typhoid]."[54] In general, however, Chadwick and his followers downplayed the concept of disease specificity and claimed decomposition was harmful because it removed

---

these levels of discourse combine to disguise medical progress during this period does not diminish their historical importance, and indeed, enhances it. The remainder of this chapter and parts of others will examine the disease-decomposition association in regard to some of these perspectives.

[53] B.W. Richardson, The Cause of the Coagulation of the Blood, being the Astley Cooper Prize Essay for 1856 . . . (London: Churchill, 1858), pp. 118-21, 345-46; and Murchison, Treatise, pp. 447-48.

[54] Murchison, Treatise, pp. 447-48.

oxygen from the atmosphere and replaced it with actively poisonous or asphxiating gases that frequently stank -- H_2S, NH_3, PH_3, CH_4, and even CO_2. If these were not the acute morbid poisons of specific diseases, they were nevertheless "predisposing causes" which by destroying the "vital principle" rendered the body more susceptible to disease.[55] Thus Chadwick's famous epigram: "all smell is, if it be intensive, immediate acute disease, and eventually we may say that, by depressing the system and rendering it susceptible to the action of other causes, all smell is disease."[56] The greater the importance assigned to predisposing causes, be they simple asphyxiants or cumulative poisons, the correspondingly less necessary was the hypothesizing of some other specific agent to account for disease. There was no categorical distinction between these predisposing poisons and the morbid poison supposed to be immediately responsible for the disease. An individual might sicken simply as a result of long exposure to a cumulative predisposing cause.[57] Chadwick regarded epidemics of cholera in this manner, as unusually severe occurrences of the endemic diarrhea that normally prevailed in cities which did not properly administer their filth.[58] It mattered little whether

---

[55] See for example "Quondam," "Water supply and disease," The Builder 12(1854): 58.

[56] In evidence to S.C. Met. Sewage Manure, Q 1503.

[57] Pelling, Cholera, pp. 21-22; Southwood Smith, Treatise on Fever (Philadelphia: Carey & Lea, 1830), p. 382; and E. Chadwick in disc. of E. Byrne, "Experiments on the removal of organic and inorganic substances in water," MPICE 27(1867-68): 34.

[58] Lewis, Chadwick, p. 190.

these poisons were inhaled or ingested, and in regard to drinking vitiated water Chadwick wrote: "all matters dissolved in it are taken up, and this without regard to their quality, the most poisonous being soaked up by the blood-vessels of the alimentary canal as readily as the most harmless."[59] Here the gut (the lungs were understood to act similarly) was portrayed as a sponge, a passive organ which would absorb automatically whatever came along. If the material absorbed were "decomposed organic matter, principally of an animal character" it would be "precisely the agent calculated to induce relaxation of the bowels, a most pernicious thing during epidemic cholera."[60]

This view of cholera reveals Chadwick's characteristic neglect of the details of pathology and disinterest in the phenomena of specific diseases in favor of an emphasis on the quality of the urban environment. All types of organic matter generated pretty much the same products of decay, and these were effective not so much as a result of the morbid chemical natures of some among them, but in proportion to their concentration and the length of time the victim was exposed to them. Although Chadwick regarded an atmosphere vitiated by the products of organic decomposition as the main health threat to city dwellers, this atmosphere could be taken up by water, and its poisons remained potent in solution. Where there were uncovered reservoirs too near a city, or where a non-continuous water supply required the

---

[59] General Board of Health, Report on the Epidemic Cholera of 1848 and 1849, P.P., 1850, 21, [1273.], p. 91.

[60] Ibid., p. 94.

use of household cisterns, the water would absorb the impurities of the atmosphere. Similarly, if the processes of decomposition occurred in a river, as they did most dramatically in the Thames during the dry, hot summers of 1858 and 1859, the poison might be transferred from the river to the atmosphere near the river, and the water therefore constituted a health hazard even if it were not drunk. No air could be good where water was bad, nor would water exposed to a bad atmosphere remain pure.[61]

Armed with this persuasive concept that decomposition made cities unhealthy Chadwick attempted to reform a variety of sanitary practices and administrative institutions, ranging from the techniques to be used in applying sewer manure to the governance of London. The view that decomposition in general was responsible for disease strengthened two precedents which became primary characteristics of the Victorian public health movement: the practice of equating danger to health with repugnance to the senses and the harnessing of scientific and medical ideas as justifications of political, administrative, financial, and aesthetic decisions. These precedents were not incompatible with subsequent versions of the relationship between filth and disease.

In 1842 William Farr introduced the concept of zymotic disease,

---

[61] Ibid.; Report by the General Board of Health on the Supply of Water to the Metropolis, P.P., 1850, 22, [1218.], pp. 16, 34; and Stevenson Macadam, "On the contamination of the water of Leith by the sewage of Edinburgh and Leith," in RCST, 3rd Report, P.P., 1865, 27, [3472.], Appendix 5, pp. 6, 20-21, 36-37; also RCST, Second Report, Appendices, pp. 44-45, for an attempt by S. Smith and Holland to show the health effects of living near the Manchester rivers.

which found its chief basis in Liebig's fermentative pathology. In drawing an analogy between disease and decomposition, Farr was recognizing a class of diseases in which the morbid poison was increased in the victim's body during the course of the disease, just as the quantity of yeast increased during fermentation. In small pox the body produced a quantity of variola capable of producing the disease in others, just as Liebig's ferments converted organic matter to their own mode of decomposition and instilled in this organic matter the ability to transmit the same change to further "fresh" organic matter.[62] The zymotic analogy also suggested that the pathological processes of these diseases were decompositions; matter undergoing a certain mode of decomposition outside the body transferred that process to the blood or other tissues when swallowed or breathed, just as Liebig had suggested.[63] For Farr and others who followed Liebig, the emphasis was therefore on the process of decomposition rather than its products. As an example, in 1858 Henry Letheby, the Medical Officer of the City of London, described the morbid effects of sewer gas: "it is matter in an active state of decomposition; . . . matter in this condition has the power to disturb the equilibrium of the organic molecules, and to propagate its own state of decay. When this occurs in the living body, it is productive of the most terrible consequences."[64] A.H. Church, a chemistry professor at the agricultural

---

[62] Eyler, Farr, pp. 102-4; Pelling, Cholera, pp. 102-3; B.W. Richardson, Field of Disease, pp. 817-18.

[63] Eyler, Farr, pp. 102-4; Pelling, Cholera, pp. 102-3.

[64] Letheby, "Sewage and sewer gas," p. 288.

college at Cirencester, spoke similarly of the effects of putrefying matters in drinking water: "They are themselves decaying or changing, and they cause changes, like putrefaction, sometimes of a hurtful kind, when they are introduced into man's stomach."[65] Like Chadwick's products of decomposition, these changing matters could be transmitted through the air as the "effluvia of decay" -- "the demons of an adulterated atmosphere" -- or by water, although there was some doubt that their unorganized condition could withstand for long the solvent action of water.[66] Again the transfer of morbid poisons between water and air, in either direction, was eminently plausible.[67]

Although one can detect, beginning in the late 1850s, a growing tendency to explain outbreaks of disease in terms of zymotic processes rather than zymotic products,[68] the ideas of Farr were not incompatible with those of Chadwick, and for many they were complementary. As early as 1852 Lankester and Redfern pointed out that the decaying organic materials in water were only occasionally exceptionally

---

[65] A.H. Church, Plain Words About Water (London: Chapman and Hall, 1877), pp. 14-15.

[66] "Thames improvements," The Field, 1870, i, pp. 155-56. The question of their behavior was much discussed by the Royal Commission on Water Supply (P.P., 1868-69, 33, [4169.], in passim). See also Henry Letheby in testimony to Select Committee on the East London Water Bills, P.P., 1867, 9, (399.), Appendix 11, Q 702; and Murchison, Treatise, p. 452.

[67] General Board of Health Medical Council, Report of the Committee for Scientific Inquiries in Relation to the Cholera Epidemic of 1854, P.P., 1854-55, 21, [1980.], p. 48.

[68] Ibid., p. 48 and in passim. Another influential study was A.W. Hofmann and Lyndsay Blyth, Report on the Chemical Quality of the Water Supplied to the Metropolis, P.P., 1856, 52, [2137.], p. 5.

dangerous, while their products -- nitrates and phosphoretted and sulphuretted hydrogen -- invariably had a "depressing" effect on the body.[69] The differences lay chiefly in emphasis: for Chadwick a concern for predisposing causes, for Liebig and Farr a recognition of disease specificity and an interest in the identification of the agents responsible. Writers who explained disease in terms of matters in a state of change often pointed out that an atmosphere full of the concentrated products of decay could not fail to act as a cooperating factor.[70]

---

[69] Edwin Lankester and Peter Redfern, Reports made to the Directors of the London (Watford) Spring Water Co. on the Results of Microscopical Examinations of the Organic Matters and Solid Contents of Waters supplied from the Thames (London: n.n., 1852), p. 3; and Jabez Hogg, "River pollution with special reference to impure water supply," JRSA 23(1874-75): 587.

[70] Notably John Simon, "The filth diseases and their prevention," Supplement to the Annual Report of the Medical Officer of the Privy Council and the Local Government Board, n.s. 1, P.P., 1874, 31 [1066.], pp. 9-10; William Procter, The Hygiene of Air and Water: Being a Popular Account of the Effects of the Impurities of Air and Water, Their Detection, and the Modes of Remedying Them (London: Hardwicke, 1872), pp. 11-12; W.H. Corfield, A Digest of Facts relating to the Treatment and Utilisation of Sewage, 2nd ed. revised and corrected (London: MacMillan, 1871), p. 17; and Robert Angus Smith, "On disinfection and disinfectants," in R.C. Cattle Plague, Third Report, P.P., 1865, 22, [3656.], Appendices, p. 156. Of course they were correct that ingesting products of putrefaction may be quite dangerous. In 1879 Jeremiah McCarthy and W.S. Greenfield published a study distinguishing the problem of septic intoxication -- like ptomaine poisoning -- from septic infection ("Report to the Pathological Society of London by the committee appointed to investigate the nature, causes, and prevention of those infective diseases known as pyemïa, septicemia, and purulent infection," in Ninth Annual Report of the Local Government Board. Supplement containing the report of the medical officer for 1879, P.P., 1880, 27, [C. 2452.], Appendix B4, p. 218). See also John Burdon Sanderson, "Report of further investigations of the properties of the septic ferment," in Report of the Medical Officer of the Privy Council and the Local Government Board for 1875. Scientific Investigations in Pathology and Medicine, P.P., 1876, 38, [C. 1608.], p. 12.

Liebig's notion that matters experiencing a certain mode of fermentation induced that same type of fermentation in new organic matter lent itself to the idea of the specificity of diseases far more readily than had Chadwick's ubiquitous products of ubiquitous decay, and it raised hopes that the fermentations of particular diseases -- the "hurtful kinds of decay" -- might be identifiable. In 1855 the General Board of Health's Committee for Scientific Inquiries in Relation to the Cholera Epidemic (of which John Simon was a member) expressed its dismay that there was no complete study of the "chemistry of organic decomposition during the epidemic prevalence of cholera -- especially into the successive transformations of animal refuse at such times."[71] Simon, who as Privy Council Medical Officer directed one of the rare endeavors at government-sponsored scientific research in nineteenth century Britain, eventually assigned the task of discovering the particular "fermentative state" in which cholera discharges were virulent to J.L.W. Thudichum in the wake of the next cholera epidemic in 1866.[72] Thudichum's largely spectroscopic studies of cholera materials did not prove fruitful, and in general sanitarians found it difficult to refine the "zyme" concept. After the mid-1850s nitrogenized or animal organic matter was commonly regarded as the most dangerous type of organic matter on the

---

[71] General Board of Health, Sci. Inq. Cholera, 1854, p. 37.

[72] J.L.W. Thudichum, "Cholera chemically investigated," in Ninth Annual Report of the Medical Officer of the Privy Council, P.P., 1867, 37, [3949.], Appendix 10, p. 486. Also see Simon's views on the goals of these studies in Tenth Annual Report of the Medical Officer of the Privy Council with appendices, P.P., 1867-68, 36, [4004.], pp. 21-22.

grounds that its putrefaction, being the most rapid, was likely to be the most virulent.[73] Farr recognized the existence of nitrogenous "zymads" for each zymotic disease, but even his naming of them ("cholerine" for instance) reflected no very specific conception of their nature.[74]

In 1865 three scientists studying the outbreak of cattle plague in Britain (the King's College, London physiologist Lionel Smith Beale, the Manchester sanitary chemist Robert Angus Smith, and the practicing chemist and editor of Chemical News William Crookes) came to the conclusion that this disease was caused by some tiny, mobile, and elusive bit of semi-living matter which they designated a germ, since it contained the ability to germinate into a case of the disease. In the studies of the 1866 cholera epidemic the name became widely used, in particular by chemists frustrated by their inability to detect the morbid poison of that disease.[75] This germ theory that emerged in Britain in the mid-1860s was effectively a metamorphosis of Farr's zymotic theory, bearing a far closer resemblance to it than to the bacterial germ theory of the 1880s. It was a germ theory whose

---

[73] Hofmann and Blyth, "Chemical quality," p. 5; Procter, Air and Water, p. 60; and Cornelius Fox, Sanitary Examinations of Water, Air, and Food. A Handbook for the Medical Officer of Health (Philadelphia: Lea & Blakiston, 1878), p. 13.

[74] Eyler, Farr, pp. 105-7. B.W. Richardson, who believed that morbid poisons were pathologically-altered glandular secretions, believed he had isolated the morbid poison of peritonitis as a crystalline alkaloid (The Field of Disease, p. 842).

[75] RCWS, Evidence, QQ 2810, 2838, 2861, 3114, 3169, 3898, 5451, 6244, 6401-3, 7011-14, 7137; Select Committee on East London Water Bills, Evidence, P.P., 1867, 9, (399.), QQ 2353-56, 2366; Appendix 11, QQ 639, 693, 702.

germs had not yet been discovered, but were only "figures of speech," indicating a belief that the agents of disease were specific tiny and resistant entities which had eluded the searches of chemists and microscopists.[76] In the tradition of John Snow the early germ theorists pointed out that the phenomena of the conveyance and specificity of diseases were consistent with the activity of a biological entity.[77] A biological entity seemed a more plausible explanation of a particular disease process than were Liebig's unorganized ferments, despite the supposedly distinct mode of fermentation each of them was believed to propagate. A biological entity could be expected to reproduce in a suitable nutritive medium, just as the zymotic morbid poison did in the victim's body: in Beale's view contagia were "very minute particles of matter in a living state, each capable of growing and multiplying rapidly when placed under favourable conditions."[78] By virtue of its vitality a living zyme

---

[76] RCWS, Evidence, QQ 2810-12. See also R.C. Cattle Plague, Third Report, p. vi, footnote; Charles Meymott Tidy, "Processes for determining the organic purity of potable waters," J. Chem. Soc. 35(1879): 53; Richardson, Field of Disease, pp. 820-21. This hesitancy to define germs is evident in America. See "An inquiry into the causes of typhoid fever in Massachusetts," pp. 164-65; and C.D. Folsom, "The disposal of sewage," in Seventh Annual Report of the State Board of Health of Massachusetts for 1876 (January 1876), (Boston: Wright and Potter, 1876), pp. 283-84.

[77] John Snow, "On continuous molecular changes, more particularly in their relation to epidemic diseases," in Wade Hampton Frost, Snow on Cholera (New York: Commonwealth Fund, 1936), pp. 156-57. See also J.K. Crellin, "The dawn of the germ theory: particles, infection, and biology," in F.N.L. Poynter, ed., Medicine and Science in the 1860s, Proceedings of the 6th British Congress on the History of Medicine, University of Sussex, 6-9 September 1967 (London: Wellcome Inst., 1968), pp. 57-76.

[78] L.S. Beale, "On the nature of the morbid poison -- contagium of

could resist the ravages of chemical and physical forces: "outside the body they are capable of resisting for very long periods the influences of conditions, which if not restrained by organic action, would produce chemical decomposition."[79] These germs were not regarded as bacteria, but something smaller, and less alive, "matter of a kind which will always be undiscoverable to the microscope."[80] It was only gradually during the 1870s and 1880s, and largely as a result of continental work, that these "germs" began to be recognized as fungi or bacteria, and the concepts of germs and zymes became distinct.[81]

Much of the talk of living germs by British scientists in the late 1860s was a consequence of inadequacies they perceived in Liebig's explanation of fermentation, and a feeling that perhaps Liebig's sweeping reductionism required of purely chemical processes results they could not produce. Thus Beale, whose cattle plague germs resembled protoplasm more than cells, felt that only a living zyme could direct the body's tissues to follow the same process of

---

the cattle plague," in R.C. Cattle Plague, Third Report, p. 152.

[79] J. Burdon Sanderson, "Intimate pathology of contagia," p. 255.

[80] R.C. Cattle Plague, Third Report, p. vi; Beale, "Contagium of cattle plague," p. 150; William Farr, Report on the Cholera Epidemic of 1866 in England. Supplement to the 29th Annual Report of the Registrar General of Births, Deaths, and Marriages in England, P.P., 1867-68, 37, [4072.], p. lxvii; Hogg, "River pollution," p. 589; Blyth, Dictionary of Hygiene, s.v. "bacteria" and "germ," pp. 71, 252.

[81] Burdon Sanderson's annual reports, published in Simon's annual reports of the medical officer of the privy council during the early 1870s, which reviewed and sometimes repeated continental work, appear to have been instrumental.

fermentation as the zyme was undergoing.[82] Likewise, William Crookes could not understand how purely chemical substances could multiply.[83] More than anything these criticisms reflect their authors' preferred modes of explanation; Liebig had confronted exactly these issues and had argued analogically that there was no reason chemical processes could not perform these functions.

Thus, even for those who believed germs were legitimate organisms and not marginally-living products spontaneously generated from decomposing filth, it was hard to keep the concept of germs distinct from the process of decay. Pettenkofer's theory that the specific agents of cholera, and by implication of typhoid, required a nurturing "nidus" of decomposing matter in which to develop was taken up by Simon and became popular among British sanitarians.[84] Equally well-known were the experiments of Karl Thiersch, repeated at Simon's request by Burdon Sanderson, which showed that minute bits of cholera evacuation did not become virulent until they had fermented for a few days.[85] Even if germs were discrete organisms and were the ultimate causes of disease it was only germs in matter in a certain condition

---

[82] Beale, "Contagium of cattle plague," p. 150.

[83] William Crookes, "On the application of disinfectants in arresting the spread of the cattle plague," in R.C. Cattle Plague, Third Report, p. 187.

[84] Pelling, Cholera, p. 283; also R.A. Smith, "Disinfection and disinfectants," p. 156; and Blyth, Dictionary of Hygiene, s.v. "putrefaction," pp. 475-76.

[85] Pelling, Cholera, pp. 236, 242-43; J. Burdon Sanderson, "On the experimental proofs of the communicability of cholera," in Ninth Annual Report of the Medical Officer of the Privy Council, P.P., 1867, 37, [3949.], pp. 438ff; and J.L.W. Thudichum, "The relation of the microscopic fungi to great pathological processes, particularly the

of decomposition that one had to worry about. Among the early British germ theorists Edward Frankland and John Tyndall were unusual in recognizing the independence of germs from decomposing organic matter.[86]

No matter what form the morbid poisons of zymotic diseases took, the pathological processes in these diseases continued to be understood as putrefactions of the living tissue. After the mid-1850s it is likely that pathology and not epidemiology was the key element that linked disease with decomposition. John Simon in particular, in his yearly reports during the 1860s and 1870s, frequently presented the pathological processes of the zymotic diseases as fermentations. Septicemia and erysipelas, which closely resembled the putrefaction of dead organic matter, exemplified diseases which spread inward from a lesion on the body's surface, while typhoid exemplified decomposition of the gut, and tuberculosis of the lung.[87] The definition of fermentation was sufficiently loose that phenomena in a considerable variety of diseases could be used to support the contention that they were fermentations within the body. De-oxidation, for example, was a characteristic of putrefaction. Although John Simon had rejected Liebig's zymotic pathology in his 1850 General Pathology,

---

process of cholera," Monthly Microscopical Journal 1(1869): 16-20.

[86] Frankland in Evidence to R.C. Water Supply, Q 6372; John Tyndall, "Observations on the optical deportment of the atmosphere," BMJ, 1876, i, pp. 121-24. But see Frankland's statement of the more common view (E. Frankland, "The water supply of London and the cholera," Quarterly J. Science 4(1867): 315).

[87] John Simon, "Filth diseases," pp. 12-16; Pelling, Cholera, p. 283; Murchison, Treatise, p. 386.

he had nevertheless pointed to the de-oxidized condition of the excretory organs as a characteristic common to many diseases and had drawn the parallel between this pathological condition within the body and the unhealthy and de-oxidized condition of cities that contained too much decomposing matter.[88] Thudichum, in his investigation of the particular choleraic decomposition, pointed to a similar occurrence, a failure of oxidation during the "essential" stage of cholera.[89] Inflammation might be conceived as the element common to all zymotic diseases. Burdon Sanderson showed that pus, which caused a surgical septicemic inflammation when applied to a wound, produced peritonitis when injected into the peritoneal cavity, and pointed to the similarity of this inflammation with the appearance of the ulcerated intestine of a typhoid victim.[90]

Even B.W. Richardson, who rejected both the zymotic theory of Simon and Farr and the germ theory, accepted fermentation a la Liebig as a fundamental part of the pathological process. In 1858 he suggested the alkalinity of the blood in many diseases might be due to the ammoniated atmosphere of cities, which might sometimes be absorbed by bodily fluids.[91] Later (and not to the exclusion of his

---

[88] John Simon, General Pathology. A Course of Lectures delivered at St. Thomas's Hospital (London: Renshaw, 1850), pp. 282-84; Pelling, Cholera, p. 78; and Royston Lambert, Sir John Simon 1816-1904 and English Social Administration (London: McGibbon and Kee, 1963), pp. 47, 270-79.

[89] Thudichum, "Cholera chemically investigated," p. 460.

[90] J. Burdon Sanderson, "Reports of an experimental study of infective inflammations," in Reports of the Medical Officer of the Privy Council and the Local Government Board, n.s. 6. Report of Scientific Investigations . . . in Pathology and Medicine, P.P.,

ammonia theory) Richardson emphasized zymotic diseases as alterations of the normal fermentations of physiology which could spread by contact to other fluids in the body. Richardson's description echoes Liebig: "a particle of any one of the poisons [an altered secretion of a glandular fermentation] brought into contact either with the blood of the living animal or with certain secretions of the living animal, possesses the property of transforming that secretion into a substance like itself." He added: "The process of change is catalytic."[92]

As the central position of Joseph Lister in the development of the germ theory in Britain reveals, the actions of germs were also understood in the context of this septic pathology. In introducing his hypothetical cattle plague germs, Crookes noted that their pathological action was to initiate "a decomposition caused by the act of nutrition of the living cell."[93] John Simon, in his famous report on the Filth Diseases (1874) presented the "common" septic ferment (B. termo at this time) as being the cause of typical zymotic disease. In the bowels its decomposing activity resulted in the diarrhea endemic in places where decomposing filth was always present. In small

---

1875, 40, [C. 1371.], Appendix 2, pp. 63-64.

[91] Richardson, Cause of Coagulation, pp. 348, 369-70.

[92] Richardson, The Field of Disease, pp. 840-49, esp. pp. 844-45; idem., Clinical Essays (London: Churchill, 1862), I, pp. 96-99.

[93] Crookes, "Application of disinfectants," p. 187. Also see R. Angus Smith, Air and Rain. The Beginnings of a Chemical Climatology (London: Longmans, Spencer, 1872), p. 520. Jabez Hogg wrote: "These atoms [bacteria] are the active agents in putrefactive changes; . . . having gained access to certain internal organs . . . they proceed to leaven the whole mass of blood" ("River pollution," pp. 588-89).

doses it might be the cause of tuberculosis.[94] A similar septic ferment produced the putrefactive changes of typhoid, another of cholera. Since the normal process of putrefaction was thus conceived as fundamentally malign, there was in this view of zymotic diseases no basis for making a distinction between the bad microbes that caused diseases and the good ones required for the cycling of matter in nature. H. Hoffert observed that bacteria were normally content with decomposing the organic detritus found in nature, but if one allowed them to multiply by conserving filth these same bacteria would turn on living humans and zymotic diseases would result.[95]

These views about the causes and processes of disease combined to form a popular medical concept in which the phenomena and processes of decomposition represented the paradigmatic malignant influence, but they failed to stimulate research by British sanitary scientists to discover how the process of decay -- admitted to be necessary for the continuance of life by both Pasteur and Liebig -- sometimes became the cause of disease. As has been suggested, the various ways in which decomposing matters might be considered harmful had in some degree evolved from one another, and rarely were they considered mutually exclusive. At the same time there were elements in the sanitary enterprise that effectively worked against a resolution of this issue.

---

[94] Simon, "Filth diseases," p. 16.

[95] H. Hoffert, A Guide to the Sewage Question for 1876, treated from a Sanitary, Economical, and Agricultural Point of View (Weymouth: Sherren & Son, 1876), pp. 8-9.

A central factor here was the contemporary legitimacy of multi-factoral explanations of disease. To admit a new mode of disease transmission did not require abandoning previously-accepted modes. Many accepted John Snow's conclusion that cholera could be spread by drinking water specifically polluted with the excreta of cholera sufferers but few agreed with Snow that this was the only way the disease was spread.[96] Typhoid was believed to arise from a variety of sanitary errors which vitiated air or water. In 1874 John Simon, who at the time believed in the existence of a specific and living typhoid germ, compiled a list of typhoid outbreaks his staff had investigated, and each of these was ascribed to its particular local cause.[97]

While legitimate medical theory in a more or less pure form permitted a considerable range of acceptable explanations of the outbreaks of disease, inestimable further confusion resulted from the activity of the great number of sanitary experts who, without medical training, assumed the authority to speak competently on medical questions. These included lawyers (Chadwick, C.N. Bazalgette), engineers (Robert Rawlinson, Thomas Hawksley), chemists (A.S. Taylor, J.A. Wanklyn), politicians (Lord Robert Montagu), and clergymen (Charles Kingsley, Henry Moule). A little sanitary science was a dangerous thing, and the sanitary literature is full of eclectic perversions of medical theories. To the engineer William Eassie the common under-

---

[96] Pelling, Cholera, pp. 63, 227.

[97] Simon, "Filth diseases," pp. 43ff.

standing that decaying organic matter supported the growth of germs appeared as the notion that gases (presumably from putrefaction) supported germs: "When they [germs] meet with hurtful gases, or units with more organised bodies in a diseased state, they generate and multiply to an alarming extent, even to the producing of an epidemic."[98]

Moreover, sanitary science was rarely discussed in abstract, and statements on the relations of disease and decomposition usually appear in support of or in opposition to specific sanitary measures such as proposed sewage farms or particular sources of water. If one were promoting one's patent purifying system, it was well to point to its effectiveness against all the modes in which decomposing organic matter might be supposed to exert its malign influence. Thomas Spencer, inventor of the "Magnetic Carbide" water and sewage filter, assured the public that his invention would remove the gaseous products of putrefaction, any putrefying matters, and any germs in the water.[99] Similarly Frederick Krepp, writing to support the Liernur system of sewage disposal (in which a huge vacuum cleaner sucked sewage from houses through air-tight pipes to an air-tight cart), applied whichever theory of disease seemed suitable as he sequentially proved

---

[98] Wm. Eassie, "The effects of growing vegetation upon human health," SR 7(1877): 232. The paper was given at the Sanitary Institute Congress at Leamington in 1877.

[99] [Thomas Spencer], A Brief Description of Spencer's Patent Magnetic System of Purifying Water (London: Magnetic Filter Co., 1869), p. 20.

unhealthy all other methods of sewage disposal.[100]

In light of their inability to isolate morbid poisons, it is perhaps not surprising that British sanitarians, when they spoke in general terms, thought it wiser to be safe and inclusive, to consider all possible dangers manifest in decomposition, rather than trying to prove the safety of decomposition under certain circumstances. Henry Letheby's attempt to cover all possibilities is typical: "Organic matter could only be hurtful when it was in certain states of change or decay, or when it was endowed with specific germs of morbific action, as in the case of the germs of disease, and perhaps also where, as it would seem, the excrements were hurtful to the animals or vegetables excreting them, but not to other species."[101] Writers of manuals of hygiene like Charles A. Cameron, medical officer of Dublin, and George Wilson, medical officer for mid-Warwickshire, listed distinct conceptions of the nature of morbid poisons and their relation with decomposition, apparently feeling no obligation to choose among them. For Cameron, proteinaceous materials in a condition of "retrograde metamorphosis" might cause typhoid; worse yet, if these materials contained the evacuations of typhoid

---

[100] Frederick Krepp, The Sewage Question: Being a Review of all Systems and Methods hitherto employed in various cities for draining cities and utilising sewage: treated with reference to Public Health, Agriculture, and National Economy generally (London: Longmans, Green and Co., 1867), pp. 35, 50, 61, 67, 87, 105, 110.

[101] In discussion of E. Byrne, "Experiments on the removal of organic and inorganic substances in water," MPICE 27(1867-68): 36-37; and Letheby, "Composition and quality of metropolitan waters in July, 1866," CN 14(1866): 83.

patients they might be "pregnant with the germs of that disease."[102] Even the doctrine that the only dangerous filth was that specifically polluted by the excreta of a victim of one of the intestinal zymotic diseases did not, in practice, serve to refine the filth - disease relationship since there was no way of distinguishing this dangerous filth from ordinary safe filth. Contrary to the fears of some, that the doctrine of specificity would undermine sanitary improvement, the opposite was often the result, and among those who advocated the strongest measures for protecting the public from zymotic diseases were those, like Frankland, who argued in favor of many levels of safeguards on the grounds that one could never be sure when the undetectable contagia were present or what purifying influences they could withstand.[103]

Indeed, on a practical level, to the local inspectors of nuisances who went about disinfecting stinks indiscriminately and to the medical inspectors who explained outbreaks of disease, it made little difference what form the morbid poisons took, or what the exact

---

[102] C.A. Cameron, A Manual of Hygiene, Public and Private, and Compendium of Sanitary Laws (Dublin: Hodges, Foster, 1874), pp. 63-64; George Wilson, A Handbook of Hygiene and Sanitary Science, 4th ed. (London: Churchill, 1879), pp. 60, 68, 194-95; and Procter, Air and Water, pp. 11-12, 46, 51, 60, 67, 73-74. For a similar American view see W.R. Nichols, in Albert Buck, ed., Treatise on Public Health and Hygiene (London: Sampson, Low, Marston, Searle, and Rivington, 1879), I, pp. 218-19. See also Hogg, "River pollution," pp. 581, 584, 587; W.H. Corfield, Water and Water Supply (New York: D. van Nostrand, 1875), pp. 25-26; and E. Lankester, A School Manual of Health (London: Groombridge, 1868), pp. 20, 47-50.

[103] See Frankland in Evidence on the Cheltenham Corporation Water Bill, 1878 (House of Lords Record Office, Minutes of Evidence, House of Commons, 1878, Vol. 5, QQ 2910-11); Simon in R.C. Water Supply, Evidence, Q 7142; and Pelling, Cholera, p. 282.

relations between decomposition and disease were. John Simon, who more than anyone else attempted to make British public health medicine scientific by sponsoring the researches of Burdon Sanderson on the nature of contagia and of Thudichum on the chemistry of pathology, effectively ignored these researches when it came to the practical task of administering towns' sanitary efforts. Like Chadwick, Simon recognized that attending to the proximate cause of disease (decomposing matter) would substantially improve the public health.[104] To the medical inspectors who worked under Simon the existence of filth in places where it ought not to be was sufficient explanation of the occurrence of zymotic disease, and certainly where filth prevailed everywhere, there was frequently little basis for demonstrating particular routes of disease transmission by epidemiological techniques. Dr. Bristowe asserted in his report on Grantham: "Knowing as we do that typhoid fever has been shown over and over again to arise from the effluvia of accumulated human excrement, and that . . . such effluvia have prevailed, there remains . . . scarcely a doubt that the disease must be attributed to this cause."[105] George

---

[104] Simon's "Filth diseases" illustrates this point. In the first half of the report he argues that diseases are caused by semi-specific germs. In the second he advocates the removal of all types of nastiness from towns. The connection is the "common" septic ferment, which Simon believes exemplifies the character and habits of zymotic germs. Compare pp. 9-10 with pp. 37-38. There was also some feeling that the exact form of the morbid poison of zymotic diseases made little difference therapeutically (E.J. Syson, "The antiseptic treatment of zymotic diseases," SR 7(1877): 339-40).

[105] In Seventh Annual Report of the Medical Officer of the Privy Council, P.P., 1865, 26, [3484.], Appendix 9, pp. 518-19. For other examples see Simon in Tenth Annual Report of the Medical Officer of

Buchanan, Simon's eventual successor, wrote of Bridport: "Bridport must be purified from the mass of putrid matter that lies in its privy vaults, and poisons the air by its decomposition. So long as the system lasts of storing up decaying filth for months and years, the town will have a high death rate and the epidemic diseases will find an extensive number of victims."[106]

Thus, what emerged from this melange of theories of disease, appearing in a multitude of different contexts, serving a multitude of purposes, and expressed by all sorts possessing varying degrees of medical knowledge, was simply the relegation of decomposition to the position of a process fundamentally antagonistic to the human enterprise. In the popular sanitary literature decomposition and all the phenomena associated with it -- stinks, stagnancy, slime, aquatic invertebrates and vegetation -- assume the same malign status. Statements like Samuel Homersham's condemnation of reservoirs are frequent. They "would represent the cesspool of a large district of the country around. . . . it would be impure and contaminated with the droppings of animals, decaying vegetable matter, the exuviae of fish, and living organisms, and was altogether unfit . . . for domestic

---

the Privy Council, 1867-68, pp. 10-12; and the inspectors' reports (Appendix 2, pp. 29, 33; Appendix 4, p. 45). Beginning with the annual report of 1874 these local reports are listed as "Abstract of medical inspections made in year with regard to matters of local sanitary administration," as in Reports of the Medical Officer of the Privy Council and the Local Government Board, n.s. 4, Annual Report for 1874, P.P., 1875, 40, [C. 1318.], pp. 99-105.

[106] Seventh Annual Report of the Medical Officer of the Privy Council, p. 525.

purposes."[107] Another writer sickened at the idea of sewage "festering, . . . fermenting, and decomposing, . . . exhaling offensive and poisonous gases continually about our homes, and filtering incessantly both natural and diseased animal juices down our springs and wells."[108]

V. Decomposition in a Benign World?

It is not surprising to find in the sanitary literature considerable sentiment that decomposition ought to be prevented. Since by "decomposition" most writers meant the nuisances and health hazards associated with putrefaction, they often accepted slow combustion, in which no nuisances were engendered, as exemplifying the process of decay in a healthy natural environment. Many argued that putrefaction or fermentation were not necessary stages in the natural transformations of substances, and as will become clear in the next chapter, some denied the necessary existence of any process of decomposition whatever. The more common view, here expressed by Philip B. Ayres in a defense of cesspools, was that nature working properly would prevent the development of nuisances: "In the beneficent process of creation, such conditions have been imposed on animal matter, that when deprived of life it shall be as little injurious as possible to living

---

107 In discussion of Thomas Beggs, "The water supply of London as it affects consumers," JRSA 15(1866-67): 214.

108 Francis Taylor, Human Manure, Its Collection and Conversion to Guano (London: Churchill, 1861), p. 6.

beings; it either forms food for other animals and plants, or, if not adapted for these purposes, it is endowed with peculiar properties that retard its decomposition, and thus admit of such great dilution of its emanations with atmospheric air as to render them practically innoxious."[109]

Often the explicit goal of sewage treatment was to prevent putrefaction, either by supplying excess oxygen to make the transition from organic to inorganic rapid and inoffensive, or by adding antiseptic chemicals which would simply stop decomposition from occurring. The concept of a "septic" form of sewage treatment was fundamentally absurd in this context although some, notably Robert Angus Smith, recognized that the eventual result of the process of putrefaction was a purified (inorganic) condition.[110] Even for Smith, however, decomposition was not the ideal mode of purification; instead, an antiseptic mode of treatment would insure that the dangerous process of putrefaction with its dangerous products never occurred: "The problem . . . of preserving meat, or of preserving the entire animal from corruption, and the problem of preserving sewage or feces from decomposition, become entirely the same."[111] In

---

[109] Philip B. Ayres, "On the disposal of the fecal and other refuse of London on rational principles in contrast with the schemes of sewerage now before the public," J. Public Health & Sanitary Review 3(1857): 20.

[110] R. Angus Smith, "On the examination of water for organic matter," CN 19(1869): 279, 281; same title, Proc. Manchester Lit. and Phil. Soc., 3rd series 4(1871): 47.

[111] R. Angus Smith, "On disinfection and disinfectants," p. 159; and Gibson and Farrar, "Robert Angus Smith," p. 249.

1859 A.W. Hofmann and Henry Witt rejected all chemicals proposed for treating London's sewage because while many of them could temporarily prevent putrefaction none of them could ultimately stop decomposition.[112] Similarly, Smith rejected charcoal as a disinfectant because it had "no power of preserving organic substances."[113] Others spoke of the need to find substances which would "mummify" excreta, or which were capable of "arresting corruption."[114]

Generally, however, the goals of purification technology had the same ill-defined character as the association of disease with decomposition. A.W. Blyth's description of the variety of tasks included in the concept of disinfection shows why so great a range of substances could be sold as disinfectants. According to Blyth, disinfectants had to deal with inorganic products of decomposition (H_2S), unoxidized intermediate organic products of putrefaction ("compound stinking ammonias"), organic matters dead but in "a state of change," "little bits of pus, dry and hard without but soft within," "living and growing cells," and finally "soft easily destructible bodies."[115]

---

[112] Hofmann and Witt, "Report on chemical investigations relating to the main drainage of the metropolis," in D. Galton, J. Simpson, and T. Blackwell, Report Relating to the Main Drainage of the Metropolis, P.P., 1857, 10, (233.), p. 40. Also see RCST, Second Report, p. 14; Appendices, p. 65; and F. Crace-Calvert, "Carbolic or phenic acid and its properties," JRSA 15(1867): 731.

[113] R. Angus Smith, "On disinfection," JRSA 5(1856-57): 337, 339.

[114] J.L.W. Thudichum, "On an improved mode of collecting excrementitious matter, with a view to its application to the benefit of agriculture and the relief of local taxation," JRSA 11(1862-63): 443. Also see J.T. Way, "On town sewage as manure," J. Royal Agricultural Society 15(1854): 137; and Corfield, Digest, p. 208.

[115] A.W. Blyth, "Disinfection" in his Dictionary of Hygiene, pp. 187-88.

Usually, the prevention of the sensible nuisances of putrefaction was judged to be the measure of effective disinfection, hence the frequent and ancient practice of equating deodorization with disinfection. Andrew Fergus, a Glasgow physician who was one of the most adamant advocates of an antiseptic system of sewage treatment, admitted to Frankland during the Rivers Commission hearings that what he was really concerned with was the prevention of sensible nuisances, and that the destruction of refuse by oxidation was as safe a means of preventing "decomposition" as an antiseptic treatment.[116]

The propriety of decomposition was also a matter of great importance with regard to ideas about the harmonious workings of nature and civilization. Edwin Chadwick believed decomposition should be prevented both because it caused disease and because putrefaction volatilized ammonia, a valuable fertilizer.[117] Other sewage farming advocates tried to prove that the processes of decomposition were not so active as to dissipate all the valuable elements in sewage en route from cities to fields.[118] Thus when Thomas Hawksley, M.D., spoke of

---

[116] A. Fergus, Excremental Pollution, a Cause of Disease: with hints as to Remedial Measures (London: n.n., 1872), pp. 3, 10; idem., The Sanitary Aspect of the Sewage Question, with remarks on a little suspected, and not easily detected source of typhoid and other zymotics (Glasgow: Porteous, 1872), pp. 1, 13; Fergus in testimony to RPPC, 1868, Fourth Report, Evidence, QQ 2786-89; Fergus, "The sewage of towns and the disposal of organic refuse," SR 1(1874): 264; and idem., "Preventative or state medicine," SR n.s. 1(1879-80): 108.

[117] S.C. Met. Sewage Manure, Evidence, Q 1503. This concern becomes more understandable when we realize that many believed NH_3 and not NO_3 was the proper state of fertilizing nitrogen. In that case it was important that the processes of resolution not proceed to their ultimate stage of nitrification.

[118] See Select Committee on the Sewage of Towns, First Report,

decomposition as a process in which organic matter was "perverted into a curse which blights vitality" he had in mind both the direct pathological effects of decomposition and its eventual effect on the nutrition of the populace.[119] Others conceived that the Creator had seen fit to enforce sanitary laws through a carrot and stick approach. By making excreta and refuse disgusting and productive of disease He encouraged us to get rid of it; by giving it fertilizing potency He encouraged us to get rid of it on the soil where it belonged.[120] The concept was embodied in Palmerston's ecological maxim that "dirt is valuable matter in the wrong place."[121] The occurrence of putrefaction therefore signaled an imbalance in the relationship between civilization and nature. One writer worried that if the sewage of a town were allowed to decompose, insufficient oxygen would remain for the respiration of the town's inhabitants.[122] An 1872 Builder editorial presented putrefaction as the consequence of the Faustian

---

P.P., 1862, (160.), p. 14; Evidence, QQ 254, 656, 844, 1261.

[119] Thomas Hawksley, M.D., Matter -- Its Ministry to Life in Health and Disease; and Earth as the Natural Link between Organic and Inorganic Matter (London: Churchill, 1866), p. 3. Readers of sanitary literature should take care to distinguish Hawksley, M.D., from Hawksley, C.E.

[120] Lloyd Stevenson ("Science down the drain") has allued to a religiousity which is a central part of the "sanitarian syndrome." This is very much in evidence in many discussions in which an attempt is made to include the phenomena of decomposition within a benign world view. See Krepp, The Sewage Question, in passim; and "The 'Cloaca Maxima' of the 'Metropolis Magna'," The Builder 33(1875): 1073; and Kingsley, Yeast -- A Problem (New York: J.F. Taylor, 1903), pp. 279-80.

[121] The statement was made in a speech to the Romsey Farmers Club (Taylor, Human Manure, title page).

[122] John Astle, "The utilisation of sewage -- Stroud sewage," The Field 33(1869): 397.

bargain of urbanization.[123]

When nature was working properly, the fate of sewage and refuse was what may be called utilization, as distinct from decomposition, and which will form the subject of the next chapter. Briefly, this concept emphasized the assimilation of dead organic matter into the food chain, either in a direct form by animals and plants or, for those who recognized that the humus theory was no longer tenable, in a direct form by animals, and by plants only after the occurrence of a salutory form of decomposition -- a slow combustion -- in an environment containing an excess of oxygen.

Unfortunately, this smugly optimistic version of the natural cycles on which civilization was based, with its notion of preserving the refuse of animals for the use of plants, contradicted fundamentally what most Victorians accepted as the most nearly correct version of the workings of the organic world: the organic chemistry of Liebig. For the natural machine to function -- the analogy was not uncommon -- decomposition, including putrefaction, had to occur, and this was the other horn of the decomposition dilemma. Liebig pointed out that plants obtained their food in an inorganic form, their phosphorous and potassium from the soil, the carbon dioxide (exhaled by animals) and ammonia (from decomposing matter) from the air.[124] Even

---

[123] "The exact relation between the food of plants and the refuse of towns," The Builder 30(1872): 818.

[124] Liebig was overly optimistic in his estimation of the abilities of plants to get their nitrogen from the air. The more common English view, that plant nitrogen came as ammonia or nitrates (there was great disagreement on which) from decaying matters in the soil, made the conservation of combined soil nitrogen important, and made even more

a wholesome Victorian conception of nature, to the extent it accepted Liebig's views, required some process for transforming dead and cast off organic substances into the food of plants. In Liebig's chemistry putrefaction was an essential first step in this transformation. For Liebig, this was a "grand natural process," this "dissolution of all compounds formed in living organisms."

> The compounds formed in the bodies of animals and plants, undergo, in the air, and with the aid of moisture, a series of changes. . . . Thus their elements resume forms in which they can again serve as food for a new generation of plants and animals. . . . Death, followed by the dissolution of the dead generation, is the source of life for the new one.[125]

Not only was Liebig emphatic that plants ate inorganic food; equally he stressed the principle of conservation of the fertilizing elements in agriculture. To grow the same crop again in the same place, the farmer had to return to the soil those same elements which had been removed in harvesting the crop. Agriculture, and by inference civilization, had very short futures when this was neglected. This raubwirtschaft, the failure to return to the soil the precise mineral elements necessary to maintain productivity, was especially characteristic of English agriculture in Liebig's view.[126] The adop-

---

imperative the recycling of sewage. See R. Haines, "Notes on nitrification," CN 4(1861): 259-60; "Metropolitan sewage," BMJ, 1859, i, p. 70; and esp. W. Gregory, "Review of Liebig's Organic Chemistry in its Application to Agriculture and Physiology," Quarterly Review 69(1842): 343-44.

[125] Liebig, Letters, 1851, #15, pp. 180-81; A.W. von Hofmann, The Life-Work of Liebig. The Faraday Lecture for 1875 (London: Macmillan, 1876), pp. 18-19.

[126] J.L.W. Thudichum, The Discoveries and Philosophy of Liebig

tion of water closets meant England was throwing into the sea its capital of phosphates.

> In the large towns of England the produce both of English and foreign agriculture is largely consumed; elements of the soil indispensable to plants do not return to the fields, -- contrivances resulting from the manners and customs of the English people, and peculiar to them, render it difficult and perhaps impossible, to collect the enormous quantity of phosphates which are daily, as solid and liquid excrements carried into the rivers.[127]

Implicit also was the warning that unless these minerals were returned to the land in a form available to plants they would be entirely useless.

If Liebig was responsible for posing this dilemma, he was also guilty of circumventing it. To be sure Liebig had a great deal to say in his popular works about the danger of decomposing matter, and in the same books, but invariably in different sections of them, he had an equal amount to say regarding the mineral theory of manures and the importance of returning to the fields the elements that had been taken from them. Further, Liebig typically emphasized the degradative role of animal digestion in converting organic material into an inorganic form. He presented excreta as the "smoke and soot," or the "ashes" of the bodily furnace, forgetting to mention that these products were still capable of unpleasant, and by his own admission, dangerous changes.[128] Although he served as a consultant

---

(London: Trounce, 1876), p. 18; Hofmann, The Life-Work of Liebig, pp. 18-19.

[127] Liebig, Letters, 1851, #30, p. 473.

[128] Liebig, Letters, 1851, #33, pp. 499-500; #34, p. 514; Animal

for schemes to recycle London's sewage, Liebig was not much bothered with the practical problem of recycling sewage safely; his functions were to lend his prestige to the schemes and to harangue the press with attacks on competitive schemes and with frequent reminders of the importance of sewage recycling for the future prosperity of profligate England.[129]

Nevertheless, the dilemma existed and the resolutions British scientists and sanitarians offered for it have a great importance in bringing about their eventual acceptance of the processes of decomposition as truly necessary parts of the economy of nature, and later in their designing of technologies which would artificially facilitate these processes -- i.e., sewage treatment plants. The dilemma and its resolutions are equally important in understanding the reaction of British scientists to Louis Pasteur's statements that, in effect, microbes were what made the world go around. Their central resolution of the dilemma, the conception of a process of biological utilization that opposed and made unnecessary the processes of decomposition, forms the subject of the next chapter.

<u>Chemistry</u>, 1842, p. 250. See also W.A. Miller's testimony to the S.C. Met. Sewage Manure, <u>Evidence</u>, Q 519.

[129] Volhard, <u>Liebig</u>, II, pp. 59-65; Select Committee on Sewage (Metropolis), <u>P.P.</u>, 1864, 14, (481.), <u>Appendices</u>, pp. 347-48, 422.

IV. BIOLOGICAL PURIFICATION: 1850-1880

> The law of decay is to the organic world what the law of gravitation is to the physical.
>
> --Earle B. Phelps*

The origins of the idea that aquatic organisms remove sewage from rivers are commonly found in the 1890s, years of great excitement about the activities of the newly discovered saprophytic and nitrifying bacteria, and when enhancing the activities of these bacteria became the chief desideratum in the design of sewage treatment plants. In different form, however, the concept of the biological purification of polluted streams existed fifty years earlier. British scientists and natural theologians interested in the circulation of matter recognized in the 1840s and 1850s that aquatic organisms were enormously important in disposing of the organic pollutants dumped into rivers. In Britain interest in the activities of these organisms peaked in the 1850s, in part due to widespread interest in the chemistry of natural transformations, in part due to the popularity of aquaria, and finally as a result of the application of the concept by scientists consulted on issues of water quality during this period when standards of water quality and techniques of water analysis both were in a transitory and confused state. During the 1850s there was an active and sometimes experimental interest in the "habits and economy" of aquatic organisms. Although the

---

* Earle B. Phelps, Stream Sanitation (New York: John Wiley, 1944), p. 38.

concept still appears in the sanitary literature of the 1860s and 1870s, by this time it had become entrenched in sanitarian rhetoric, a scientific principle used either to demonstrate the harmony of creation or to defend pollution. Bound by these contexts, the accumulated scientific credibility of the concept of biological purification vanished.

This chapter examines the benevolent mid-Victorian outlook from which the concept of biological purification emerged, the development of the concept by Liebig, the application of the idea of biological purification as a principle of natural theology and a defense of pollution, its experimental elaboration through the development of aquaria, and its technological application in sewage farming, and concludes by considering the failure of the biological purification concept to inspire research and the reasons for its rejection by British scientists during the 1870s and 1880s.

I. Providence and Cycles: The Natural Theology of Mutability and Transformation

A world in which the essential, inevitable, continual, and ubiquitous process of organic decomposition was regarded as invariably offensive and harmful, and frequently lethal, was unacceptable to many British sanitarians. Mid-Victorian sanitarianism was an actively religious enterprise, and sanitary rhetoric consistently reflects the natural theology of William Paley, the belief that the phenomena of nature are the contrivances of the Creator and exhibit His universal benevolence. As Gladstone put it in an address to the Bolton Liberal Association on the actual and ideal quality of urban life, "God made this world to be pleasant to dwell in."[1] Improving the quality of life clearly required great expenditure of human energy and ingenuity, but the sanitarian natural theologians could also point to various aspects of the creation as processes whose apparent purpose was to purify the world of organic, and particularly of human refuse, and thereby to safeguard the public health.

Many early nineteenth century discoveries in organic and biochemistry, and in physical geography seemed to indicate the Creator's concern for the quality of urban life and were recognized in this regard as contrivances for purification. The discovery of the reduction of carbon dioxide to form respirable oxygen by photosynthetic plants -- and the complementary production of the carbon dioxide needed by plants in the process of animal respiration -- showed splendidly that the respective departments of life were adapted to purify one another's exhaled gases. There was likewise great celebration of the chemistry of transformations and the conservation of matter. Natural theologians traced the movements of matter through successive

---

[1] Quoted in SR 7(1877): 96. R.M. Young distinguishes between the natural theology of Paley, which stressed the benevolence of the Creator, and that of Thomas Chalmers, who portrayed God as a strict and unforgiving enforcer of the laws of creation (Young, "Malthus and the evolutionists: the common context of biological and social theory," Past and Present, #43(May 1969): 114-25). See also Lloyd G. Stevenson, "Science down the drain -- on the hostility of certain sanitarians to animal experimentation, bacteriology, and immunology," BHM 29(1955): 2.

stages of nature and marvelled at the useful operations in which it participated at every stage. The sanitarian natural theologians were delighted with oxygen, a wonderful material that both supported animal vitality and by safely combusting de-vitalized organic matter freed the world of impure animal and plant refuse.[2]

This fascination with the natural theology of chemical transformation is very much in evidence in the sewage recycling literature where it is constantly reiterated that only through recycling could the intended function of refuse organic matter be fulfilled. The continual series of transformations was also the means adopted by nature to purify the world, and specifically to free it from the effects of decomposing matter. Corpses and sewage became flowers and grain; ill-starred lovers found ultimate consummation in the intertwining rose and briar. As one poet expressed it:

> If life no more can yield us what it gave
> It is still linked with much that calls for praises, --
> A very worthless rogue may dig the grave,
> But hands unseen will dress the turf with daisies.[3]

When it worked properly nature hurried putrid matter on to new and useful manifestations. This faith in the benevolent course of the

---

[2] See C.G.B. Daubeny, "On ozone," J. Chem. Society 20(1867): 1-28. For those with faith in this chemical version of natural theology the bloody world of Darwin posed no threat. Charles Kingsley, who was both a sincere natural theologian and a Darwinist, appears not the least bit bothered when, in Water Babies, his personified insects are devoured by his equally personified trout. Nature's chief administrator is simply assigning the molecules used in the insects' bodies to a new and equally praiseworthy department of nature (Kingsley, Water Babies [New York: Dutton, 1949], p. 61).

[3] F.L., "A human skull," Cornhill Magazine 2(1860): 718.

chemistry of natural transformations is so fundamental a part of the mid-Victorian sanitary outlook that it is worthwhile to consider briefly the works of a few who typify this theme.

In the Bridgewater Treatises, which form a compendium of early Victorian natural theology, the issue of chemical transformations -- of the cycling of matter -- is dealt with in the section on digestion in William Prout's eclectic contribution to the series, Chemistry Meteorology and the Function of Digestion considered with reference to Natural Theology (1834). Prout, a physician and physiological chemist, pointed to the interdependence of animals and plants as an example of "the extraordinary skill manifested in the disposal of the various parts of the organized system," but his proof of design here rested entirely on the phenomena of gas interchange and was less sophisticated than the treatments of later authors.[4] Prout also found

[4] The phenomenon of plant respiration proved to be a considerable problem for those who were trying to show that plants and animals were in every way complementary and interdependent. Prout is one of the few who acknowledged the process, but he maintained that the amount of oxygen released during photosynthesis was greater than the amount of CO_2 produced in plant respiration and therefore the net effect of vegetable existence was to purify the exhalations of animals (Prout, Chemistry Meteorology and the Function of Digestion considered with Reference to Natural Theology [London: Wm. Pickering, 1834], pp. 451-53, 543). In 1848 W.T. Brande also argued that the overall effect of plant life was to purify, and furthermore maintained that some plants (the fig tree was his only example) put out oxygen even in the dark (Brande, Manual of Chemistry, 6th ed., 2 vols. in 3 [London: J. Parker, 1848], Vol. II, Pt. iii, "Organic Chemistry," p. 1863). W.A. Miller in 1862 insisted that the emission of CO_2 by plants at night was merely incidental. He thought that they were merely letting off CO_2 which they had taken in during the day and not got around to reducing (Miller, Elements of Chemistry, 2nd ed., 3 vols. [London: J. Parker, 1860-62], Vol. 3, p. 828). See also J. Reynolds Green, A History of Botany 1860-1900 (Oxford: Clarendon Press, 1909), pp. 419-22.

in the chemical transformations of matter en route through the food chain an indication of the efficiency with which the natural economy had been designed. Like most of his contemporaries Prout believed that animals incorporated certain proximate principles built up by plants -- albumen is an example -- in a more or less direct fashion into the tissue of their bodies. Therefore the lower orders of the food chain appeared, by "a wise arrangement," to serve the higher orders by making nutritive matter more easily assimilable for the apex of creation. "By the same means," Prout went on, "that accumulation of dead animal remains, which would soon be overwhelming, is entirely prevented."[5] This illustration, in which the same process that provides the nutriment for higher orders of life also purifies the world of their refuse, is very much in evidence in the purification-as-utilization arguments of Victorian sanitary philosophers. Prout concluded that transformation, and necessarily mutability, were wonderful aspects of nature. Without death there could be no birth. When "we contemplate the repeated employment of the same materials" we see "that the greatest possible effect is everywhere produced by the simplest possible means" -- truly the mark of a Deity with an eye toward efficiency.[6]

Eight years after the publication of Prout's Bridgewater Treatise, the concept of animal/plant interdependence received a considerably more extensive treatment by the French chemists J.A. Dumas and

---

[5] Prout, Chemistry, Meteorology, Digestion, p. 544.
[6] Ibid., pp. 544, 547.

Auguste Cahours. Besides citing the oxygen/carbon dioxide interchange between animals and plants, Dumas and Cahours pointed out that plants decomposed water and ammonia, while animals produced these; that plants reduced and emitted oxygen, while animals oxidized and absorbed oxygen; that plants absorbed heat and electricity while animals emitted these; and finally that plants were stationary while animals moved. This table of oppositions that Dumas and Cahours had assembled became standard fare in British chemistry texts, usually appearing in those introductory or concluding sections where the authors wished to illustrate the "admirable adjustments" of organic chemical processes in nature.[7]

The themes adumbrated by Prout received a far more emphatic enunciation and a considerably more detailed scientific justification in the early 1850s by the Scottish agricultural chemist James Finlay Weir Johnston, reader in chemistry and mineralogy at the University of Durham from 1833 until 1855. Johnston held an M.A. from Glasgow and had done further study with Berzelius about 1830.[8] In 1853 he published an article on "The Circulation of Matter" in Blackwood's Magazine, in which he expanded on some of the themes of his Chemistry of Common Life (1850). Johnston was completely enraptured with the

---

[7] J.A. Dumas and A. Cahours, "Sur les matières azotées neutres de l'organisation," Annales de la Chimie et de Physique, 3rd series 6 (1842): 386. The table is found in Brande, Manual of Chemistry, Vol. II, Pt. iii, pp. 1883-84; and Miller, Elements, III, p. 822. Its essential components are given verbally by Stevenson Macadam (The Chemistry of Common Things [London: T. Nelson, 1866], p. 171).

[8] Sir E. John Russell, A History of Agricultural Science in Great Britain 1620-1952 (London: George Allen and Unwin, 1966), p. 130.

cyclic chemical transformations that nature effected. After quoting from Hamlet regarding the destiny of "imperial Caesar's" dust (used to stop up chinks in a drafty wall) and Alexander's remains (used to stop up beer barrels), Johnston happily expatiated on the fate of the flesh generally:

> We need scarcely concern ourselves, therefore, with the destiny of the organic part -- the tissues and blood of our bodies. Its fate is decided by fixed and unchanging laws. When it has served our purpose new and immediate uses await it. We attempt in vain to detain it from pre-determined labours, or, by the arts of the embalmer, to compel it to perpetuate a loved or honoured form. No need to wait, . . . for the body to crumble into dust. The fluids and tissues decompose rapidly, and are dissipated, so that what is now part of the body of a Caesar or a Venus, may literally within a week become part of a turnip or a potato.[9]

Johnston was also greatly impressed with the efficiency with which all the cycles (he considered the carbon, nitrogen, and phosphorous cycles) were integrated. Doubtless inspired by the gears, belts, and shafts that ran the cotton mills of the industrialized north of England, Johnston spoke of the "restless activity" of particles of matter moving according to the "motions of all the wheels"; wheels on whose "perpetual movement" the continuation of life depended. The immense importance of what had seemed to be insignificant parts of nature -- like the tiny concentration of CO_2 in the atmosphere -- impressed him.[10] Our senses would not perceive its absence yet life

[9] J.F.W. Johnson, "The circulation of matter," Blackwood's Magazine 73(1853): 550, 556.

[10] Ibid., pp. 558, 560.

could not exist without it. So thoroughly was Johnston concerned with the natural theology of chemical transformations that he took time to consider the Christian doctrine that "the dead shall be raised incorruptible," pointing out that in regard to "the knowledge of the time" this statement clearly required some exegetical attention.[11] As had Prout, Johnston recognized that plants purified the air for animals (the photosynthesizing plant was not simply "amusing itself"), and that they organized inorganic matter into animal food. Unlike Prout, Johnston realized that animal refuse had to be mineralized before it could be useful to plants.[12]

The themes Johnston explored are common in the sanitary literature. In their discussions of purification, sanitarians were concerned with the condition and destination of the impure matter involved; and the fate of this matter was frequently considered in regard to the intentions of the Creator. Natural phenomena such as rivers, soil, and sewage were believed to have purposes in the design of the world, and the sanitarian's job was to see that these purposes were fulfilled. Even the products of human artifice, as for example the wastes of chemical factories, were viewed by some as intended for the purification of polluted rivers, since many harsh chemicals acted antiseptically and temporarily deodorized sewage-polluted streams. One finds these concerns about the cycling of

---

[11] Ibid., pp. 558-59.

[12] Ibid., pp. 553, 555, 559-60; idem., Chemistry of Common Life, 2 vols. (New York: D. Appleton, 1855), II, pp. 292, 349-50.

matter, in this context of natural theology, even in what are otherwise practical guides for sanitary improvement. In the city of Hygeia, Benjamin Ward Richardson's sanitary utopia, not only would the sewage be recycled on well-operated sewage farms, but the dead would be buried in shrouds or wicker-baskets so that their bodies would be quickly recycled, and "the economy of nature" thereby preserved.[13]

II. Liebig and Biological Purification

Liebig's works on organic chemistry were among the principal scientific works that provided the raw material for the sanitarian natural theologians. While there is ample evidence in the works on organic chemistry and the Familiar Letters of Liebig's own teleological convictions, his British disciples, and in particular his translators Lyon Playfair and William Gregory, paid special attention to the teleological messages that could be found in Liebigian chemistry.[14]

---

[13] Benjamin Ward Richardson, Hygeia -- A City of Health (London: MacMillan, 1876), pp. 40, 43. A company was formed to put Richardson's plan into action ("Hygeia," JRSA 24(1875-76): 914).

[14] Wm. Gregory, "Review of Liebig's Organic Chemistry in its Application to Agriculture and Physiology," Quarterly Review 69(1842): 333-34; L. Playfair, "Address to the Health Section of the Social Science Association," SR 1(1874): 258, 263. Compare the latter with Liebig, Familiar Letters on Chemistry in its Relations to Physiology, Dietetics, Agriculture, Commerce, and Political Economy, 3rd ed., revised and much enlarged (London: Taylor, Walton, and Maberly, 1851), #15, pp. 180-81. See also Otto Sonntag, "Religion and science in the thought of Liebig," Ambix 24(1977): 159-69.

Liebig provided British sanitarians with the specific version of the purification-as-utilization argument that dealt with the question of what happened to the sewage dumped into rivers. Both in the 1851 edition of the Familiar Letters and the 1852 edition of the Animal Chemistry Liebig devoted a good deal of space to opposing the views of Schwann, Kutzing, Caignard-Latour, and others who attributed fermentation (and putrefaction) to the vital activities of various microscopic organisms. In the Animal Chemistry in particular, Liebig went so far as to issue a lengthy and vehement diatribe about the proper methods of scientific reasoning, in which he pointed to the biological explanation of fermentation as a classic example of the fundamental error of confusing cause and effect.[15] Following this extended methodological chastisement Liebig described the "universal cause," the catalytic or contact process of change that he believed was responsible for a great many sorts of combinative and decompositive chemical changes. Finally, Liebig turned to the principal evidence that appeared to support the thesis he opposed: the presence of microorganisms in decomposing matter. He argued that it was unphilosophical to consider these organisms as the causes of processes to which they were themselves subject. After all, the substances that made up microorganisms were just as complex as those from which the bodies of higher organisms were constructed. "How is it possible then," Liebig wondered, "to regard fungi and

---

[15] Justus Liebig, Animal Chemistry, or Chemistry in its Application to Physiology and Pathology, edited from the author's manuscript by Wm. Gregory. From the 3rd London edition, revised and greatly

infusoria as causes of those processes, when they themselves, these supposed causes, putrefy, ferment, and decay, so that nothing is left of them but their inorganic skeleton?"[16]

Liebig still had to explain what these organisms were doing at the site of putrefaction. This was no problem for Liebig: "The presence of animalculae, which are often found in prodigious numbers in putrefying matters, cannot itself be considered wonderful, since these animals find there the conditions of their nutriment and their development combined."[17] Liebig regarded these creatures as expediters of decomposition. They ingested the already-decomposing materials, and by "their own processes of nutrition and respiration, they are the accelerators of the process of resolution." He went on:

> Since, with the conversion of the elements of organic beings into carbonic acid and carbonate of ammonia, all putrefactive processes are at an end, it is plain that the time necessary for this purpose must be singularly shortened, when the putrefying body becomes the abode of infusoria, and millions of these animals are most industriously engaged in causing the constituents of the organic beings, by means of their digestive processes, to be resolved into these ultimate products.[18]

---

enlarged (New York: John Wiley, 1852), pp. 127-28.

[16] Ibid., p. 144; Liebig, Letters, 1851, #19, pp. 237-38.

[17] Liebig, Letters, 1851, #19, p. 238.

[18] Liebig, Animal Chemistry, 1852, pp. 144-45. Owing to the vast number of editions of Liebig's works that appeared in the 1840s I have been unable to discover when he first introduced this argument. The earliest version I have found is in the 1847 French edition of the Familiar Letters in which Letter #19 of the English 1851 edition appears as Letter #16 (Liebig, Lettres sur la Chimie considérée dans ses applications à l'industrie, à la physiologie et l'agriculture, nouvelle ed. Française pub. par M. Charles Gerhardt [Paris: Masson, 1847], pp. 181-87).

It will be well to make some comments on Liebig's view. First, it is apparent that Liebig accepted microbes as the means in practice, if not the cause philosophically, of much eremacausis (oxidative decomposition), one stage in the process of decomposition. After all, in terms of the initial materials ingested and the final products ejected, animal metabolism and eremacausis were identical processes. Pasteur also recognized this similarity, and a comparison of Liebig's views with an extract from an abstract of an 1862 lecture by Pasteur should help illustrate one of the reasons British sanitarians failed to recognize the revolutionary changes Pasteur was effecting in the theory of decomposition.

> "If," says M. Pasteur, "microscopic beings were to disappear from the earth, its surface would soon be encumbered with dead organic matter of every kind, animal and vegetable. These microscopic beings are the chief agents through which oxygen receives its properties of combustion. Without them, life would be impossible, because the work of death would be imperfect. After death, life appears in a new form, and gifted with new properties. The germs of microscopic beings are everywhere present; commence their evolutions, and by their aid, and the curious power they possess (and which M. Pasteur points out in his paper), oxygen fixes in enormous quantities on the organic substances which are pervaded by these beings and gradually operates their complete combustion." This is the power which, according to M. Pasteur, the organic cells possess; viz., transporting the oxygen from the air into all organic matters, and of so effecting their complete combustion with the development of much heat. And, he adds, we have here a faithful image of respiration, and of the combustion which is its consequence, in the action of the organic globules, which the blood incessantly carries into the pulmonary cells, where they are brought to take up the oxygen of the air, and whence they convey it into all parts of the body, in order to burn up there the different principles of the body. What these corpuscles effect, is likewise effected in

fermenting fluids by the mycodermic cells present in them.[19]

This passage contains nothing that is not in Liebig's Animal Chemistry of 1852 and even adopts the tone of admiration of natural transformations that typified British chemico-theology in the 1840s and 1850s. But while Pasteur went on to recognize that putrefaction might be understood as a similar metabolic process carried out by animals that could live in an anaerobic environment, Liebig would admit no such thing. He recognized that de-oxidizing fermentations went on *in vivo* in the process of digestion, but he did not regard the results of these fermentations as the normal metabolic end-products. Moreover, Liebig stressed that both the process of putrefaction and its products were highly deleterious to all life, including microscopic organisms.

Liebig, in his comprehensive view of organic chemistry, believed that a great many types of organic reactions could be subsumed by his contact theory. The processes of digestion, for example, were effected by decomposing nitrogenous materials within the body, like rennet and pepsin. Microbes were therefore simply sites, readily multiplying to meet the needs of the job, and agents for accomplishing the eremacausis of putrid matter, and the contact process of eremacausis that took place within the microbe, and not the

[19] BMJ, 12 April 1862, p. 390; Pasteur, "Études sur les mycodermes. Role des ces plants dans la fermentation acétique," in P. Vallery-Radot, comp., Oeuvres de Pasteur, 7 vols. (Paris: Masson, 1922-39), Vol. 3, pp. 10-12.

microbe itself, was the cause of the decomposition. The only unusual characteristic of scavenging animalculae that distinguished them from higher animals was their ability to make use of putrefying food whose vibratory decomposition would have overwhelmed the vital resistance of the higher forms. Since contact decompositions went on in roughly the same way both inside and outside the body, it was of no great significance that eremacausis sometimes happened within animalcules, for the identical process was going on, if more slowly, wherever oxidizable matter was in contact with matter undergoing oxidation. Scavenging organisms, by virtue of their resistant constitutions and the special enzymes at their disposal, vastly increased the rate of this process while at the same time confining and controlling its dangerous intermediate stages.

The hygienic implications of this process were not lost on Liebig. He called these infusoria "the true enemies and destroyers of all contagions and miasms." The theological implications were also apparent: "The infusorial animalculae cannot be the cause of putrefaction -- of the production of poisonous matter deleterious to plants and animals, -- but an INFINITELY WISE INTENTION designs them to accelerate the transition of the elements of putrefying substances into their ultimate products."[20]

Liebig's understanding of the relationship between microorganisms

[20] Liebig, Animal Chemistry, 1852, p. 145; idem., Letters, 1851, #19, p. 242. For a similar view see "Review of A. Starr, Discourse on the Asiatic Cholera," Lancet (1848), ii, p. 69.

and decomposition could be interpreted in two different ways. Liebig preferred to think of these infusoria principally as oxidizers and viewed purification as the achievement of complete oxidation. He observed that they performed an extensive amount of metabolism which led neither to growth nor to reproduction, but simply released energy to sustain vitality. For Liebig these scavenging animalcules, with their cast iron stomachs, were simply combustion chambers for burning putrescent or putrescible matter. Other microorganisms (also "animalculae") helped by emitting some of the oxygen that supported this combustion.[21]

Most British sanitarians, who were equally concerned with reconciling the evil of putrefaction with the benevolence they attributed to the Creator, emphasized a complementary aspect of animalcular activity. As well as destroying putrid matter, microorganisms utilized it, converting some of it into new tissue. The putrid matter, instead of regressing to a less useable condition, was continually being reintroduced into the food chain as the scavening fauna "rescued" organic material from its inorganic destiny and rejuvenated it by assimilating it into their own tissues. The same process was believed to occur in plants, which in some obscure manner were observed to live on sewage. The question of whether plants assimilated the organic matter in sewage directly or only after it had been somehow mineralized was usually conveniently ignored. Both Liebig

[21] Liebig, Letters, 1851, #19, pp. 239-41.

and the British, however, regarded animalcular purification as the prevention of the offensive and possibly infectious process of putrefaction, and both also recognized in this process the transformation of dangerous matter into new and useful stations of existence.

III. Purification-as-Utilization in the British Sanitary Literature 1850-1880

The microscopical studies of Thames water by Arthur Hill Hassall in 1850 and 1851 serve as an excellent beginning for a survey of the sanitarians' use of the purification-as-utilization concept. Hassall (1817-1894) had become interested in microscopy while a medical student in Dublin in the early 1840s. Before returning to London in the late 1840s he had completed excellent studies of the littoral microorganisms of the nearby Wicklow coast. In London he completed his medical education, taking a London University M.B. in 1848 and M.D. in 1851.[22]

Early in 1850 Hassall published _A Microscopic Examination of the Water Supplied to the Inhabitants of London and the Suburban Districts_, beautifully illustrated with color plates showing the organisms living in the various waters supplied to Londoners. Edwin Chadwick, in the midst of a campaign to reform the London water supply administratively, qualitatively, and quantitatively, was impressed by Hassall's work, and Hassall provided extensive microscopical

---

[22] E.G. Clayton, _Arthur Hill Hassall, Physician and Sanitary Reformer_ (London: Balliere, Tindall, & Cox, 1908), pp. 2-5.

evidence to support Chadwick's proposals which was appended to the Report on the Supply of Water to the Metropolis, issued by Chadwick's General Board of Health later the same year. Thomas Wakley, the reforming editor of the Lancet, also took notice, and Hassall produced a short paper for the Lancet in February 1850 and then, writing as the Lancet Analytical Sanitary Commission, a much more extensive illustrated report in February 1851. These studies established Hassall's reputation as a water analyst and the public health authorities called upon him periodically during the rest of the decade to examine various waters and other materials.[23]

For Chadwick and Wakley Hassall's microscopy served to demonstrate, far more starkly than anyone had previously done, the dreadful condition of the London water supply. The organisms that crowded his drawings might be beautiful examples of the Creator's handiwork, but as the constant inhabitants of the domestic water they were repulsive. Owing to Hassall's active participation in the campaign to prove London water impure, it is a little surprising to

---

[23] Hassall, A Microscopic Examination of the Water supplied to the Inhabitants of London and the Suburban Districts (London: S. Highley, 1850); Hassall in App. 3 of the Report by the General Board of Health on the Supply of Water to the Metropolis, P.P., 1850, 22, [1283.], pp. 31-58; idem., "Memoir on the organic analysis or microscopical examination of the water supplied to the inhabitants of London and the suburban districts," Lancet (1850), i, pp. 230-35, and editorial, p. 246; "Review of Hassall, A Microscopical Examination," Lancet (1850), i, p. 448; Lancet Analytical Sanitary Commission [Hassall], "Records of the results of microscopical and chemical analyses of the solids and fluids consumed by all classes of the public," Lancet (1851), i, pp. 187-93, 216-25, 253-56, 279-84; Clayton, Hassall, pp. 5-6; and Hassall, The Narrative of a Busy Life. An Autobiography (London: Longmans, Green & Co., 1893), pp. 58-70.

find him eloquently espousing the view that the animalculae he had discovered in the Thames-derived supply had the purpose of purifying the river of the organic detritus it contained. That this argument should appear in all three major versions of Hassall's London water study, and that it should be dealt with by Charles Kingsley and F.O. Ward, who relied on Hassall's studies in writing the condemnatory articles on the London water supply they published in the literary press, is evidence of the pervasiveness of this concept of biological utilization.[24]

Hassall's treatment of the concept of biological utilization appears in its most complete form in the 1851 Lancet report, where he headed one section with the title "Uses of Vegetables and Animals in Impure Water." He began with words that echo Johnston and Prout: "In the existence in impure water of different kinds of organic productions, we recognize the fulfillment of wise and beneficial purposes." Hassall went on: "If all the organic matter present in some waters were to be removed by the ordinary processes of putrefaction, decomposition, and the formation of offensive and deleterious gases and compounds, incalculable mischief would constantly ensue." So, "to obviate this, nature has ordained that some of the organic matter of impure water, in place of undergoing decomposition, should be imbibed by other and living forms, and these dying, that other

---

[24] Charles Kingsley, "The water supply of London," North British Review 15(1851): 241; F.O. Ward, "Metropolitan water supply," Quarterly Review 87(1850): 493-94.

generations should take their place, and fulfill a similar important office." Finally, "The purposes fulfilled by living vegetable and animal productions in water, are then of an eminently useful and preservative character; . . . particularly . . . the vegetable productions, . . . for these not only remove the organic matter dissolved in the water, by absorbing it into their own tissues for appropriation, but they still further purify water by the effect of their respiration."[25]

This is a splendid example of the purification-as-utilization argument common among British sanitarians during this period, and it will be useful to extract from it some themes characteristic -- though sometimes only implicitly -- of the argument generally. The first is that the biological utilization of pollutants is avowedly part of the plan of a benevolent Creator for keeping the world pure. Second, this process of utilization is understood to oppose the process of decomposition. The organic matter in water may either decompose or it may be utilized. Organisms do not decompose it. This is the crux of the difference between Liebig and the British on one hand, and Pasteur on the other. Third, the emphasis is on what we would now consider to be a sub-cycle in the normal pattern of cyclical transformations of matter. The dominant cycle is not from inorganic to plant to animal and back to inorganic. Instead, the matter remains in an organic condition. Whenever it becomes de-vitalized it is

---

[25] Lancet Analytical Sanitary Commission, "Records of the results," p. 189.

quickly reinserted into the food chain by being ingested by scavenging plants and animals which will serve as food for higher animals. Fourth, purification is explicitly understood to be "preservation," the maintenance of the matter in a salubrious, useable, and in this case organic, condition. Finally, Hassall's treatment reveals, and more importantly relies on, an incorrect understanding of plant physiology. The belief that plants, in some obscure fashion, are able to assimilate organic matter directly means that an inorganic stage is unnecessary. This belief is common in the sanitary literature, and we shall see several examples of it further down the line. Hassall's peculiar version of plant respiration (confusing it with photosynthesis) is only one of several modes of confusion. As indicated earlier (footnote 4) plant respiration caused a considerable disturbance of the sanitarian world view.[26]

Hassall's discussion of the principles of the biological utilization of sewage is remarkable in its detail. More often statements of the purification-as-utilization concept appear only incidentally in the sanitary literature. In 1852 Edwin Lankester, another highly reputed microscopist, observed (in another report on the deplorable state of the London water supply) that, "although there is no doubt that these beings [aquatic microorganisms] are wisely adapted to take up those matters which would be more injurious were they not present," nevertheless these scavengers were not to be trusted to do this

---

[26] See also Hassall, A Microscopical Examination, pp. 25-26; and Hassall in GBH, Water Supply Report, App. 3, pp. 33, 40.

reliably.[27]

Along with Lankester, many of those who admitted the process existed, and who regarded it as an admirable natural operation, still failed to attribute to it any substantial effect. In 1864 Henry Acland, F.R.S., Regius professor of medicine at Oxford, testified before the Commons Select Committee on the Sewage of Towns (Metropolis) regarding the presence of fish at sewer outfalls: "it is well known that there are arrangements in nature of that kind for the supply of food of various animals, which then prey upon one another." Asked if this operation would tend to purify the river, he replied, "if they convert, in the ascending series of life, those putrefying excrements into healthy life, of course they must do so."[28] Just as Hassall had done, Acland emphasized that these fish purified the river by assimilating the organic refuse rather than by oxidizing it. There is also a hint here of Liebig's conception of scavengers as organisms with such strong constitutions that they could halt the putrefactive vibrations and, having overcome them, rely on the formerly decomposing material for their nutrition.

Other Oxford dons shared Acland's perspective. When the first Royal Commission on Rivers Pollution (the Rawlinson Commission) held

---

27 E. Lankester, "Report . . . on the organic contents found by the microscope in water supplied from the Thames and other sources," in Lankester and Peter Redfern, _Reports made to the Directors of the London (Watford) Spring Water Company on the Results of Microscopical Examinations of the Organic Matters in the Waters supplied from the Thames and other sources_ (London: n.n., 1852), p. 17.

28 Acland in Select Committee on the Sewage of Towns (Metropolis), _Evidence_, _P.P._, 1864, 14, (487.), QQ 3482-83.

hearings at Oxford during its investigation of the condition of the upper Thames (1-8 Nov. 1865), professors Rolleston (physiology), Daubeny (botany), and Brodie (chemistry), as well as Acland, turned up to offer their opinions. Rolleston believed considerable purification was accomplished by "the american weed" (Anarchis). Accumulations of this plant presented "a mechanical obstacle" to stream flow, allowing the plant a better opportunity for oxygenating the stream and "subducting other matters from the fluid passing through it." Rolleston maintained that "anybody with or without any preconceptions as to the workings of vegetable life can see that the water which leaves is clearer and cleaner than that which enters such a bed of weeds." Rolleston also commented on the uptake of sewage through a food chain that climaxed with fish, though he did not indicate what the intervening stages were.[29] Daubeny, though he doubted that water plants could effectively purify streams, understood the question in relation to the usual view of plant/animal complementarity. Putting animal wastes into a stream would stimulate plant growth, and "a stream so highly impregnated with animal matter as to be highly injurious to animals would be most favourable to vegetation." He added that "nature appears to have adopted that means for lessening the evil."[30]

Occasionally, the doctrine of biological utilization or the

---

[29] RPPC/1865/1st Rept. (River Thames), P.P., 1867, 33, [3634.-I], Evidence, statement, pp. 73-74.

[30] Ibid., Evidence, QQ 918-24.

axiom of animal/plant complementarity took the form of assumptions from which verifiable biological predictions could be deduced. W.B. Carpenter, for example, argued that a multitude of yet undiscovered marine plants must exist in the south polar ocean in order to provide food for the animals that lived there and to "purify" the water of the excess carbon dioxide produced by the respiration and decomposition of these animals.[31] The physiologist and sanitary reformer B.W. Richardson relied on a law of the conservation of useful matter to console a pro-sewage recycling audience in 1866. Even if sewage could not always be returned to the soil (its proper fate), it was still utilized in streams for the benefit of all the organisms involved. Richardson proclaimed: "If we cannot give the sewage to the land, the sea will have it, and the fish will have it, and we shall have the fish. Whatever man may do, he cannot disturb the motion of the earth, or change the balance a single grain. Nature, almighty conservator, from age to age ever keeps up one continuous supply and demand; and from the womb of death there springs in line the most perfect form of life."[32]

These views of Richardson's are almost as charming and as exemplary as those of Hassall which began this section. For Richard-

---

[31] W.B. Carpenter, The Microscope and its Revelations (London: Churchill, 1856), p. 334.

[32] Benjamin Ward Richardson, "On the poisons of the spreading diseases, their nature and their mode of distribution," in John Hitchman, ed., The Sewage of Towns. Papers by various authors, read at a congress on the Sewage of Towns, held at Leamington Spa, Warwickshire, Oct. 25th and 26th, 1866 (London: Simpkins, Marshall, 1867), p. 107.

son the conservation of matter, which he expressed as "nature ordains that nothing ever was or will be lost," meant that nature was so well organized that nothing ever ended up in a condition in which it was not useful. This is evident in the economic, "supply and demand" metaphor he uses to describe nature, in the implicit social contract in nature by which the distribution of matter is regulated for the use of all types of beings, and in the appearance of the "great chain of being" concept, here being used to prove that nature is organized to provide for the needs of higher beings.

Frank Buckland's observations on the utilization of organic pollutants by aquatic organisms, published in 1878, represent a suitable conclusion for this survey of the purification-as-utilization argument. Buckland was a journalist, "fisherman-zoologist," natural historian, and ex-army surgeon, who served as an inspector under the salmon fishery acts from 1867 until his death in 1880.[33] Buckland was a vehement anti-Darwinist and, like his father the geologist and clergyman William Buckland, viewed nature as the creation of a benevolent God (and one with an appreciation of English country life and field sports). As a fisheries inspector Buckland was continually fighting pollution -- pushing for passage of stronger anti-pollution legislation, threatening polluters with legal action, and advising them on pollution-control technologies. His paper on "The Pollution of Rivers and its Effects upon the Fisheries and the Supply of Water

---

[33] Roy McLeod, "Government and resource conservation: the salmon acts administration, 1860-1886," J. British Studies 7, pt. ii (1968): 126-38.

to Towns and Villages," given in October 1877 to the first congress of the Sanitary Institute of Great Britain, was intended to portray in the gravest of terms the harm done by polluted rivers. Yet, like Hassall, Buckland found himself impressed with the "admirable provision" built into nature for the prevention of the evil effects of river pollution:

> It is probably not generally known to the public the reason why these banks of foetid mud are not more pernicious than they are. Below the exit of the sewer at Canterbury, . . . the bed of the river is a mud bank instead of pure gravel. Having taken samples of this mud home, I found that, by a wonderful balancing power of nature, which ordains that nothing in creation should be lost, when it had settled, this mud bank was the habitation of thousands of different creatures of three or four different species, whose food consisted of the (to us, invisible) portions of putrid animal and vegetable matter, which they by converting them to their own uses, prevented from increasing to such an extent as to become directly injurious to human health.
>
> This same admirable provision of nature may be observed in nearly all stagnant waters. The domestic rain-water butt and horse-pond are familiar examples. This marvellous phenomenon of nature assisting us to dispose of our sewage should be observed by all, as all are really interested.[34]

It is not surprising that statements such as Hassall's, Richardson's, and Buckland's did not lead to a science of aquatic ecology. They did not report scientific accomplishments, nor did they sketch out programs of research to be done. Instead, they were celebrations of extrapolated laws of created nature. They were statements of

---

[34] F. Buckland, The Pollution of Rivers and its Effects upon the Fisheries and the supply of Water to Towns and Villages (London: C.L. Marsh, 1878), p. 16.

praise and wonder rather than of skepticism and inquiry. It is also not surprising to find "serious" scientists like Huxley, Frankland, and Tyndall rebelling against this happy valley of a world where nature so reliably put right whatever went amiss and supported the existing order.

IV. Trusting Nature: Scavengers as a Defense of Pollution

The argument that rivers were purified biologically had great utility for those who wished to uphold the safety of polluted water supplies or for those who contended that the sewage they put into a river was quickly rendered harmless by cooperating natural agencies. The defense of pollution forms a major context for this concept of purification-as-utilization and in chapter 10 the discovery in Britain of the bacterial processes active in the self-purification of rivers and in modern sewage treatment technology will be traced to the utilization argument in this context.

Once again, the London water controversy of 1850-52 proves a good starting point. As we have seen, even Hassall, who strongly opposed use of the filthy supply the London water companies derived from some of the worst reaches of the Thames and the Lea, argued that the microorganisms he discovered in the water acted to purify it. Both Hassall and the chemists Alfred Swaine Taylor and Arthur Aiken (both of Guy's Hospital), who represented the West Middlesex water company in particular and the rest of the companies generally, followed Liebig in regarding the organisms in water as sanitary allies whose natural function was to improve the quality of the water. It was only in the implications of this understanding that Hassall and the chemists disagreed. In 1851 both parties interpreted the evidence for the benefit of the Commons Select Committee on the Metropolitan Water Supply. Hassall argued that the presence of microorganisms in water indicated the presence of their food, decaying organic matter, and the more organisms present, the more dangerous decaying matter the water contained. Despite their numbers, it could not be assumed that microorganisms processed the organic matter sufficiently rapidly to keep dangerous products from forming. Moreover, it was possible that the microorganisms themselves were harmful. Taylor and Aiken, on the other hand, believed microorganisms could be found in all fresh water, and even in streams devoid of organic contamination. These organisms ate the dangerous organic matter rapidly, and if anything their presence in large numbers indicated purity, for they could not tolerate an environment of concentrated putrefactive products. They could be swallowed with impunity. In Taylor's view these animalcules also had the convenient physiological capability both of consuming organic matter for their own use and of emitting oxygen to burn up more organic matter and other contaminating compounds: "Animalcules are a wise provision of nature to remove a quantity of organic matter contained in the water. . . . there is organic matter there, and these little animals are constantly consuming that organic matter which might otherwise be injurious to health; they are setting free oxygen and therefore tending to aerate the water."[35]

The opposing views were institutionalized in this form. Whenever sanitary reformers attacked water pollution by revealing the hordes of tiny plants and animals polluted waters contained, the defenders of the status quo trotted out this war-horse of an argument, asserting that these organisms accomplished purification and indicated purity. The argument defended the use of Thames water as the London supply in the major inquiries of 1850-52, 1867, and 1893-94 (on the last occasion in newly-fashionable bacterial dress), and it was used to justify the dumping of London's sewage into the lower Thames in an important investigation in 1882-83. It also contributed to the defense of pollution in innumerable provincial inquiries.

It is easy to take the view that Taylor, Aiken, and men of their ilk were unscrupulously marketing their scientific skills and reputations. Taylor's smug satisfaction with the demonstrably appalling London water supply is blatant in his testimony to the 1851 and 1852 select committees on the issue. The issue is not that simple, however.[36] The appearance of the purification-as-utilization argument in support of institutionalized pollution and its existence as an illustration of Paleyan natural theology are neither incompatible nor

---

[35] Taylor in Evidence, Select Committee on Metropolis Water Supply Bills, P.P., 1852, 12, (395.), QQ 550-52; Select Committee on the Metropolis Water Supply Bill, Evidence, P.P., 1851, 15, (643.), QQ 4062, 12200-11; Lancet Analytical Sanitary Commission, "Records of results," p. 188; and Hassall, A Microscopical Examination, pp. 25-26.

[36] See for example R.A. Lewis' treatment of the career of Thomas Hawksley, C.E., and others (Lewis, Edwin Chadwick and the Public Health Movement [London: Longmans, Green, & Co., 1952], pp. 120, 132-33, 274).

unrelated. In an age when sincere people believed the laws of laissez-faire political economy were among the created laws of nature, it was hard to ignore a scientific principle which supported the unregulated effects of free enterprise -- as with the water supplied by the London companies -- as well as the constitutional unwillingness of rate-payers to support sanitary reform. This relationship between the biological utilization argument as a defense of pollution and as an axiom of natural theology is well illustrated in an 1858 report by William Odling (1829-1921; later F.R.S. and Oxford chemistry professor) to the Vestry of St. Mary, Lambeth, for which he was the medical officer of health. The report is on the condition of the lower Thames and opposes Joseph Bazalgette's plan for the main drainage of London.

Odling's 1858 report came on the heels of the "great stink" of the Thames during the summer of that year. Owing to a lack of rainfall and to uncommonly hot weather, the sewage component of Thames water was unusually great, and the London section of the river was undergoing an especially noisome putrefaction. The stink was indeed a monumental event; the pages of _Punch_ for the first three weeks of July contain little besides stink-related satires and poems. Edwin Chadwick's identification of the volatile products of putrefaction as the causes of zymotic diseases was still accepted by many, and the stink therefore led to a considerable panic in parliament which was then in session on the bank of the river. In this political climate the members quickly passed Bazalgette's plan for dealing with the sewage of London in a manner that would prevent a recurrence of such a dangerous nuisance. Bazalgette proposed a series of trunk sewers parallel to the Thames and at various distances from it on both sides of the river. These would intercept local sewers and take all London's sewage several miles below the metropolitan area where it would be dumped raw into the estuary.

The Bazalgette plan was the product of a decade of administrative chaos. The London main drainage problem had been studied by seven previous Metropolitan Commissions of Sewers beginning in the late 1840s and there were several alternatives to Bazalgette's plan. Each of these had been blocked by continual factionalism, personal antagonisms, and technical and financial disputations, and even in an atmosphere of panic the adoption of Bazalgette's plan required considerable political pushing and shoving. Decades after its completion many still regarded it as a huge sanitary error, if an impressive piece of engineering.[37]

Bazalgette's great sewer, therefore, which seems to us in retrospect as a necessary, obvious, and progressive sanitary reform, was by no means so obvious to Odling and his contemporaries. The plan was expensive, scientifically and hygienically suspect, the events of its adoption included some of the lowest and ugliest varieties of political behavior, and it had been railroaded through parliament during a

---

[37] A. Sayer, Metropolitan and Town Sewage, their nature, value, and disposal . . . (London: Calder, 1857), in passim; RCMSD, First Report, P.P., 1884, 41, [C.-3482.], Commission report, pp. xi-xxxv.

sanitary crisis. Pioneers of urban sanitation like Chadwick and Bazalgette could not agree on the principles which ought to guide their practice. It is not surprising, therefore, to find Odling looking to the natural processes of the river for practical sanitary guidance. In telling his employers not to waste their money on this sanitary reform, Odling tells them that the Thames can take care of itself, and his tone is reminiscent of Prout and Johnston:

> Vital development -- the generation of countless forms of animal life, the preying of animals upon animals, is perhaps the greatest and most effective check upon the process of putrefaction. . . . Reveling in the midst of putrefaction, thriving in the sewers, and fattening in the charnel houses, do we find animal life in endless profusion and variety. These animalculae, however, do not develop themselves from out of those portions of organic matter that are already semi-oxidized or decomposed, but in the midst of decomposition, they build up their organism out of the non-decomposed portions, which they as it were rescue from putrefaction. Then, in other conditions or localities they form the prey of animals of a different grade, which could not have resisted the foul atmosphere in which the original stercorine and carrion forms were generated. We are unable to form any conception of the amount of putrefaction prevented by this wonderful provision of nature in its almost unheeded supply of scavenger animals. It must be remembered that animals of every grade, no matter how-much-soever they grow or multiply do not add one iota to the existing quantity of organic matter. Throughout their lives they are constantly consuming more organic matter than they yield, which excess organic matter they convert into mineral matter by oxidation; and at their deaths they must either furnish food for other animals, or else undergo a more violent continuance of the oxidation to which they were subjected during their lives, and become entirely converted into innocuous mineral matter.[38]

---

[38] Wm. Odling, Report on the Effects of Sewage Contamination upon the River Thames (Lambeth: The Vestry/G. Hill, 1858), pp. 7-9.

Here the context of natural theology is again evident in the phrase "wonderful provision of nature." In comparison with most other versions of the purification-as-utilization argument Odling's treatment reveals a clear recognition of the oxidative function of animal metabolism (Liebig's emphasis), as well as an understanding of the assimilation and continuation of organic matter through the operation of the food chain.

Odling's defense of the Thames' salubrity was based on other factors as well. Since refuse matter was so often soluble or capable of aqueous suspension, rivers necessarily acted as drains, he argued. There was proof that an effective purification was occurring. Epidemiological studies showed no increase of mortality or zymotic disease during the great stink. Odling could find none of the presumably dangerous hydrogen sulfide the Thames ought to have exuded if it were insufficiently oxygenated. He believed the main problem was the accumulation of mud on the river's banks and that the construction of embankments to regulate the scour of the river would solve that problem far more effectively than Bazalgette's huge sewers.[39]

---

[39] Ibid., pp. 3-4, 9-10, 17-19. See also "Westminster" to Times, 16 June 1858; W. Odling to Times, 17 June 1858; W. Ord to Times, 6 July 1858. The failure of the great stink to generate an epidemic of zymotic disease was a key piece of evidence in demolishing the simplistic identification of the products of putrefaction as the causes of disease (Wm. Budd, "Observations on typhoid or intestinal fever: the pythogenic theory," BMJ (1861), i, pp. 485-87; Wm. Ord, "Report on the Thames nuisance of 1858-59," in Second Annual Report of the Medical Officer of the Privy Council for 1859, P.P., 1860, 29, [2376.], pp. 54-56; and John Simon, "The filth diseases and their prevention," in Medical Officer of the Privy Council and Local Government Board, Annual Report to the Local Government Board, n.s. 1, P.P., 1874, 31, (1066.), p. 9).

If Odling's magnificent vision of the biological process that kept the Thames pure remained hypothetical, his version of the process is at least interesting in its detail and the philosophy of nature it represents. In many later appearances during the 1860s and 1870s the concept is present simply as a rhetorical assertion that nature can be trusted to purify rivers effectively. The concept rarely appears in detail, little effort is devoted to investigating the actual effects of organisms on polluted waters -- either to support or to oppose the assertion -- and the defenders of pollution no longer wax philosophical on the beauty of a process that so admirably served civilization. Perhaps the most frequent user of this argument during the later 1860s and early 1870s was Henry Letheby (1816-1876), who held M.B. and Ph.D. degrees from London University and served as medical officer for the city of London and professor of chemistry at the London Hospital. In 1864 Letheby began a program of monthly analyses of the London waters, ostensibly for the Association of Metropolitan Medical Officers of Health but in fact paid for by the associated water companies, for whom he was effectively a permanent consultant and advocate.[40] During the late 1860s, when the London water companies were experiencing a period of particularly intense

---

[40] "Henry Letheby," DNB 11, p. 1010; RCWS, Evidence, QQ 3861, 3909. See also Frankland's comments on Letheby in a letter to Tyndall (Tyndall Papers, Royal Institution, Cat. 9E3, typescript 3960, 19 Jan. 1871): "Letheby is at the present moment a most serious obstacle in the way of sanitary reforms. You will always find him on the side of the joint stock companies and against the public -- companies pay him well and the public does not pay."

criticism, Letheby brought forward the purification-as-utilization argument frequently. Before the Commons Select Committee on the East London water bills (1867), Letheby testified "all the matters that are capable of undergoing oxydation are eaten by the fish, and plants are very greedy in taking up all that is in solution; and then comes the power of atmospheric oxygen; so that between all these operating circumstances the sewage is quickly gone."[41] In the same year he appeared before the Royal Commission on Water Supply, which was seriously considering alternative water schemes that would put the London companies out of business: "Not but that I am quite ready to admit that the discharge of sewage into a river is a most improper thing, but considering the powerfully oxidizing influence of water upon sewage, the many agencies which are at work destroying it, the using of it up by vegetables and aquatic plants and by fish, and above all the power of oxydation, I think that none of the sewage discharged into the Thames can at the present moment be discovered at Hampton."[42] In 1869 Letheby brought up the argument in a paper given to the Association of Medical Officers of Health on "The Method of Estimating Nitrogenous Matter in Potable Water" which was mainly a polemic against Frankland. Here he claimed that the organic matter in water "assumed a malignant form" only rarely, "and that even then nature was always careful to hasten it on to more wholesome

---

[41] Select Committee on the East London Water Bills, P.P., 1867, 9, (399.), App. 11, Evidence, Q 633.

[42] RCWS, Evidence, Q 3891.

conditions, as by molecular changes of all kinds, by oxydation, and by organic appropriation."[43]

The purification-as-utilization argument also filtered down to a popular audience where it became one of the mid-Victorian layman's principles of aquatic ecology. When the first Rivers Pollution Commission visited upper Thames towns in the autumn of 1865 with the objective of cajoling local authorities to act more responsibly and keep their sewage out of London's water supply, it was met by the response that the Thames flora and fauna could be trusted to protect the Londoners. Henry Darvill, the town clerk of Windsor, told the commission that the sewage of his town was absorbed by weeds in the river for use as fertilizer. Frederick Gould, a Kingston-on-Thames dentist and corporation member, claimed that their sewage was food for the fish. Both Darvill and Gould reflected the predominant sentiments of their communities in being reluctant to undertake sanitary reforms solely for the benefit of downstream Londoners.[44]

In fact the argument that scavenging was a reliable means of river purification was so pervasive, and apparently so commonly used as an excuse for not treating sewage, that the earlier Sewage of Towns Commission had attempted to circumvent the impact of the argument, though they did not deny its validity. In their preliminary report (1857) the commissioners appealed to the public's sensibility: "We are aware that it is said that the evil of sewage pollution is

---

[43] "Processes of analysis of potable water," <u>BMJ</u> (1869), i, p. 379.
[44] RPPC/1865/<u>1st Rept.</u>, <u>Evidence</u>, QQ 3990-92, 5925-39.

really much less than it seems; that there are natural causes at work, as the influence of the air, aquatic vegetation, fish, etc., which naturally diminish the quantity of offensive matters thus mixed with the water. Granting that such causes may to a certain extent mitigate the evil, we still say that as a matter of common sense and public decency it is not to be tolerated that the sewage of one town shall flow through and still less be the water source of another."[45]

Occasionally in the arguments of those who rely on this principle of biological utilization to defend acts of water pollution there are statements which in detail and character seem like precursors of modern aquatic ecology. Henry Letheby, in the midst of his testimony before the Select Committee on the East London water bill, casually issued a detailed plant saprobiensystem, a description of the sequence of plants found below a source of pollution, each species tolerating a different degree of foulness, and each species or group accomplishing a distinct purifying operation. According to Letheby,

> the first plant to grow in it would be the sewage fungus. They would effect a great purification. Then there would be the river ranunculus, and then plants like water-cresses, and then the flags which you see growing on the river banks, and then the American weed; and we know step by step, as those plants make their appearance, what is the purity of the water and in the end there would be nothing [no pollution] there.[46]

---

[45] RCST, Preliminary Report, P.P., 1857-58, 32, [2372.], p. 10. See also their Second Report (P.P., 1861, 33, [2882.], p. 8).

[46] S.C. East London Water Bills, App. 11, Evidence, Q 736.

To regard this statement of Letheby's as incipient ecology would be a mistake. The context of the statement itself tells us a great deal about why the multitude of studies of polluted rivers in Britain during the second half of the nineteenth century failed to produce a science of aquatic ecology. Letheby's observation here is incidental. Letheby was interested neither in discovering the habitat requirements of different aquatic plants nor in studying their physiological responses to their polluted environment. The problem at hand was to use his considerable experience as a scientific observer of polluted rivers to convince a group of politicians that polluted rivers quickly regained their purity. The repeated reliance on this biological utilization argument by defenders of pollution like Letheby effectively deprived it of its credibility. Wherever it popped up it was either being used to excuse acts of pollution or to celebrate a natural order that science had proved did not exist, and whose perpetuation was an obstacle both to scientific and social progress. It is not surprising that Frankland, in his detailed studies of various aspects of river pollution, totally ignored the biological aspects of the question.

V. Society in Miniature: Warington, Lankester, and the Natural Theology of Aquaria

The closest that the purification-as-utilization argument came to producing a science of aquatic ecology came with the invention of the aquarium in 1849 by Robert Warington, sr. Warington (1807-1867) was the chemical operator for the Society of Apothecaries. He had been a brewery chemist, managed a citric acid factory for J.B. Lawes, and was active in the anti-food adulteration campaign.[47] In May 1849 Warington placed two gold fish and one aquatic plant (Vallisneria spiralis) in about six gallons of spring water in a twelve gallon tank and covered the top with a muslin cloth to allow the air to come in but to keep out the London soot. Warington hoped to observe "the adjustment of the relations between the Animal and Vegetable Kingdoms"; those complementary functions of animals and plants which were such a topic of current interest. W.T. Brande, in his Manual of Chemistry, had noted that both animals and plants were necessary to maintain the purity of ornamental ponds, and this observation was apparently Warington's inspiration for the project.[48]

Warington began his experiment with only the fish and the plant in the aquarium but soon observed a slime forming on the sides of the aquarium which he attributed to the leaves of the plant, which had lost their "vitality and begun to decompose."[49] In accord with the contemporary medical theory that such a process of decay must harm the animals in the environment, Warington attempted to prevent it by

---

[47] "Robert Warrington," DNB 20, p. 844; Russell, Ag. Sci. in G.B., p. 160. See also Philip F. Rehbock, "The Victorian aquarium in ecological and social perspective," in M. Sears and D. Merriman, eds., Oceanography in the Past, Proc., 3rd International Conference on the History of Oceanography, 22-26 Sept. 1980, Woods Hole (New York: Springer, 1980), pp. 522-39.

[48] Brande, Manual of Chemistry, II, pt. iii, pp. 1857, 1863.

[49] Warington, "Notice of observations on the adjustment of the relations between the animal and vegetable kingdoms, by which the functions of both are permanently maintained," J. Chem. Soc. 3(1850):

adding scavengers -- snails -- to the aquarium. He understood the snail's activity in terms of Liebig's view of scavenging. "By its vital powers" the snails converted "what would otherwise act as a poison into a rich and fruitful nutrient, again to constitute a pabulum for vegetable growth." Warington's snails quickly put matters right, all the inhabitants of this little ecosystem thrived, and the biomass of the system increased. The plants and animals purified each other's exhaled gases, the fish fattened on a diet of baby snails, and the snails did well on the proliferation of Vallisneria, whose numbers had gone from one to thirty in six months. In March 1850 Warington happily told the Chemical Society that he had perfected "the balance between animal and vegetable inhabitants" which enabled "both to perform their vital functions with health and energy." Warington continued the experiment for at least five and a half years without changing the water.[50]

This modest experiment of Warington's is important for several reasons. First, it launched an aquarium fad which swept Britain during the 1850s, to the point of endangering some marine species. More important here are Warington's views of the process of decomposition that occurs in the aquarium. While Warington recognized the principal components of the table of animal/plant complementarity in a general fashion and emphasized the exchange of gases, there are

---

53.

[50] Ibid., pp. 53-54; idem., "Memoranda of observations made in small aquaria, in which the balance between animal and vegetable organisms was permanently maintained," Annals and Mag. of Natural History, 2nd series, 14(1854): 367.

curious omissions in his discussion which reveal both the influence of Liebig and the peculiar purpose of the experiment itself. For example, it is noteworthy that Warington was uninterested in the fate or effects of cast off animal matter, apparently assuming that this was either excreted in a purified, inorganic form or that it was produced in an organic form which plants could assimilate without an intervening process of decomposition. His statement that the plant "feeds on the rejected matter, which has fulfilled its purpose in the nourishment of the fish and the snail" does not indicate whether he is thinking of excreta as Liebig's "ashes" of metabolism or as unmetabolized organic matter.[51] Warington shares this vagueness with British sanitarians generally, who were not much troubled with the processes, if any were necessary, which transformed the refuse of animals into a condition suitable for plant nutrition.[52]

Like Liebig and Hassall, Warington regarded scavenging as a process that stopped decay. The snails used their vital force to convert putrescent or putrescible materials into more snail stoffe and into gases or other undescribed materials plants required. With this assimilation of decaying matter by the snails, there is apparently no decomposition whatever going on in the aquarium, so truly had

---

[51] Warington, "Notice," p. 54.

[52] By contrast, scientists working on the same general group of problems in France were very much concerned with this question (R. Aulie, "Boussingault and the Nitrogen Cycle," Diss. Yale U., 1968, in passim; and Theophile Schloesing, "Assainsement de la Seine," Annales d'Hygèine Publique et Médecine Légale, 2nd series, 47(1877): 211-15).

it become a sanitary Eden.

Throughout most of the decade Warington continued to work with aquaria, making observations about the effects of light, food, and temperature; experimenting with marine aquaria; and trying to find groups of aquatic species that lived harmoniously together. In designing his marine aquarium Warington checked the oxygenating capacities of different marine plants, seeking the species that would most effectively support animal life. Despite their unsophisticated nature these experiments show Warington on the verge of transforming concepts of chemico-theology into an experimental science of aquatic ecology. But Warington's experiments had a distinct direction, and he is better understood as an experimental natural theologian than an experimental scientist. It is significant that in assembling his aquaria he did not attempt to duplicate natural communities of organisms, and thereby to discover their interactions by systematically varying chemical, physical, and biological parameters. Instead, Warington's goal was to create artificial biological societies whose chief characteristic was equilibrium.[53] His interest was not in discovering how natural communities were set up but in demonstrating an axiom which had more importance in natural theology than as a principle of biology. This is made clear in the opening paragraph

---

[53] Warington, "On preserving the balance between animal and vegetable organisms in sea water," Ann. and Mag. of Natural History, 2nd series, 12(1853): 319; idem., "Memoranda," in passim; idem., "On the aquarium," Proc. Royal Institution 2(1854-58): 403-8; idem., "On the injurious effects of an excess of want of heat and light on the aquarium," Ann. and Mag. of Natural History, 2nd series, 16(1855): 313-15.

of his 1850 paper to the Chemical Society: "This communication will consist of a detail of an experimental investigation . . . which appears to illustrate . . . that beautiful and wonderful provision which we see everywhere displayed throughout the animal and vegetable kingdoms, whereby their continued existence and stability are so admirably sustained, and by which they are made mutually to subserve, each for the other's nutriment, and even for its indispensable wants and vital existence."[54]

To Warington, and to the Victorian reader, the ideal aquarium represented a perfectly-functioning, hygienic miniature society in which the vital operations of each member produced a maximum beneficial result not only for itself, but for all the other inhabitants too. Warington's delight in this mutual utility is evident in his description of the equilibrium workings of the aquarium: "the snails . . . thrive wonderfully, and besides their function in sustaining the perfect adjustment of the series, afford a large quantity of food to the fish in the form of young snails, which are devoured as soon as they exhibit signs of vitality and locomotion."[55] Note that the fish do not simply eat the snails; instead the snails serve as food for the fish, their societal duty being the sacrifice of some among them to the fish in order to sustain the equilibrium of the community. Later Warington states that the snail acts "the

[54] Warington, "Notice," p. 52.
[55] Ibid., p. 54.

important part of a purveyor to its finny neighbours," and the image of snails satisfying the seignurial rights of the fish, their natural masters, is even stronger.[56] It is not surprising to find later writers advising the study of aquaria for the moral lessons they could teach.[57]

Warington, as a chemist, had a special interest in experimenting with the chemical interactions of the various organisms. Later writers on aquaria were usually less concerned with these chemical processes, and they tend to deal with the subject less comprehensively and philosophically than Warington did. Instead, their books tend to be devoted to the minute details of aquarium-keeping, and much

---

[56] Ibid., p. 54. Describing Warington's experiment, F.O. Ward spoke of "young snails, whose tender bodies serve as succulent food for the fish" (Ward, "Metropolitan water supply," p. 494).

[57] J.E. Taylor, The Aquarium, its Inhabitants, Structure, and Management, 2nd ed. (London: Bouge, 1881), pp. 24-25, 97, 137-38. To the Victorian readers who kept up with Warington's later papers or who tried to maintain aquaria themselves it must have seemed that aquatic society was more characteristically Irish than English, for the organisms showed an alarming reluctance to live together harmoniously. Warington had great difficulty managing his aquaria, and the apparently smoothly-running system described in his paper to the Chemical Society and described in an 1853 Builder article ("Maintenance of the balance between animal and vegetable kingdoms," Builder 11(1853): 232-32) is an illusion. Snails were a particular problem since some species were overly-voracious and quickly consumed still-living plants, while other species were consumed by the fish. Running the marine aquarium proved even more difficult and Warington appears to have given up the idea that he could maintain it in equilibrium. Some, apparently inspired by Warington's philosophical treatment of the aquarium in the 1850 paper, assumed that a properly-working aquarium would require no additional input of food, a suggestion Warington found ridiculous. Compare Warington's Chemical Society and Builder papers with those published in the Annals and Magazine of Natural History, especially his "Sea water," pp. 321-22; and his "Observations on the natural history of the water-snail and fish kept in a confined and limited portion of water," Ann. and Mag. of Natural History, 2nd series, 10(1852): 274. On his opinions on aquarial

attention is given to describing the different species of organisms which can be kept in aquaria. J.G. Wood, in his The Fresh and Salt Water Aquarium (1868), did not even portray the aquarium as a demonstration of the complementarity of animal and plant life. In Wood's aquaria most of the oxygenation was accomplished by diffusion from the atmosphere. Wood regarded plants as unnecessary if the fish bowl were only shallow enough and mentioned their oxygenating capacity only casually and in passing. Not surprisingly, most of the inhabitants of Wood's aquaria died prematurely, but he considered this as no great tragedy since the inquisitive young naturalist could profit by examining their corpses under the microscope.[58]

One who did not overlook the microcosmic character of the aquarium and considered it even more thoroughly than Warington was the journalist, physician, sanitary reformer, and microscopist Edwin Lankester (1814-1874). Lankester's book The Aquavivarium contains a complete, though not an experimentally-based, discussion of the principles of the transformative chemistry of decomposition and purification as they appear in the working aquarium. Like Warington, Lankester observed that the aquarium was a perfect miniature society. "A vessel of water containing plants and animals must be looked upon as a little world . . . all of its inhabitants live and prosper."[59] Like

---

feeding see his "On the aquarium," p. 408.

[58] Rev. J.G. Wood, The Fresh and Salt Water Aquarium (London: G. Routledge, 1868), pp. 8-9, 16, 90.

[59] Edwin Lankester, The Aquavivarium, Fresh and Marine; being an account of the principles and objects involved in the domestic culture of water plants and animals (London: Hardwicke, 1856): p. 1.

Warington, Lankester had much to say about the biological processes that produced and removed oxygen and carbon dioxide. He also recognized that plants required their nitrogen in an inorganic form as ammonia and that in the aquarium this ammonia was derived from "animal substances in a state of putrefaction."[60] Since the aquarium was a miniature world, the maintainance of the nitrogen cycle on which all life depended required putrefaction.

Then Lankester faced the other horn of the decomposition dilemma.

> Both animals and plants die, and the elements of which they were composed are ultimately reduced to a state in which they may again become the food of plants. But before this takes place, a process of putrefaction sets in, which has a power of spreading from the dead to the dying and the dying to the healthy, so that putrefaction is a process to be avoided as much as possible.[61]

Unlike the modern reader, Lankester did not see a problem here. He went on: "In order to prevent this in the great field of the world, certain animals are formed who prefer dead to living prey, whose digestive powers enable them to convert putrefying tissue into the substance of their own bodies. Such animals are the vultures, and carrion crows among the birds; the crocodile . . . ; the sturgeons . . . ; the beetles . . . ; and the water-snails . . . these are the scavengers of nature. . . . If we would, then, avoid mortality from putrefying substances which spread cholera and fevers among our water pets, we must employ some scavengers."[62] Here Lankester's

---

60 Ibid., p. 5.

61 Ibid., p. 6.

62 Ibid., pp. 6-7.

views are typical. Purification means the prevention of putrefaction and the preservation of matter in an organic condition by means of its incorporation into fresh organisms. To the modern reader, assured that these processes of utilization and decomposition into plant food are inextricably associated with one another, this passage seems to demand some explanation from Lankester. We would like to know how the sanitarian is supposed to protect the public health and at the same time ensure that the natural cycles on which life admittedly depends are completed.

It is clear that Lankester believed in looking to nature for guidance in sanitary matters. In discussing the utility of scavenging animals he remarks, "if man imitated nature more closely, we should find a larger number of scavengers in our great towns than we do at present."[63] This statment is interesting for it reveals the means Lankester and his fellow sanitarians used to escape the decomposition dilemma. To Lankester, scavenging meant utilization, and he does not make a categorical distinction between a biological utilization in which putrefying wastes were incorporated into the bodies of carrion crows or animalcules and one in which they were carted into the country and thrown on agricultural soil. The conflict is avoided by switching from a philosophical to a pragmatic mode of discourse. The process that makes wastes into plant food, if one is necessary, occurs reliably somewhere where people are not and does not generate

---

[63] Ibid., p. 7.

those sensible putrefactive nuisances about which Victorian sanitarians were so worried. In the benevolently designed world some process of utilization, whether by animals or plants, was the chief means of purification, and the sanitarians devoted far more attention to praising the design than they did to investigating its details or enhancing its efficiency.

Lankester was confident in the benevolence of the creation. Even if putrefaction were necessary, other natural processes operated to ensure that it did not become a nuisance. Aquaria worked according to these natural laws and modeled nature. Their study would lead to "contemplation of a universe governed by His wisdom and His love."[64] In common with many sanitarians and with Paley, Lankester could contemplate unhealthy and unsightly circumstances like stagnant ponds and still see in them evidence of the benevolent creation. In a later work he gave the following advice to young microscopists in search of specimens:

> If, now, we go to a very dirty pond indeed into which cesspools are emptied, and dead dogs and cats are thrown, we shall find abundant employment for our microscope in the beautiful forms of animalcules which are placed by the Creator in these positions to clear away the dirt and filth, and prevent its destroying the life of higher animals.[65]

One can easily imagine the mothers of children who owned Lankester's book being less than enchanted with this advice, despite its pious

---

[64] Ibid., pp. 70-71.

[65] Edwin Lankester, Half Hours with the Microscope, illustrated by Tuffen West, 3rd ed. (London: Hardwicke, 1863), p. 64.

moral.

VI. Harnessing Providence: Technologies of Biological Utilization

Since biological utilization was recognized as one of the principal agencies by which nature purified refuse of cities, and since Victorian sanitary engineers were continually urging one another to follow the guidance of nature in solving urban sanitary problems, it is not surprising to find biological sewage treatment schemes in the sanitary literature. One "Hygeia" wrote to the Lancet in 1850 to suggest that "ponds and lakes on estates" might be biologically purified. Hygeia called for the surface of the pond (he was particularly worried about the much-polluted Serpentine in Hyde Park) to be covered with aquatic plants, "and if this vegetable life be abundant enough, not a bubble of putrid gas will be exhaled."

> All will be the food of plants! If to this be added an appropriate number of fresh water mollusca, batrachian tadpoles, and fish, the circulation between animal and vegetable ingesta and egesta will be complete, and the water, though stagnant, will be as clear and pure as crystal.[66]

Hygeia added that "even trout might live in these ponds, if a little fountain were made to play in their center."

The date of Hygeia's letter (22 June) and its content suggest the influence of Warington's paper to the Chemical Society (4 March) and possibly also of Hassall's microscopical study of the London

---

[66] "Hygeia," "On the purification of ponds and lakes on estates in general, and of the Serpentine in particular," Lancet (1850), i, p. 764.

water supply which had appeared late in the winter. As was the case in those treatments, Hygeia's letter is full of celebration of the wisely-arranged harmony exhibited in the scavenging process. Hygeia simply assumed that so long as the plants and animals were present in proper numbers they would purify one another's wastes and regarded the details of these operations as unimportant.

Henry Acland, the Oxford professor of medicine whose views of the purification-as-utilization concept have already been mentioned, suggested in 1857 that artificial means be used to increase the amount of animal life in sewage-polluted rivers. Acland predicted that "it may hereafter be found to be quite feasible, and perhaps also commercially and economically advantageous, to take measures to raise largely the stock of edible aquatic animals for the purification of the rivers of England."[67] It was necessary that the "organic rejectamenta" participate in the "ascending series" of chemical transformations, for this process, by keeping matter in an organic condition, would prevent putrefaction.

In 1865 Acland's colleague T.D. Rolleston suggested a purification scheme in which sewage would be conducted through shallow weed-filled ditches.[68] Seven years later the engineer C.E. Austin amplified Rolleston's suggestion, envisioning "a pond stocked with fish

---

[67] Acland, Notes on Drainage, with special reference to the sewers and swamps of the Upper Thames (London: Murray and Parker, 1857), p. 7; and in S.C. Sewage of Towns (Metropolis), Evidence, Q 3503.

[68] RPPC/1865/1st Rept., Evidence, statement, pp. 73-74.

and water plants, such as canes, reeds, rushes, and water cresses which . . . growing organisms would entirely absorb the original organic matter."[69] It was of no importance whether some intervening process must occur before these organisms could absorb the sewage. Like most sanitarians Austin was basing his suggestion on the observation that somehow when sewage was applied to farmland plants grew and the sewage disappeared. To him it seemed more sensible to grow aquatic plants rather than land plants since the fertilizer dissolved in sewage was so exceedingly dilute.

In 1877 W.E. Buck, the medical officer for Leicestershire and Rutland, wrote to the Sanitary Record suggesting that the sewage of small towns might be purified by running it through osier beds. This plan would provide gainful employment for the local poor who were to harvest the sewage-fertilized osiers and use them to weave wicker baskets. Buck's letter too betrays a lack of interest in the physiological processes through which plants purify and utilize sewage: "The osier is a greedy plant, liking both sewage and water, and having the peculiar property of purifying the sewage."[70] One Lewis Thompson suggested that sewage might even be utilized by raising edible shellfish, a suggestion which appalled a Chemical News reviewer -- and quite rightly.[71] Thompson also believed water

---

[69] C.E. Austin, On the Cleansing of Rivers (London: R.J. Mitchell, 1872), p. 20.

[70] W.E. Buck, "Purification of sewage by osier beds," SR 7(1877): 152-53.

[71] "Review of Lewis Thompson, Remarks on the Purification of Water and other Things," CN 34(1876): 19.

cress helped purify rivers by extracting nitrogen compounds from the river water. He was convinced that some simple natural remedy would be found to purify water and that the water companies with their complicated filters were "blindly hunting by expensive methods to perform a really inexpensive operation." "Nature's remedy" was "staring them in the face."[72]

It is not clear that any of these schemes for using aquatic ecosystems to purify sewage was ever realized. Victorian sanitarians did acquire immense experience in biological technologies of sewage purification, however, for the second half of the nineteenth century was an heroic age for sewage irrigation farms and patent sewer manures. Sewage irrigation in particular was regarded by many as the only effective means of sewage treatment, and by the early 1870s the Local Government Board was insisting that sufficient irrigation land be included in any municipal sewage plan for which the central government was to sanction a loan. Sewage farming literature shared with other forms of biological utilization rhetoric a preoccupation with the natural theology of recycling and a willingness to appeal to whatever scientific principles would show irrigation projects or dry manure patents safe and beneficial.

Throughout the second half of the nineteenth century both the practices and the scientific principles of agricultural sewage utilization were topics of continual debate. In regard to sewage

---

[72] Lewis Thompson, "The presence of nitrites in the water of the Thames," CN 34(1876): 82.

irrigation there was disagreement about the amount of land per capita necessary, about the optimum concentration of sewage and the optimum dosage, about which crops and soils were best suited to sewage farming, and about whether purification happened on the surface of the soil or beneath it. Likewise, sewage manure patentees went to great lengths to prove that the temporarily-clear effluents they could usually obtain were meaningfully pure; that the manures they produced were excellent fertilizers even though chemical analyses showed them to be devoid of the usual fertilizing constituents; and that they could run their processes successfully without producing an awful stench. Large amounts of money were involved, and whenever these issues were debated at town gatherings or the meetings of sanitary societies the principles of natural science and natural theology were invariably called into service on all sides.

In common with those who wrote on the uptake of sewage by aquatic plants, sewage farmers as a group could not agree whether plants utilized sewage directly or required it to have undergone an intervening process of decomposition. Nor did many of them care, being wholly caught up with admiration at this wonderful transformation that changed animal poisons into lush and succulent vegetation. With Walt Whitman, they were

> . . . terrified at the Earth, it is that calm and patient
> It grows such sweet things out of such corruption.[73]

---

[73] Quoted in Dorothy Hartley, Water in England (London: George Allen and Unwin, 1964), p. 258.

What science there was to explain the purification that occurred during sewage irrigation had been developed by J.T. Way and Robert Angus Smith in the late 1840s and early 1850s. Way, who served as the Royal Agricultural Society's consultant chemist from 1847 to 1857 and was professor of chemistry at the Royal Agricultural College at Cirencester, suggested in 1854 that the selective chemical affinities of clay minerals attracted the ammonia in the sewage and in this way held the sewage in the soil until it could be absorbed by the plant. In 1861 the surgeon Francis Taylor presented Way's thesis in the sort of language that is all too typical of tracts of sanitary natural theology.

> Whether the [silicic] acid is combined with lime, potash, or soda, or magnesia, in the earth, it leaves any of them on the approach of ammonia: -- it is divorced instantly and becomes married to its favourite, and a silicate of ammonia is formed. Volatile sal becomes through her union with an acid husband a fixed and fertile matron, most useful to the earth.[74]

While Way was interested in how sewage was taken up by plants, R. Angus Smith was concerned with the process of purification that went on in the soil. In studies carried out for Chadwick in the late 1840s Smith observed the impressive oxidizing and nitrifying abilities of porous soils. Although he recognized that humus catalyzed these reactions, Smith attributed the action of the soil mainly to

---

[74] J.T. Way, "On the use of town sewage as manure," J. Royal Agricultural Society 15(1854): 163; in testimony to S.C. Sewage of Towns (Metropolis), Evidence, QQ 4728-30; Francis Taylor, Human Manure, Its Collection and Conversion to Guano (London: Churchill, 1861), p. 5.

the immense specific surface its manifold particles supplied as sites for the reaction. Smith concluded: "The action of air and water on surface is then a powerful one, and probably is capable of doing many marvelous things with the substances given it to treat."[75] Despite the fact that in this process oxygen destroyed the organic substances in sewage Smith did not regard it as incompatible with the philosophy of purification as preservation, nor categorically different from those treatments intended to halt putrefaction. Smith's goal was to ensure that sewage did not sensibly putrefy. If decomposition happened in the soil -- and here Smith was stressing a salubrious, largely oxidative form of decomposition -- any dangerous stages were hidden, and therefore inconsequential. Thus Smith suggested that sewage first be prevented from decomposing by mixing it with carbolic acid, and then encouraged to decompose in the soil. Happily, the soil could apparently overcome the effects of antiseptics:

> substances which will not oxidize in the air at all [as those treated with carbolic], and will not decompose in the air, will decompose completely in the soil, because the soil is a porous substance, which has the power of absorbing oxygen to a very large extent, and this oxygen is transferred by the soil to the organic substance there, so that oxydation is forced as it were.[76]

In this manner Smith could reconcile the sanitary imperative of preventing putrefaction with the requisites of Liebigian agriculture.

---

[75] Smith in General Board of Health, Report on Water supplied to the Metropolis, App. 3, p. 93.

[76] Select Committee on the Sewage of Towns, First Report, Evidence, P.P., 1862, 14, (160.), Q 220.

The sanitarian might safely use antiseptics to stop putrefaction temporarily and trust the soil to decompose the treated sewage into an inorganic form that plants could use. Smith echoed the natural theologians, conceiving this process of soil oxidation as "a very important provision of nature for the prevention of the evil consequences of putrefaction; it is the complete destruction of all dangerous gases and the perfect purification of the most impure substances. . . . Whatever goes through this ordeal is made pure."[77]

Although Edward Frankland emphatically confirmed Smith's view of the oxidizing capability of the soil in 1868, and even went so far as to show how properly managed soil might become a virtual oxidizing machine, there was in Britain little further detailed study of the processes of purification and utilization that occurred on sewage farms during this period. The failure to investigate this transformation was not due to a lack of research on sewage farming; both the Royal Commission on the Sewage of Towns and the British Association Sewage Utilization Committee extensively studied sewage farming on plots at Rugby and Romford, respectively, and thorough investigations were also done by private individuals and corporations. The research done under these various auspices was usually only empirical, directed toward discovering optimum doses of sewage, or which crops grew best on sewage. The processes that made sewage available for plants received relatively little attention. Way, who

---

[77] Smith in General Board of Health, Report on the Water supplied to the Metropolis, App. 3, p. 92.

managed the experiments for the Sewage of Towns Commission, defended the low priority given the issue of how sewage nitrogen was dealt with by the soil. Way recognized that the nitrogen had to be in an inorganic condition to be used by plants but considered that "in an agricultural sense, refinements of this kind are in the present state of our knowledge scarcely admissable, and we may be satisfied to consider the value of sewage as a manure to depend, so far as nitrogen is concerned, on the total ammonia which it may yield."[78]

Other factors helped to divert the attention of English scientists during this period from the processes of soil/plant/manure chemistry. Even as late as the 1870s some respected British sanitary scientists still believed plants assimilated organic matter directly. Others, perhaps more significantly, continued to speak in the traditional terms of the humus theory, assuming that since decomposition happened automatically and was simply a condition of devitalized organic matter, it therefore made no difference whether one spoke of organic matter or its oxidized products as plant food.

It is possible to find the humus theory brazenly stated and predominantly displayed in British scientific literature well after Liebig had presumably disproved it. The article on "Zoology" in the Encyclopedia Metropolitania (1845) contains the assertion: "it is generally admitted that plants are able . . . to assimilate other organic matter to their own."[79] Sewage either underwent decomposition

---

[78] Way, "Deodorization experiments," in RCST, Second Report, App. 6, p. 67.

[79] J.L. South, "Zoology," Encyclopedia Metropolitania, or

or it was absorbed by plants; both processes did not occur. According to the first Rivers Pollution Commission, of which J.T. Way was the chemist: "The organic matter it [sewage] contains is held in temporary union with the active soil to be afterwards absorbed by the roots of plants or decomposed by the air, so that in a short time, varying according to the activity of the vegetation and decomposition no impurity whatever remains."[80] The Commission also believed that rapidly-growing plants could purify (absorb?) sewage simply by coming in contact with it as the sewage flowed across the surface of the soil.

More commonly these statements of the humus theory are put forward by those who were either unaware that there was a point of agricultural chemistry at issue here or who regarded the issue as unimportant. For example, W. Allen Carter's 1877 statement that "the wastes of the human body are the most deadly enemies it has to struggle with in its efforts for preservation. . . . Here nature steps in, in the most providential manner, and we find that those wastes so deadly to the animal kingdom are in a state exactly suitable for the maintainance of the vegetable kingdom"; appears to be simply a rhetorical reference to the sanitary truism that animals and plants are complementary. Carter was an engineer speaking to engineers. Elsewhere he alluded to a controversy about the necessity of

---

Universal Dictionary of Knowledge (London: Fellowes . . . , 1845), Vol. 7, p. 111.

[80] RPPC/1865/3rd Rept. (Rivers Aire and Calder), 2 vols., P.P., 1867, 33, [3850.], Vol. 1, p. xv.

decomposition, but treated it as a matter of no great importance.[81] Similarly, James Duthie, clerk of the Preston Board of Works, maintained that "soils and the roots of plants have a great power of abstracting the poisonous impurities from sewage water," but it is likely that Duthie's language here is simply vague and that he is making a general observation on what effectively happens in sewage irrigation.[82] Edwin Chadwick, whose opposition to science and readiness to sacrifice principles of science and medicine in the name of sanitary expediency is often transparent, spoke in 1867 of "the peculiar powers of vegetation for taking up animal and vegetable matter."[83] William Menzies, who managed the Windsor castle estate and designed sewage treatment and drainage schemes, wrote in 1869 that "vegetation is the great purifier of all refuse," since plants built "into some useful and safe shape, the polluted matter which is discharged from human dwellings."[84] Again it is likely that Menzies is making a general statement here, rather than consciously affirming the humus theory.

Way's decision to postpone investigation of the process of sewage

---

[81] W. Allen Carter, "Methods of sewage removal," in Sanitary Papers read to the Edinburgh and Leith Engineering Society and the Royal Scottish Society of Arts (Edinburgh: Bell and Bradfute, 1877), pp. 3-4.

[82] James Duthie, A Treatise on the Utilisation of Towns' Sewage particularly with reference to its application to Preston (Preston: Dobson, 1870), p. 18.

[83] Chadwick in disc. of E. Eyrne, "Experiments on the removal of organic and inorganic substances in water," MPICE 27(1867-68): 35.

[84] Wm. Menzies, "The sanitary treatment of the refuse of towns and the utilization of sewage," Builder 27(1869): 325-26.

decomposition in the soil, made on the grounds of scientific priorities, complemented the tendency of many sanitarian natural theologians to portray the transformative processes of the soil as miraculous events illustrating the Creator's goodness whose workings were better celebrated than investigated. Charles Kingsley spoke of the "fruitful all-regenerating grave of earth" into which sewage should be put. Thomas Hawksley, M.D., expatiated on "The Ministry of Earth to Life" at the Leamington Sewage congress, and Lord Robert Montagu lectured the House of Commons on the natural economy of sewage farming, both men pointing to the mechanisms articulated by Smith and Way as illustrations of design. James Copland, M.D., F.R.S., marvelled at the metamorphosis of noxious effluvia into "floral aroma."[85]

The rhetoric of the sanitarian natural theologians is not always distinguishable from that of operators of sewage farms or sewage manure patentees who defended the safety and efficacy of their operations with the principles of natural theology and natural science. They too took advantage of the disinclination of British agricultural scientists to inquire closely into the purifying process that occurred in the soil, sometimes stretching the meaning of purification to

---

[85] Charles Kingsley, Yeast -- A Problem (New York: J.F. Taylor, 1903), pp. 279-80; T. Hawksley, M.D., "The power for good or evil of refuse organic matters," in Hitchman, ed., Leamington Congress Papers, pp. 13-18; Montagu in debate on River Waters Protection Bill of 1865, Hansard's Parliamentary Debates 177(8 March 1865): 1323; J. Copland, The Drainage of London and Large Towns, their evils and their cure (London: Longman, 1857), p. 30.

include whatever their processes did, elsewhere stretching the principles of soil science and plant physiology to allow the plant and the soil hygienically-significant functions. Sanitarian natural theologians often spoke of the action of the soil on sewage as magical, and this willingness to adopt a magical or sometimes metaphorical mode of discourse was advantageously used by these special pleaders.

The case of Alfred Carpenter is illustrative. Carpenter, M.D. (London), was a respected sanitarian and a staunch defender of the controversial sewage farm at Croydon, where he was medical officer of health. Between 1869 and 1875 Carpenter carried out experiments that seemed to him to prove that the rye grass grown on the Croydon farm preferred its fertilizer in organic combination. Plants fed a solution of organic matter showed a growth spurt which was not seen in plants fed inorganically.[86] These experiments confirmed Carpenter's observation that the farm was safe and effective and explained why. Epidemiological studies had shown an improvement in the mortality rate of the area surrounding the farm since it had been built, and chemical analyses had proved the effluent was pure, but now Carpenter had scientific justification for these observations. To the argument that sewage farms must be unhealthy because they were sites of decomposing matter Carpenter could now reply that decomposition did not occur on

---

[86] A. Carpenter, "The power of soil, air, and vegetation to purify sewage," SR 2(1875): 287. Few details of the experiment are given but there appear to be problems with its design. Using three similar plants, Carpenter did not feed one, gave another a solution of organic matter, and gave the third a solution of NH_4NO_3. This last would not supply the complete needs of the plant.

the farm, and the farm could not, therefore, be a health hazard. To defend the farm Carpenter explicitly denied Liebig's doctrine of mineral manures.

> The true principle of sewage utilisation is, for . . . oxidation not to take place at all, but that the organic matter that is arrested by the soil should be taken up into its natural storehouse, made to revert into the formed material of plant life at once, rather than to change into a chemical salt which has to be decomposed again by the vital power of the plant, reabsorbed into its juices, and so brought back to organic life by a roundabout process, instead of that direct one which Nature provides in her agricultural laboratory.[87]

Unfortunately for Carpenter, by 1875 the proof that decomposition had been prevented was no longer a completely acceptable proof of salubrity. There were germs to worry about, and Carpenter maintained that his Croydon rye grass would deal with any germs in the sewage as well as imbibing the organic matter it contained. He had no experiments to support this belief but it seemed to make sense. Epidemiology proved the farm removed the disease-causing agents from the sewage; the aqueous medium in which the hypothetical germs were suspended would insure that they would be exposed to plant roots; germs were surely small enough to be easily assimilable by the roots; and they were made out of nitrogenous organic matter, which was one of the substances plants absorbed from sewage. Carpenter concluded: "No germs of disease will pass by without being made to stand and deliver up all the nitrogenous matter they contain, and which exists

[87] Ibid., p. 285.

in everything likely to develope germs of mischief."[88] Given the contemporary uncertainty which permeated both plant physiology and the question of the nature of germs, the suggestion is by no means as absurd as it seems in retrospect. Among those who took part in the discussion of Carpenter's paper some, like the agricultural chemist A. Voeckler, took issue with his denial of the mineral theory of manures, but many others agreed with Carpenter that plants probably did absorb sewage directly, and no one questioned the reality of his germ-eating plants.[89]

Occasionally this manipulation of the principles of agricultural science to suit the purposes of sanitary argument produced what initially appear to be remarkably cogent observations. For example Henry Moule, a country clergyman who had invented -- and frequently defended in print -- the "dry earth closet" which used baked soil to deodorize excreta, was one of the few sanitarians to recognize that the process which sewage underwent in the soil was in fact a digestion. In trying to show why it was so important that dry earth, and not water or some group of chemicals, be used to treat excreta Moule wrote "the earth in fact digests and assimilates the manure for the plant."[90]

---

[88] Ibid., p. 287.

[89] Ibid., pp. 311-13. A year earlier, Adam Sçott, the English agent for the Liernur Pneumatic Sewage Co., had maintained that plants could eat germs, but advised against relying on a planted field to deprive sewage of the germs it contained (Scott, "The sewage problem: Capt. Liernur's improved system of town drainage," SR 1 (1874): 361).

[90] Moule, Town Refuse the Remedy for Local Taxation, 2nd ed.

Lifted from its context and admitting some allowances for the mode of expression this appears an accurate statement of the process that makes organic matter available to plants. Moule, however, was not thinking of this digestion as a vital process of organisms, nor does his statement reflect Liebig's shrewd observation that complete oxidation is the universal characteristic of digestion. Instead, Moule was speaking metaphorically. He drew a series of parallels between animals on one hand and the plant/earth system on the other. Together, the earth and the plant fulfilled all the functions characteristic of the animal, and Moule, who had drunk too long at the fount of transcendental anatomy, conceived that plants were effectively animals with external digestive systems. This set the basis for Moule's extended discussion of the providential and magical processes of the soil and explained why A. Voeckler's discouraging analyses of Moule's dry earth manure must be wrong. According to Moule Voeckler had failed to consider this wonderful process of digestion which the dry earth performed on excreta, and he had forgotten -- and here Moule went all the way back to Jethro Tull for his agricultural authority -- that the size of nutriment particles, and not their quantity, determined their value to the plant. Moule's tendency to grope among the trash-heap of cast-off scientific concepts for analogies or metaphors which could lend momentary credibility to his invention is characteristic of sewage manure patentees

---

(London: Wm. Ridgway, 1872), pp. 17-22, esp. 17-18.

who, in common with quacks of every variety, found the assumption of the trappings of science a useful sales technique.

VII. Conclusion -- The Eclipse of Biological Utilization

In the 1870s and 1880s the idea of purification-as-utilization lost much of its attractiveness. As it became apparent that micro-organisms were responsible for disease some feared that the purifying organisms might themselves be a problem. It was not clear, for example, that there were distinct species of bacteria which could be discriminated by their physiological activity, appearance in culture media, or reaction to different dyes. Many believed the various forms in which bacteria appeared were simply morphological variations of a single pathogenic species. Moreover, Simon's view that the normal septic ferment was the cause of a variety of diseases undermined the distinction between pathogens and saprophytes. There was, therefore, a degree of plausibility in suggestions like Hoffert's (Chapter III, p. 126) that epidemics of zymotic diseases were simply manifestations of normally placid saprophytes on a rampage.[91]

---

[91] Bulloch has shown that many leading proto-bacteriologists in the 1870s accepted bacterial polymorphism or pleomorphism to some degree. These included Hallier, Lister, E. Ray Lankester, Huxley, Billroth, Nageli, and H. Buchner (Bulloch, The History of Bacteriology [London: Oxford University Press, 1938], pp. 188-203). In 1878 the chemist C.T. Kingzett wrote of pathogenic organisms: "The yeast plant and Bacillis subtilis are the representatives of a large class of organisms which may be grouped under the generic term of saprophytes, and these are the kind of organisms which are found to be associated with infective inflammations and contagious fevers" (Kingzett, "The chemistry of infection, or the germ theory of disease from a chemical point of view," JRSA 26(1877-78): 313). See also H. Hoffert, A Guide

At the same time as purifying organisms became suspect, the concept of purification became more obscure. Hassall and his contempocraries had been confident that purification could be achieved by preventing the putrefaction of refuse organic matter. As long as organic matter continued to be incorporated into the bodies of healthy organisms no dangerous putrefaction would occur. In the late 1860s and early 1870s Frankland, Tyndall, and others of high reputation began to assert that this easily measurable organic matter was not the dangerous impurity in sewage. Instead the morbid impurities were so elusive and obscure that there was no way of proving they had been removed from water and little theoretical basis for assuming that some biological or chemical process in rivers and ponds would be likely to consume them.

More significantly, the concept of purification-as-biological utilization lost its credibility as a principle of serious science. In several ways it had been integrated into an ideology that threatened the existence of a rigorous and progressive British biological establishment. The purification-as-utilization concept had remained a loosely-defined axiom of Paleyan natural theology, a view of the world antithetical to Darwinism and one which subjected scientific knowledge to theological conclusions. The concept appeared most prominently as a defense of sanitary inaction, and in support of

---

to the Sewage Question for 1876, treated from a Sanitary, Economical, and Agricultural Point of View (Weymouth: Sherren & Son, 1876), pp. 8-9; and "The exact relation between the food of plants and the refuse of towns," Builder 30(1872): 818.

blatant acts of pollution. Frankland, who believed that the techniques of Victorian engineering working under the guidance of science could lead to a healthy and productive society, found the reforms he thought necessary and obvious -- the acquisition of a non-polluted water supply for London is an example -- opposed by those asserting that in the activities of the scavenging animals there was proof that nature supported the status quo, making reforms unnecessary. Similarly, the germ theory, which to men like Frankland and Tyndall seemed a medical breakthrough with tremendous potential for improving the public health, was denied on the grounds that it was inconsistent with the benevolent Creation. In 1870 James Morton rejected Lister's theories because he could not believe that nature was "some murderous hag," while B.W. Richardson admitted in 1883 that he could not accept the germ theory because it held out such a pessimistic prospect for humankind.[92]

Finally, sanitarians like Richardson, again led by their faith in the benevolence of the world, mounted a major campaign to prevent physiological vivisection. To many members of the scientific elite this seemed to threaten fundamentally the potential for a thriving biological science in Britain, and in particular the ability of biological research to contribute significantly to public health, for the attack was directed against much bacteriological research as

[92] Morton, quoted in Richard Fisher, Joseph Lister 1827-1912 (New York: Stein and Day, 1972), pp. 179-80; Benjamin Ward Richardson, The Field of Disease. A Book of Preventative Medicine (London: MacMillan, 1883), pp. 861-62.

well.[93] It is not surprising (and will be considered in detail in Chapter 8) to find Edward Frankland denying the existence of the main mechanisms, biological utilization included, that supposedly purified streams, or Tyndall presenting the world seen by the microscope as an innumerable host with which mankind must (and could successfully if guided by science) do battle. Nor is it surprising to find the physiologists T.H. Huxley and Michael Foster writing articles which undermined the concept of animal/plant complementarity that figured so centrally in the Paleyan world view.[94]

It is important to consider briefly the reaction to Pasteur's discoveries by British sanitarians. Pasteur does not figure prominently in the British sanitary literature, and the past two chapters have dealt with some of the reasons for this lack of interest. British sanitarians gave to Pasteur's work a fundamentally different meaning than that which it has since come to have. The modern view of Pasteur the scientific hero is based on what are usually presented as his two distinct spheres of accomplishment. First we recognize

---

[93] Stevenson, "Science down the drain," pp. 6-9. Also see Richard French, Antivivisection and Medical Science in Victorian Society (Princeton: Princeton University Press, 1975), in passim, and esp. pp. 27, 32, where he shows that Richardson was more amenable to compromise with the physiologists than were most antivivisectionists.

[94] Frankland's opposition forms the bulk of Chapter 8. For Tyndall see his "Fermentation and its bearings on surgery and medicine," in his Essays on the Floating Matter of the Air in Relation to Putrefaction and Infection (New York: D. Appleton, 1888), pp. 273-74. See also [Michael Foster], "Animals and plants," Quarterly Review (Am. ed.) 126(1869): 138; and T.H. Huxley, "On the border territory between the animal and vegetable kingdoms," MacMillan's Mag. 33(1876): 375.

his studies of the 1860s and early 1870s which showed how microbes keep the earth's surface free of dead organic matter by carrying out various types of fermentations, some of which are even industrially useful. We consider separately his elaboration of the germ theory of disease in the 1880s, and while we may recognize how Pasteur's interests led him from the first sphere to the second, it remains difficult for us to recognize the unity these two courses of research had for mid-Victorian sanitarians.

To British sanitarians in the early 1860s what was new in Pasteur's work was not true and what was true was not new. Pasteur's "discovery" that oxidizing microbes resolved organic refuse into a harmless form was well-known to them through the work of Liebig, Hassall, Lankester, Odling, and others who had written on biological scavenging. Unlike Pasteur, however, they did not commonly treat this oxidation as a process of decomposition, emphasizing the organisms that were supported by this metabolic process rather than the inorganic products of it. As we have seen this resulted partly from the considerable vagueness regarding the details of the cycles elements followed. When Pasteur told them that putrefaction was caused by microbes they at first ignored him. Putrefaction was not a metabolic process, its products were poisonous to all forms of life, and, more importantly, putrefaction did not need a cause, being simply a manifestation of the absence of vital force. Later, in the mid 1870s when British sanitarians began to admit the existence of a living septic ferment that was responsible for putrefaction, the vanguard had come to consider the action of this septic ferment as the quintessentially malignant natural process, and they had become distrustful of the ideology which emphasized the providential activities of scavenging microorganisms. Progressively-minded sanitarians promoted a world view in which nature was no longer seen as a servant of humanity and the rectifier of human sanitary errors. Both the Darwinian thesis and the germ theory suggested that microorganisms looked after their own interests first; and the rhetoric of the germ theorists in particular fostered the view that microbes were fundamentally anti-human. As microorganisms became the agents of dreaded diseases it was forgotten that they were also responsible for the processes that kept matter circulating smoothly through the departments of nature. Insofar as Pasteur was responsible for the development of the germ theory in Britain, it was a result of his early work on fermentation, and not of his studies of the specific microbial diseases of beer, wine, and silkworms.

In this regard the epiphany of Joseph Lister is exemplary. Lister saw that the carbolic acid the citizens of Carlisle used to prevent their sewage from going through the usual cyclic transformations which organic matter experienced might do the same thing when applied to the wounds of his surgical patients. In either case the action of the septic germ was illegitimate. To British sanitarians Pasteur's proof that microorganisms were the causes of necessary decompositive transformations did not alter the fact that these dangerous transformations ought to be avoided, and nature forced to use alternate routes for cycling matter.

V. FROM SENSIBILITY TO NIHILISM: WATER QUALITY AND SELF-PURIFICATION 1850-1860

Between 1850 and 1870 the transmission of intestinal zymotic diseases by specifically contaminated water was recognized and became widely accepted. This recognition brought reassessments of the definition of pure water and the mechanisms of purity. Traditional, "sensible" criteria such as color, smell, taste, clarity, or the ability to support fish, and simple chemical criteria, such as quantities of lime or magnesia, were shown to be hygienically meaningless, but it was not immediately clear what was to replace them. Some, therefore, became "purificatory nihilists," denying the existence of any natural or artificial processes that reliably removed the zymotic morbid poisons from polluted water. Often they were also "analytical nihilists," insisting that no diagnostic techniques could distinguish safe from virulent water.[1] Thus, whatever it was that constituted purified water became increasingly obscure during this period. Self-purification, which according to traditional sensible standards of water quality had held the status of observed fact, became an hypothesis requiring proof, but one for which many leading scientists would accept no proof.

The widening of the gap between the answers society required for

---

[1] I use the term "nihilism" in roughly the same way historians of medicine speak of the therapeutic nihilism of early nineteenth century French clinicians. Analytical and purificatory nihilists specified the limitations of scientific knowledge, yet developed a practice built on the admission of scientific inadequacy.

making public health policy and the competence of science to supply answers is apparent in the attempts during the early 1850s to harness science to solve London's water supply problems and during the latter part of that decade to deal with that city's sewage. Scientists' involvement in these issues reflects a pattern of contradiction and circumlocution on such questions as the nature of morbid poisons or the existence of self-purification that became commonplace during the second half of the century and that bears much of the blame for the ineffectiveness of British water-and-sewage science during the period.

One of those sensible standards that was carefully examined and then discarded during the 1850s and early 1860s concerned the presence of microorganisms in water. Sanitarians faced a real paradox with regard to life in water. On the one hand, as documented in the last chapter, they recognized that animalculae removed filth from water, but on the other they were revolted by the presence of locomotory beings in drinking water, and concerned that these reflected the presence of dangerous filth, or might even be harmful themselves. This "scavenger paradox" was not resolved during this period, and it lost much of its importance with the rise of the belief that the matters of real concern in water were obscure chemical phenomena.

This chapter examines the state of water analysis in the early 1850s, the London water controversy of 1850-52, the increasing irrelevancy of biology in water-and-sewage matters, and the beginnings of a nihilistic attitude in matters of water quality during the late

1850s and early 1860s.

I. Standards and Measures of Water Quality before 1856

In 1849, when John Snow and William Budd first implicated polluted water as a means -- if not the means -- of the transmission of cholera, analytical chemistry was totally unfitted to deal with the problems of measuring the organic contamination of water in general, or of discovering the obscure cholera-causing circumstances (if any existed) specifically. Although chemical water analysis was a relatively active sphere of scientific enterprise, the analysis practiced and the concerns to which it was directed were artifacts of that peculiar eighteenth century fascination with the manifold medicinal properties of spring waters. Chemists were interested in the inorganic acids and bases in waters, chiefly those believed to have purgative or anti-purgative effects. They were also interested in hardness, which many eighteenth century physicians had regarded as the cause of calcareous deposits (bladder stones, etc.) in the organs. Organic, "extractive" matter was of little importance, except when it existed at levels that made water sensibly offensive or when it contributed substantial amounts of nitrates or other undesirable minerals to the water.

The 1850 report on the water supply of Leicester by Alfred Swaine Taylor and Arthur Aiken illustrates what happened when mineral water analysis was applied to potable waters. During the 1850s these Guy's Hospital professors were probably the most frequently consulted water analysts, and Taylor in particular was a leading practitioner of the application of mineral water procedures to potable water analysis. Taylor and Aiken were satisfied with the water that Leicester could obtain from local streams and defended their conclusions in their "Remarks on the purity of water" which they appended to their report. Although they were not the least concerned with water borne diseases, Taylor and Aiken recognized that water might be more or less wholesome. Wholesomeness was distinct from purity, however, for water could be too soft. Ideal water contained twelve to forty grains/gallon saline matter, and under three grains organic matter. Larger concentrations of organic matter might, "when stagnant under a warm temperature . . . give rise to putrescency and form a nidus for the growth of confervae [algae] and infusorial animalculae" -- the latter innocuous, though undesirable. Purgative magnesium salts were their chief concern; otherwise good water need only be colorless, tasteless, and odorless; contain as saline matters mostly lime, common salt, and gypsum; and not dissolve lead. Such a water was "wholesome and fit for the use of men and animals," whether taken from "lake, river, spring, canal, or well."[2]

---

[2] Arthur Aiken and Alfred Swaine Taylor, "Remarks on the purity of water," Appendix to their "Report on the analysis of Leicester waters," in Thomas Wicksteed, Preliminary Report on the Sewerage, Drainage, and Supply of Water for the Borough of Leicester, and a Report upon the Analysis of Sewage Water, and the Water of the streams in the neighbourhood of Leicester made by Arthur Aiken and Dr. Alfred Swaine Taylor (London: 1850), pp. 86-88. In a later report to London's West Middlesex water company Taylor and Aiken expanded on this opinion: "Water to be wholesome, and fit for dietetic use, should be free from colour, taste, and smell. It has also been considered indispensable that it be free from vegetable, animal, and

In the 1850s the kinds of tests Taylor and Aiken did for Leicester represented the facts of analysis -- the accurately determined amounts of well-defined entities. Nevertheless there was conflict between what could be factually determined and what was of concern hygienically. With Fresenius, many analytical chemists were reluctant to abandon this certainty for less accurate procedures to measure more important but less well-defined entities. In the 1846 edition of his authoritative text on analytical chemistry Fresenius made no distinction between potable and mineral water analysis. He instructed the analyst to test for a variety of inorganic bases (those of potassium, sodium, lithium, ammonia, strontium, calcium, and magnesium), as well as dissolved gases, Mulder's crenic and apocrenic acids (hypothetical constituents of humus), and finally to determine "indifferent organic substances" by ignition.[3] By 1873 Fresenius had separated analytical procedures for "fresh" water from those for mineral water but continued to pay little attention to organic matter determinations, correctly regarding the procedures then in use as unreliable. Determinations of acids, bases, and suspended

---

mineral matter; but universal experience is adverse to this view" (Taylor and Aiken, "Report on the water of the West Middlesex Waterworks Company," in Select Committee on the Metropolis Water Bill, Evidence, P.P., 1851, 15, (643.), p. 691.

[3] K.R. Fresenius, Instruction in Chemical Analysis. Quantitative, ed. by J. Lloyd Bullock (London: Churchill, 1846), pp. 466ff. See also B.H. Paul, Manual of Technical Analysis: A Guide for testing and valuation . . . founded upon the Handbuch der Technisch-Chemischen Untersuchungen of Dr. P.A. Bolley (London: Bohn, 1857), p. 221.

matter remained the essence of fresh water analysis.[4]

Co-existing with and sometimes undermining the exact, if relatively meaningless methods of mineral-water chemistry was the sanitarians' preoccupation with decomposing matter as the measure of water quality. While potable water might be tested for an impressive variety of mineral substances, it was often judged only on the basis of sensible qualities -- i.e., its degree of disgustingness. Often drinking water was dark, thick, vapid, swarmed with visible organisms, and smelt of a swamp. Abraham Booth's 1830 statement that Thames water contained "every species of filth and unutterable things" required no confirmation from analytical chemistry.[5] Likewise, the "taint" water acquired upon standing was a quality familiar to the readers of Knight's Penny Magazine.[6]

During the early 1850s this reliance on sensible qualities was consistent with the sanitarians' preoccupation with perceptible products of putrefaction as predisposing causes of disease.[7] Thus in 1851 "Quondam" maintained that the smell of the Thames -- due to putrefactive products like H_2S -- made its water unfit for drinking

---

[4] K.R. Fresenius, Quantitative Chemical Analysis, 6th English ed. (rptd. from 4th English ed.), trans. from 5th German ed., by A. Vacher (London: Churchill, 1873), pp. 543-44. See also W.T. Brande and A.S. Taylor, Chemistry (London: John Davies, 1863), pp. 132-34.

[5] Abraham Booth, Treatise on the Natural and Chymical Properties of Water and on Various British Mineral Waters, (London: Wightman, 1830), p. 168.

[6] "Domestic chemistry III (domestic waters)," Knight's Penny Magazine 7(1838): 55.

[7] See above, Chapter 3, pp. 111-14.

and cooking, while a Dr. Druitt asserted that "sweetness" -- the absence of sensible putrefaction -- determined potability; aquatic organisms being hygienically unimportant.[8] The same criteria were used to judge the effluents from newly developed sewage treatment processes. The addition of lime, for example, was deemed successful because it yielded a clear, tasteless, and odorless effluent.[9]

The potential for conflict between chemistry and sensibility is apparent. In the 1828 hearings of the Royal Commission on Water Supply (Thomas Telford, Peter Roget, W.T. Brande) this potential was realized. The pamphleteer John Wright and the cadre of medical and lay witnesses who supported him in condemning the polluted Thames water supply based their arguments on the observed condition of the water. They maintained that a public water supply taken from a tidal river near the outlets of hundreds of sewers could be assumed impure without the aid of the arcane manipulations of analytical chemistry. The commissioners were sympathetic to this argument and questioned the reformers' witnesses extensively regarding the tangible condition of the water. They heard about the shrimp-like creatures, the fish, periwinkles, and "little round black things" that were among the water's visible inhabitants.[10]

---

[8] Quondam to Builder, "The noisome reek of the Thames," The Builder 9(1851): 334; Druitt in J. Public Health and Sanitary Review 3(1857): 16.

[9] "Clarification" and "deodorization" were used synonymously with "purification." See "Results of a trial of Prince Albert's proposal for the clarification of sewage water and the preparation of sewage manure," Builder 12(1854): 254.

[10] Royal Commission on the Metropolitan Water Supply, Evidence,

Analytical chemistry undermined the significance of these observations. Claiming for themselves jurisdiction in issues of water quality, partisan chemists directed the commission's attention to inappreciable amounts of obscure materials whose presence and significance could only be determined by men of their expertise. By contrast with these subtle distinctions of which chemistry was capable, sentiment was a backward and unscientific basis for evaluating water quality. The Grand Junction Water Company (the target of Wright's criticisms) hired Drs. Phillips, Pearson, and Gardner as analysts, and these chemists took the simple but effective stratagem of allowing their samples to deposit sediment (and some spoke of accumulations of 8-9" of mud per month depositing in their cisterns), and then analyzing only the supernatant water. They justified this practice by maintaining that whatever settled out was "adventitious," and therefore not an intrinsic property of the water. They concluded that the company's water was "as perfectly harmless as any spring water of the purest kind used in common life."[11]

The reformers had difficulty incorporating analytical chemistry into their arguments and tended to discount chemical results entirely. Dr. William Lambe, who was unofficially their analyst, was unable to demonstrate chemically the presence of the putrefying organic matter he was sure made the water harmful to health.[12] John

---

P.P., 1828, 9, (267.), p. 58.

[11] Ibid., pp. 92-97.

[12] Ibid., pp. 84-92.

Bostock, the commission's official analyst, tested for $CaCO_3$, $CaSO_4$, chlorides of soda and magnesia, and organic matter, but concluded only that the water was "disgusting . . . and improper to be employed in the preparation of food."[13] Dr. William Somerville, a Wright supporter, regarded the chemists' failure to find anything objectionable in the water as meaningless, noting that they would be equally unable to find anything harmful in a water-closet pan. Instead of relying on chemistry one should consider the presence of putrefactive products and the idea of impurity in choosing waters, Somerville advised.[14] James Mills agreed: "A slender portion of common sense authorizes one to affirm that a stream which received daily the evacuations of a million human beings, of many thousand animals, with all the filth and refuse of various offensive manufactories, . . . cannot require to be analyzed, except by a lunatic, to determine whether it ought to be pumped up as a beverage for the inhabitants of the Metropolis of the British Empire."[15]

The application of mineral water chemistry to sanitary science made water analysis a confused business for decades. An incomplete survey giving examples of water analyses done before 1875 (Appendix A) shows the range of procedures different analysts thought necessary to determine whether water was potable. When the number of substances analysts tested for is multiplied by the variety of analytical

---

[13] Ibid., pp. 81-83.

[14] Ibid., p. 53.

[15] Ibid., pp. 62-63.

procedures available for determining each, and again by the distinct formats which analysts used for representing their results, the chaos becomes inpenetrable. In part, this accounts for widespread cynicism among the public toward water analysis, exemplified by the opinion that the number of distinct verdicts on the quality of a water was limited only by the number of analysts who examined it.[16]

Yet the problem analysts faced in evaluating water quality was as much medical as it was chemical. The chemists' uncertainty was

---

[16] The classic example of this was reported by J. Carter Bell in 1875. Five analysts had examined the same water from a deep well near London. The first analysis, done by a chemist, was for CaCO3, $MgCO_3$, $CaSO_4$, NaCl, "nitrates of alkalis and organic matter," hardness, and ammonia (qual.), and found the water satisfactory. A second analysis a year later by a commercial analytical firm (for $CaCO_3$, $CaSO_4$, $CaNO_3$, KNO_3, $NaNO_3$, $MgCl_2$, SiO_2, and organic matter) found the water bad. A third, by a medical man (for solids, Cl, free and albumenoid ammonia [see Chapter 8], loss upon ignition, and temporary and permanent hardness) warranted the opinion that the water not be used unless filtered. The fourth analysis, by a chemist (for total solids, Cl, free and albumenoid ammonia, nitrates, nitrites, and temporary and permanent hardness) returned a satisfactory verdict, and the fifth (for NaCl, $CaCl_2$, CaS, $CaNO_3$, $CaCO_3$, $MgCO_3$, oxides of iron and aluminum, SiO_2, temporary and permanent hardness, and free and albumenoid ammonia) found that the water was acceptable if not outstanding. It is noteworthy that none of the analysts commented on the significance of the inorganic compounds measured -- with the exception of the chlorides, the nitrates, and ammonia -- when reaching their verdict on the water's quality. Bell was not impressed ("Water analysis," Chemical News 32(1875): 246-47). Also see: W.R., "Analysts' differences," The Builder 27(1869): 271. Part of the confusion resulted from analysts' disagreement about how acid and base radicals were actually combined in water. For example, one analyst might list sulfates in combination with calcium, while another put them with magnesium ("Water analysis," Chemical News 3(1861): 285, 315). Miller's suggestion that analytical tables only list acid and base radicals, rather than salts, was not widely accepted (W.A. Miller, "Observations on some points in the analysis of potable waters," J. Chem. Soc. 18 (1865): 129).

exacerbated by the physicians' ignorance of the pathology and etiology of zymotic diseases. Without knowledge of the causes of disease, chemists could not be sure what to test for and were therefore unable meaningfully to define contamination or purification. Such uncertainty made it impossible to agree on chemical standards of water quality and allowed partisan chemists on both sides of any water issue the opportunity to couple their professional experience to almost any data and produce the expected verdict.

To the degree that there existed an hygienically meaningful and chemically measurable quality in water whose determination could reveal the progress of self-purification, that quality was organic matter -- a term used loosely to describe both living and dead materials, and sometimes the noxious inorganic products of putrefaction as well. In the years preceding the germ theory organic matter was the bane of the water drinker, and the term received the same public opprobium later directed to germs. The preeminence of organic matter resulted partly from the centrality of organic decomposition in the sanitarians' pathological theories. At the same time organic matter, again owing to its unsavory decomposition, was associated with those sensibly offensive qualities -- bad color, smell, appearance, and taste -- which had traditionally formed the basis of water quality evaluation. For several decades the interpretation of this analytical parameter remained circumscribed by reliance on a water's sensible properties as the proof of its salubrity, no matter what chemistry might say.

The vagueness of the association between organic contamination and ill health probably sustained the use of mineral standards. In testimony to the General Board of Health in 1850 the practical chemist J.T. Cooper spoke of water quality primarily in terms of mineral impurities, while at the same time considering the organic contamination of the Thames (especially from the air) as a serious problem deserving further study. In the same hearing the Aberdeen University chemist Thomas Clark admitted he could not say why the organic matter in water was unhealthy, though it was obviously unpleasant and he did his best to remove it in his lime process of water purification.[17]

The most common view of the significance of organic matter in water in the early 1850s was that it vitiated water in proportion to the amount of it present, and as a result of the gaseous products that resulted from its putrefaction, which were acute poisons when concentrated and predisposing or cumulative poisons otherwise. This had been the conclusion of the 1828 commission and remained the context in which Chadwick's water inquiries were conducted and his proposals debated. It is perhaps best illustrated in the 1851 report of A.W. Hofmann, Thomas Graham, and W.A. Miller, the "government chemists" appointed to evaluate the contradictory views on the quality of the London water supply of the General Board of Health and the

---

[17] Report by the General Board of Health on the Supply of Water to the Metropolis, Appendix III: Reports and Evidence, Medical, Chemical, Geological and Miscellaneous, P.P., 1850, 22, [1283.], p. 165. (To be cited as GBH MWS App. III.) See also the similar opinions of J.T. Cooper, ibid., pp. 182-85.

water companies. Hofmann, Graham, and Miller were principally concerned with hardness, and they were interested in organic matter only insofar as it was capable of putrefaction. They recognized the tendency of nitrogenous organic matter for particularly offensive putrefaction and concluded from an elemental analysis of Thames water that "organic" nitrogen in quantities of about 0.1 grain/gallon were present, that these were probably from animal matters, and therefore "a minute and probably unimportant portion of animal matter would be admitted to be present."[18] The sewage of upstream towns was not worrisome since the small amount of organic matter it added to the river was disposed of by natural means without putrefying within the city or near the intakes. As the upstream population grew it was possible that the river's sewage component would increase to a point where "the sense of this violation of river purity will decide the public mind" and lead to the abandonment of the river supply. Their strongest remark concerning organic matter was that the discoloration and contamination of the water by extractive vegetable matter during the fall and winter was a "serious evil."[19]

This tendency to consider organic contamination only in terms of its quantity, and to regard it as safe if in concentrations that made

---

[18] Thomas Graham, W.A. Miller, and A.W. Hofmann, "Chemical report on the supply of water to the metropolis," J. Chem. Soc. 4(1851): 380, 383.

[19] Ibid., pp. 384-85. See also J.F. Bateman, "On the present state of our knowledge on the supply of water to towns," 25th Report of the B.A.A.S. (Glasgow, 1855), (London: John Murray, 1856), Reports, p. 72.

putrescence unlikely, is dominant in the sanitary literature of the 1850s. In 1851 Taylor and Aiken spoke of the "exceedingly small" concentrations of organic matter in the Thames, by which they meant less than 2.5 grains/gallon. They advised: "It is scarcely possible to procure potable water in large quantity entirely free from organic matter: and when the proportion is so small as not to effect the colour, taste, or smell of the water, there is no reason to believe that injury to health can result from its presence."[20] In 1855 the same figure, this time expressed as "organic matter with water of hydration," was regarded as "as small as could well be expected, and if there is never more there is no danger from it."[21] Indeed, substantially larger concentrations, such as 4 grains/gallon, raised no objections from Brande, Cooper, and Taylor.[22] As the chemist William Odling remarked in 1860, drinking-water purity was a "practical question," a matter of degree. The purer the water the better, but non-putrescent organic matter raised no great concern.[23]

In the years after 1828, as urban death rates increased, cholera swept Britain repeatedly, and the pioneer public health reformers began to publicize the danger of organic decay and to affirm the need for better water supplies, the measurement of the virulence of water

---

[20] Taylor and Aiken, "Report on the West Middlesex," p. 692.

[21] "Reports on the London water supply," J. Public Health and Sanitary Review 1(1855): 420.

[22] GBH MWS, Appendix I: Returns of the Water Companies, P.P., 1850, 22, [1281.], p. 39.

[23] S.C. on the Serpentine, Evidence, P.P., 1860, 20, (192.), Q 739.

became increasingly critical. Lambe had presciently recognized the inability of chemistry to distinguish putrescence, and even the simple determination of the organic matter in water presented vexing problems. The incineration or ignition process, the usual method for determining the amount of organic matter in water, was periodically shown to be inaccurate in the years prior to its general replacement in the mid 1860s, but owing to the lack of alternatives analysts were forced to rely on its results.[24] By the mid 1850s many chemists had become skeptical of the significance of their results. In 1856 J.T. Way warned readers that the errors inherent in examining 30 or 40 grains of material in a gallon of water -- the largest convenient sample size -- were substantial, while determination of the oxidation products of organic matter -- ammonia and nitrates -- was "almost beyond the skill of the chemist."[25] A.W. Hofmann and Lyndsay Blyth,

---

[24] In the ignition process the water sample was evaporated and the dried residue ignited. The weight lost by the residue through ignition was assumed to represent organic matter, though some chemists recognized part of the loss might be due to inorganic substances or to water of hydration, while portions of the organic matter might have been driven off during evaporation. Some analysts developed procedures to correct for some of these events, but none of these were widely adopted nor were they adequate. More circumspect analysts expressed the results of the ignition process as "organic and other volatile matter" or some similar phrase. Many chemists made extensive qualitative distinctions on the smell and color of the residue during ignition, chiefly distinguishing animal and vegetable organic matter. See Paul, _Manual of Technical Analysis_, p. 223; J.E.D. Rodgers in Select Committee on Metropolis Water Bills, _Evidence_, _P.P._, 1852, 12, (395.), QQ 6677-78; Wm. Odling in S.C. Serpentine, _Evidence_, Q 2146; A.H. Hassall in ibid., QQ 3217, 3233; Miller, "Observations," pp. 119-20; E. Frankland and H.A. Armstrong, "On the analysis of potable waters," _J. Chem. Soc._ 21(1868): 77-81; C.J. Fox, _Sanitary Examinations of Water, Air, and Food. A Handbook for the Medical Officer of Health_ (Philadelphia: Lea & Blakiston, 1878), pp. 98-102.

[25] J.T. Way, "On the composition of waters of land-drainage and

in a report to the government on the quality of the London supply in the same year, concurred. After an extensive introduction in which they pointed out the difficulty and meaninglessness of existing procedures for the estimation of organic matter in water, Hofmann and Blyth presented their data (obtained by incineration) but refrained from discussing its relevance to the issue of the quality of the London water supply.[26]

II. Water Quality Standards as Political Strategies

The London water controversy of 1850-52 was a struggle for the control of the water supply, pitting rival reform groups against one another and against the eight existing water companies, whose division of the market had rightly earned them the title of "the water monopoly." The controversy was about the quantity, quality, and cost of alternative supplies. Water quality was an important issue, though not the dominant one it would later become. Quality was conceived both in terms of hardness and sensible characters.

The controversy is an excellent illustration of the selective application of science, and particularly of sensible standards, to enhance each side's case. The Board of Health, as well as its reform

rain," J. Royal Agricultural Society 17(1856): 130.

[26] A.W. Hofmann and Lyndsay Blyth, "Report on the chemical quality of the supply of water to the metropolis," in Reports to the Rt. Hon. Wm. Cowper, M.P., president of the General Board of Health on the Metropolis Water Supply, under the provisions of the Metropolis Water Act, P.P., 1856, 52, [2137.], pp. 4-6, 14-15.

rival, the Watford Spring Company, relied extensively on microscopical investigations to demonstrate the unfitness of the existing supply. The water companies made hydrogen sulfide their critical standard of impurity and claimed their water was pure since it was free of this noxious gas. Both sides selectively denied sensible standards too; the Chadwickians hinting that since the association of polluted water with disease was not well understood, it was best to avoid polluted water altogether. By contrast the companies' representatives castigated Chadwick and the Watford Company's microscopists for condemning the water on the presence of harmless, if ugly, animalculae. Each side thus constructed its case in accord with the definition of purity that best suited its interests, and each held a distinct and corresponding vision of how natural processes in rivers effected purification.

Chadwick presented his case in the General Board of Health's <u>Report on the Supply of Water to the Metropolis</u>, issued in the spring of 1850. With regard to water quality, Chadwick was mainly concerned with hardness rather than organic or sewage contamination. The report concluded of the Thames: "its inferiority as a supply for domestic use arises chiefly from an excess of hardness."[27]

Hardness had long been one of the traditional measures of water quality, but Chadwick gave it unprecedented (and unwarranted) centrality and applied available arguments indiscriminately to justify his

---

[27] GBH <u>MWS Report</u>, <u>P.P.</u>, 22, 1850, [1218.], p. 312.

concern. To be sure, there were sound objections to hard water. It wasted soap and tea (matters of great concern to the economical Chadwick). It was unsuitable for use in boilers. The famous chef Alexis Soyer maintained that hard water was inferior to soft water for cooking.[28] For Chadwick, who linked disease with squalor and dirt, these were very real concerns. Wasting soap meant wasting health, and plentiful, cheap water would ensure a healthy population.[29]

In the GBH Water Supply Report the contention that hard water was harmful to health appears as a subsidiary argument, and the evidence Chadwick presented to support it is unimpressive, coming from such eighteenth century sources as Haller and Heberden. The medical authorities Chadwick cited disagreed which injurious physiological effects were attributable to hard water, agreeing only that harm was done. Chadwick admitted that the traditional association of hard water with "the stone" (internal calcareous deposits) had been discounted epidemiologically.[30] For Chadwick therefore, water quality

[28] Ibid., pp. 50-59, 64-80.

[29] Chadwick was not alone in emphasizing quantity (and cost) over quality. See The Times, 8 May 1851; 28 May 1851; 20 June 1851. As W. O'Brien put it: "whatever may be the evils arising from impurity in the quality, those caused by a deficiency in quantity, and an intermittent distribution, are infinitely more deplorable" (The supply of water to the metropolis," Edinburgh Review 91(1849-50): 382). See also "Domestic chemistry," p. 55.

[30] GBH MWS Report, pp. 50-57. It seems likely that much of the concern directed toward hardness resulted from association of disease with the water of urban shallow wells, the hardness of which would have been largely due to urine and other cesspool drainage. Nelson Blake quotes a delightful passage from a lecture given at the New York Lyceum in the early nineteenth century: "This liquid [urine], when

principally meant hardness and was only indirectly a medical issue.

Chadwick's publicists likewise stressed hardness in discussing the "impurity" of the Thames and the Lea. F.O. Ward, a journalist known as Chadwick's spokesman, maintained that horses instinctively chose soft water. Lyon Playfair argued that hard water caused disease, was avoided by animals, and was responsible for all manner of problems in industry. Like many others Playfair spoke of water quality entirely in terms of hardness.[31] The Times wrote: "the average hardness of this water . . . does not exceed one degree; its purity, therefore" was high.[32] Samuel Homersham, the promoter of the Watford Spring supply, also highlighted the softness of his alternative against the hardness of the Thames.[33] Hofmann, Graham, and Miller, the government's impartial referees, were also preoccupied with hardness and considered the 14° of hardness in the Thames too

---

stale or putrid, has the remarkable property of precipitating the earthy salts from their solution, or in other words, it makes hard waters soft. Although the fastidious may revolt from the use of waters thus sweetened to our palate, it is perhaps fortunate that this mixture is daily taking place, for otherwise the water of this city would become, in a much shorter space of time than it actually does, utterly unfit for domestic purposes" (Blake, Water for Cities -- A History of the Urban Water Supply Problem in the United States [Syracuse: Syracuse University Press, 1956], p. 250). See also the report of Dr. Sutherland (GBH MWS App. III, pp. 3-19, esp. pp. 7-8).

31 [F.O. Ward], "Metropolitan water supply," Quarterly Review 87 (1850): 473-75; L. Playfair in The Builder 9(1851): 765. See also Bateman, "On the present state of our knowledge of the supply of water," p. 70; and Brande, Cooper, and Taylor in GBH MWS App. I, p. 39.

32 Times, 13 November 1849.

33 Homersham, Review of the Report by the G.B.H. on the Supply of Water to the Metropolis; contained in a Report to the Directors of the London (Watford) Spring Water Company (London: John Weale, 1850), in

high.[34]

The Board of Health's failure to base its condemnation of the existing supply substantially on the sewage contamination the water received deserves comment, for it is apparent that Chadwick had no particular faith in processes of self purification. Several considerations were probably involved. First, there was little basis in Chadwick's pathological or etiological theories for singling out water as the peculiar route of transmission for any zymotic disease. Even had Chadwick been inclined to isolate the effects of organically polluted water, there were few instances where its effects were unambiguously distinct from those of other insanitary conditions. As a deliberate stratagem, emphasizing hardness was wise, for hardness was not removed by filtration, and unlike organic impurity, was therefore an intrinsic and irredeemable fault in the companies' supplies.[35] Finally, it should be noted that Chadwick himself was in part responsible for the sewer construction that was causing the pollution problem. Certainly many later writers (unfairly) blamed Chadwick for the problem of river pollution.

Although Chadwick did not stress the harmfulness of organic pollution, he and many others were to some degree concerned with polluted water as a cause of intestinal disease, and particularly of the cholera of 1848-49. In 1849 John Snow and William Budd separately

---

passim.

[34] Graham, Miller, Hofmann, "Chemical report," pp. 381-83, 387-92.

[35] GBH MWS Report, p. 82. See also "Domestic chemistry," p. 54.

published the view that cholera -- sometimes if not always -- was spread by specifically polluted water. In the frequent public meetings on the London water supply in late 1849 and early 1850 the allegation that the companies' water had been responsible for the cholera (especially in south London) was often heard.[36] In the spring of 1851 the Times went so far as to remark that "if there was one fact clear it was that impure water most surely contained the seeds of death."[37] Hassall too held this view, although he did not emphasize it.[38]

According to orthodox Chadwickian medical theory, organically polluted water was a "predisposing" cause of dysentery, which became cholera during an epidemic period. In some cases it was clearly the crucial predisposer. In one cholera outbreak recorded by Chadwick 46 of 63 people using cesspool-polluted water got cholera, while none of their 22 neighbors who did not use the water were struck with the disease.[39] To someone like Snow, convinced that water was the vehicle of the specific contagion that caused cholera, this outbreak would surely have seemed an excellent example to

---

[36] P.E. Brown, "John Snow -- the autumn loiterer," BHM 35(1961): 520; Times, 23 October 1849; 11 December 1849; 12 December 1849; 13 February 1850; 15 March 1850.

[37] Times, 21 April 1850.

[38] A.H. Hassall, "Memoir on the organic analysis or microscopical examination of the water supplied to London and suburban districts," Lancet, 23 February 1850, p. 235.

[39] Report of the General Board of Health on the Epidemic Cholera of 1848 and 1849, P.P., 1850, 21, [1273.], pp. 59-62; GBH MWS Report, pp. 16-17.

support that contention. But Chadwick had equally strong reasons for limiting the effects of water to predisposition, the lowering of the body's resistance to disease. He was convinced that the poisons in organically polluted water were the inorganic products of putrefaction particularly hydrogen sulfide, and these clearly acted cumulatively to undermine health. Most of his contemporaries accepted this view and the concern with "decomposed organic matter" -- with products rather than processes -- is evident in a remark made by Edwin Lankester at a public meeting on the water question late in 1849. In cesspool-contaminated water, Lankester pointed out, "organic matter existed in a greater or less degree, and . . . by it was produced the sulphate and carbonate of lime, and large quantities of saline matter, more or less injurious to the human system."[40] Similar poisons of decomposition existed in the air, and their effects might be accentuated by all manner of insanitary conditions -- poor ventilation for example -- as well as by polluted water. It was the emphasis on these products of decomposition that justified Chadwick's amazingly broad range of public health reforms.

Still, Chadwick and his followers were opportunistic. Where they found other reasons to oppose the use of polluted water they used them. In various forums they embraced the range of available arguments for the harmfulness of organically-polluted water. F.O. Ward emphasized Liebig's zymotic theory of disease. He quoted Liebig

---

[40] Times, 23 October 1849.

extensively and told of cattle which had acquired an unspecified zymotic ailment after drinking the putrefying effluent of a starch factory. Ward believed the gluten in the effluent had communicated its own mode of putrefaction to the bodies of the cattle. But since decaying matters did not always produce acute disease Ward strengthened his case with the traditional Chadwickian dogma of the evil effects of continued exposure to predisposing causes. Bad water was only "one of a multitude of noxious influences" that caused the "premature degradation of the townsman's body." As Ward put it:

> When litmus paper is dipped in acid it turns red; when a rosy child is dipped in town air it turns white. The first change is instantaneous, the second may take months or years to accomplish. . . . The dilute impurities of even the clearest-looking Thames water, when introduced day after day into the blood, must produce a certain effect . . . of a more or less deleterious kind.[41]

Hassall, another who was associated with Chadwick's proposed water reforms, suggested that certain microorganisms in polluted water might cause disease (see Section III below). Chadwick himself observed that along with the noxious predisposing gases might be "subtle agents" in the water, small amounts of peculiar organic matter or noxious gases undetectable by chemistry, and noted significantly that chemistry was unable to follow the course of putrefactive change.[42]

In this latter argument Chadwick was implementing analytical nihilism, in much the same form and for much the same purpose as it

---

[41] [Ward], "Metropolitan water supply," pp. 483, 478-79. The metaphor is an embellishment of one Chadwick used. See GBH <u>MWS Report</u>, p. 32.

[42] GBH <u>MWS Report</u>, pp. 31-32.

would come to dominate water quality issues during the 1870s. But in the 1850s such appeals to unknown and undetectable qualities in water were exceptional, and the reformers' arguments more commonly relied on a concept of impurity in which chemically undefined but sensibly appreciable qualities formed the basis for condemning polluted water supplies. A Builder editorial of 1851, specifically concerned with reservoirs, typifies the rhetoric in which the issue of water quality was usually discussed: the existing supply was "a liquid loaded with the sewage of a large town, and stored up for days fermenting till its loathsomeness is increased tenfold by the engenderment of things of impure natures and disgusting forms."[43] Readers would presumably be able to recognize the description and they would affirm the sentiment.

To defend the quality of the existing supply the water companies combined sensibility-based arguments with denials of the relevance of other sensible standards. They held that chemical analysis reliably showed the presence of noxious materials (which were usually also appreciable to the senses), and that perceptible qualities such as taste, brightness, smell, and color were important criteria for judging water quality.[44] They completely rejected the possibility of undetectable noxious materials in polluted waters, and were concerned

---

[43] "Metropolitan water supply," The Builder 9(1851): 494. Chadwick wrote: "as yet chemistry has failed to determine the qualities of much animal and vegetable, and above all gaseous matter, that is perceptible and offensive both to taste and smell" (GBH MWS Report, p. 31).

[44] S.C. Metropolis Water Bills, Evidence, QQ 690, 692 (Cooper); 3650-60 (N. Beardmore).

neither with microorganisms nor with organic matter that was not decomposed or decomposing.

Of the several chemists and engineers the companies brought before the select committees of 1851 and 1852, Alfred Swaine Taylor was the most outspoken. As did Chadwick, Taylor regarded organic matter in water as harmful in proportion to its amount and was happy to find it present only in "infinitely small" concentrations. Taylor believed animal organic matter was more harmful than vegetable matter because animal tissues contained sulfur and therefore decomposed into hydrogen sulfide, which with phosphoretted hydrogen constituted the harmful substances in water. In "significant" quantity both were perceptible constituents of water, and we may recall Taylor's statement (note 2) that the organic matter in water was of no concern unless it affected the color, taste, or smell of the water. Taylor insisted he could tell if water were contaminated with sewage by its smell and maintained that he could discover H_2S and PH_3 in concentrations of 10 ppm. He spoke of having isolated the "noxious ingredient" causing dysentery (probably MgO) and assured the 1852 committee that if it were absent water was safe.[45] Thus the noxious parts of sewage were detectable for Taylor, if not by the nose, by the simple techniques of inorganic qualitative assay (sewage-contaminated water, treated with lead acetate, deposited a brown precipitate).[46] The

[45] Taylor in S.C. Metropolis Water Bill, 1851, Evidence, QQ 12280-87; 12273-74; Taylor and Aiken, "Report on the West Middlesex," p. 693; S.C. Metropolis Water Bills, 1852, Evidence, QQ 417, 448, 509, 519, 435, 422-23, 10268; and 625-26, 645 (J.T. Cooper); 6628 (J.E.D. Rodgers).

assertion that sewage contained chemically undetectable noxious ingredients showed only the incompetence of the chemist making it, according to Taylor.[47]

Taylor's colleagues -- Arthur Aiken, J.T. Cooper, and W.T. Brande -- echoed his remarks. Aiken pronounced as safe, water with eight grains organic matter (the sample was taken from the Thames near the tower of London) since it did not tarnish silver, having "lost the sulfur and the hydrogen, to which the bad smell is owing."[48] Faith in the meaningfulness of these sensibly and chemically detectable qualities was sufficiently strong to form the basis of the companies' denial of the epidemiological arguments that linked the 1848 cholera to the water of the Southwark and Vauxhall company. Since this company's water was chemically similar to the water of the Grand Junction company (whose service district remained relatively healthy), the Southwark and Vauxhall maintained that differences in general sanitary conditions and not water accounted for the excess mortality in its district.[49]

The companies' advocates also took care to deny the relevance of other sensible qualities of water. Taylor ridiculed the notion that the layer of soot that covered the Southwark and Vauxhall company's

---

[46] Taylor in S.C. Metropolis Water Bills, 1852, Evidence, Q 435.

[47] See especially the supporting testimony of W.T. Brande, in S.C. Metropolis Water Bills, 1852, Evidence, QQ 728-29, 743, 750; and S.C. S.C. Metropolis Water Bill, 1851, Evidence, Q 3831.

[48] S.C. Metropolis Water Bills, 1852, Evidence, Q 790.

[49] GBH MWS App. I, pp. 33-35.

reservoir could have any detrimental effect on the quality of the water.[50] The engineer Thomas Hawksley, later to become the indefatigable proponent of the inevitable reliability of the self-purification of streams, argued that if the public wished to retain the luxury of so "fastidious" an attitude toward the appearance of its water, it would have to pay extra for that luxury.[51] The same pattern of argument -- on the one hand the assertion that apparent improvements meant meaningful purification, on the other that poor appearance was a scientifically naive basis for condemnation, is evident in the companies' stands on two key water quality issues of the 1850-52 controversy: the self-purification of rivers and the significance of aquatic microorganisms.

III. Microscopical Water Assay and the Rejection of Sensibility

The microscopic life of the Thames, the Lea; of the companies' reservoirs and the householders' cisterns, was the principal sensible argument the reformers used to condemn the companies' supply. The General Board of Health published microscopical studies by Hassall and R. Angus Smith in its Water Supply Report while the Watford Spring company hired Edwin Lankester and Peter Redfern, lecturer in anatomy and physiology at King's College, to compare the microscopical host in existing supplies with its abiotic alternative. The failure of

---

[50] S.C. Metropolis Water Bills, 1852, Evidence, QQ 10249-60.
[51] Ibid., QQ 1248-50.

these studies to demonstrate convincingly any relationship between microscopic life and unwholesomeness, together with the intimate association that developed between water microscopy and partisan politics on one hand, and with reactionary natural theology on the other, led to the abandonment of biological studies in the water and sewage inquiries of the 1860s, 1870s, and 1880s. Biology and water quality were not effectively reunited in Britain until the first decade of the twentieth century, and only then through the direct incorporation of German science. The problem was not that the microscopical studies of the 1850s were poorly done -- some of them reveal ecological insights usually regarded as twentieth century discoveries -- but they were irrelevant, and the better they were the more obvious was their irrelevance. Moreover they were compromised by the unfortunate and "unphilosophical" circumstances of their context.

Hassall and his fellow microscopists attempted to link microbial life with unwholesomeness in three distinct ways. The first and most superficial of these -- and at the same time the central one -- was simply to assert that the thickly inhabited water supply was disgusting, and that justified its replacement. As Hassall put it, the "contemplation of the extraordinary forms and remarkable structure of the majority of the living creatures found in impure water, as well as the large size of many, would create . . . in the minds of most people, an insuperable repugnance to its use as a beverage."[52]

---

[52] A.H. Hassall, A Microscopical Examination of the Water supplied

Legitimizing this appeal to sentiment were more substantive reasons for being concerned with microscopic life. All the microscopists believed the quantity of life was proportional to the organic impurity, and further that the species of organisms present in a water indicated the kind of contamination it had received; for example distinguishing water polluted by sewage from water contaminated by decaying leaves. In the microscopists' view it was the ability to make this qualitative distinction that made microscopy superior to chemistry.[53] Unfortunately the utility of this argument was severely circumscribed by the fact that most organisms were unable to tolerate concentrated H_2S and PH_3, the gases both sides agreed constituted the principal noxious impurities in sewer water. Hassall himself cited experiments on birds, fish, and infusoria, showing the lethal effects of sewer gases, and arrived at the same conclusion as Liebig, that putrefaction was antithetical to life in general. Thus while organisms indicated organic impurity to a degree, Hassall had to admit that they also demonstrated a significant degree of salubrity, and that they tended to purify polluted water. Both points were emphasized by the companies' barristers when they cross examined Hassall during the 1851

---

to the Inhabitants of London and the suburban districts (London: Samuel Highly, 1850), p. 28.

[53] Ibid., p. 1; E. Lankester and P. Redfern, Reports made to the Directors of the London (Watford) Spring Water company on the results of microscopical examination of the Organic Matters and Solid Contents of the Waters supplied from the Thames (London: 1852), Introduction, p. viii; R. Angus Smith, "On the air and water of towns," 18th Rept. of the B.A.A.S. (Swansea, 1848) (London: J. Murray, 1849), pp. 17, 25; GBH MWS Report, pp. 31-32; App. III, pp. 165, 174-77, 57.

select committee hearings, and Hassall's admissions greatly weakened the effectiveness of his testimony.[54]

Finally, some of the microorganisms of the polluted Thames might be parasites. Hassall acknowledged the lack of consensus on the subject, but referred to "grave doubts and misgivings" in regard to the supposed harmlessness of microorganisms, and worried specifically about fungi "attacking the human frame" from within.[55] Despite (or rather as a result of) its complete hypotheticality this argument was unanswerable, and Hassall gave it considerable emphasis. Among these malign infusoria might be the specific cause of cholera.[56]

The basis of the microscopists' participation in the 1850-52 water controversy was the association between disgustingness and impurity and they took every opportunity to reinforce this association by appealing to public sentiment. The illustrations of the organisms in each company's water that decorated Hassall's book were presented in circular frames, giving the impression that what Hassall had drawn was exactly what any microscopist would see. In fact Hassall eventually admitted that the illustrations were composites of typical organisms found in the sediment each water deposited.[57] The drawings

---

[54] S.C. Metropolis Water Bill, Evidence, QQ 4051-62; A.H. Hassall, The Narrative of a Busy Life; an Autobiography (London: Longmans, 1893), pp. 68-70.

[55] Hassall, Microscopical Examination, p. 28

[56] Ibid., pp. 31, 57; Lancet Analytical Sanitary Commission [Hassall], "Records of the results of microscopical and chemical analyses of the solids and fluids consumed by all classes of the public," Lancet (1851), i, pp. 222-23; Hassall, "Memoir," p. 235.

[57] Ibid., p. 9; S.C. Metropolis Water Bill, 1851, Evidence, QQ 3937-38.

were effective, and sympathetic reviewers of Hassall's work accepted them as a legitimate means of promulgating meaningful (and sensible) standards of water quality among the public. Charles Kingsley recommended Hassall's book to any consumer who wished "to instill into his imagination a wholesome terror and disgust of these wonderous atomies," while the Lancet remarked that the drawings "exhibit monstrosities that are too hideous to be seen without disgust," but did not explain exactly why such "frightful" creatures were harmful.[58] Such statements certainly confirmed and reinforced the existing standards of sensibility.

The appeal to sentiment as a standard of water quality even found its way into discussions of the natural economy of aquatic microorganisms ostensibly directed to the reader's intellect. For example, Kingsley managed to describe the purifying activities of aquatic life in language that totally undermined the content of that concept:

> The supplies swarm with living animalcules, the presence of which putting aside its disgustingness [really the essence of his case], as a mere matter of feeling, must be considered as indicative of unwholesomeness. These creatures are nature's scavengers -- their food is decomposing organic matter, animal and vegetable; they attend on putrefaction, as surely as the vultures on fallen carcasses.[59]

Likewise, Kingsley's speculations about how microorganisms were harmful were scientifically dubious, but effectively conveyed the intended sensible message. Like Chadwick and Hassall, Kingsley used

---

[58] Kingsley, "The water supply of London," North British Review 15(1851): 242; "Editorial," Lancet (1850), i, p. 246.

[59] Kingsley, "Water supply of London," pp. 241-42.

a shotgun method of scientific persuasion, offering so many plausible reasons why organisms might be harmful that the reader was hard pressed to escape without accepting at least one. Kingsley argued that some microorganisms were "capable of living and multiplying within the human body -- many more of producing irritation within the intestinal canal, by the siliceous shells and spines, finer than the point of the finest needle, which envelop them; many minute fungi can propagate disease in a healthy organic tissue which has become inoculated with them."[60]

The elevation of "mere matters of feeling" to the status of medically meaningful standards was by no means done surreptitiously and indeed was one of the fundamental assumptions and one of the principal motivations of the sanitary reform Chadwick inspired. And given the dreadful condition of many urban water supplies it is likely that improving the appearance of the water improved its wholesomeness, just as Chadwick's aspecific programs for urban cleanliness did improve the health of the urban population. The centrality of sensibility, and the interacting medical, political, and economic factors which together characterize the 1850-52 controversy are well illustrated by Kingsley. Laissez-faire had left "John Bull" at the mercy of the water monopoly -- "the hard water-and-animalculae sellers."

> you are literally filled with the fruits of your own devices, with rats and mice and such small deer, paramecia and entomostraceae, and kicking things with horrid names, which you see in the microscopes at the

[60] Ibid., p. 242.

> polytechnic, and rush home and call for brandy -- without the water -- with stone, and gravel, and dyspepsia, and fragments of your own muscular tissue tinged with your own bile. . . . Oh John! John! The love of money is the root of all evil.[61]

The crusading microscopists were much taxed with reconciling the purifying activities of microorganisms with their status as indicators of impurity, if not impurities themselves. Lankester pointed out that while stream biota lived on impurities, factors such as climate might inhibit their activity and meant that there would "always be in such waters a quantity of vegetable and animal matter in a state of decay, always disagreeable, and under some circumstances likely to be highly injurious to the health of those who consume it in their diet."[62] In an inexplicable yet palpable manner, it was apparent to the microscopists that the purifying organisms frequently constituted an impurity more offensive than the sewage they removed. Stinking and stagnant reservoirs, for example, illustrated the folly of allowing nature to purify water. Organisms excreted and died, repolluting the water with decomposing substances. If it were necessary to use organically contaminated water it was best to get it to the consumer quickly, before decomposition had begun.[63]

Such reconciliation was not easily achieved. F.O. Ward devoted considerable space in his 1850 Quarterly Review article to explaining

---

[61] Ibid., pp. 246, 253.

[62] E. Lankester, "Report . . . on the organic contents found by the microscope in water supplied from the Thames and other sources," in Lankester and Redfern, Reports to the Watford company, pp. 2-3.

[63] Hassall, Microscopical Examination, pp. 26-27.

why the traditions of natural theology, specifically the revelations of Warington, were not relevant to water supply technology. Ward was impressed with Warington's experiments on aquaria and gave them so enthusiastic a description that effective rebuttal was difficult. He decided, however, that at best organisms had "a mixed influence" on water purity. Warington's achievement showed what could be done when nature was skillfully managed, but "in natural streams these aquatic scavengers, especially the microscopic tribes, do give to existing impurities another, and often a more objectionable form; animalcular swarms being more odious to the sense than even the filth they clear away." But just when his argument seemed to be hanging solely on the principle that what was ugly was harmful, Ward changed course and offered the plausible (and to the Victorian audience, highly persuasive) opinion that nature could not be trusted to manage a waterworks. In words that tell much about the Victorian view of nature Ward described the biological dynamics of rivers:

> the green weeds, the eels, and the countless microscopic forms of life, tend, by the assimilation of feculent matter, to clarify the water which, by their presence and their excretions, they contribute also to infect. It is, indeed, the reciprocal intermixture and interference of natural processes, consequent on the very exuberance and multiplicity of the forces to which they are due, that chiefly calls for the corrective intervention of human art. The river is often dissolving one kind of sediment at the same time as it is depositing another; the breeze, which yields its deodorizing oxygen, drops also on its surface myriads of infinitesimal spores -- germs of the very taint which oxygen serves to neutralize; and the living generations are less nimble to purify than their dead predecessors are to pollute it.[64]

---

[64] [Ward], "Met. water supply," p. 494.

Nature, it seemed, was not in a Warington equilibrium.

Ward's discussion reveals how central standards of sensibility were in judging water quality. Water that teemed with life did acquire a vapid taste, became discolored, stank, and was clearly unfit to drink. This observed association of life with apparent impurity formed the basis for all the microscopists' assorted and ad hoc explanations of the significance of their examinations of water. Judged by the standards of sensibility organisms were impurities. Therefore, when the microscopists tried to associate a distinct "insalubrious" flora and fauna with impure water, their arguments became tautological, since animals and plants were among the sensible qualities upon which the verdict of disgust depended. Thus R. Angus Smith, who regarded the changing microbial population along the Thames as the best measure of the quality and quantity of the organic pollution it received, spoke of animalcules not just "larger," and "fatter," but also "uglier than any preceding" as proof that the Thames became less pure during its course through the center of London.[65] Similarly, Hassall's introductory remark that any river whose banks were covered with "carcasses of dead animals, rotting, festering, swarming with flies and maggots" did not require the microscope to detect its impure condition, ensured that the microscopical studies done anyway would necessarily confirm this verdict of sensibility.[66]

---

[65] GBH <u>MWS Report</u>, p. 41; <u>App. III</u>, pp. 94-95.

[66] Hassall, <u>Microscopical Examination</u>, p. 6.

To counteract the microscopists' testimony the water companies argued that microscopy provided merely a "sentimental" and scientifically meaningless basis for condemning water. The microscopists' attempts to associate organisms with unwholesomeness were groundless. A.S. Taylor knew of "no writer, chemist, physician, or physiologist who had ever adduced a single fact to show the injury arising from the use of water containing these microscopic animals: and I believe we should eat nothing and drink nothing if we used the microscope before hand to settle that point."[67] In the view of the companies' chemists recent research (Liebig) demonstrated that animalcules emitted purifying oxygen.[68] While they might be present in proportion to organic matter, microorganisms could not tolerate the really dangerous impurities and were "instantly killed by sewage water."[69] Taylor pointed out that the "prevailing chemical opinion now" -- especially so after the work of Warington -- was to regard organisms as purifiers.[70]

In an age inspired by Liebig's organic chemistry, the companies' representatives were correct in presenting scientific progress as the replacement of biological by chemical modes of explanation. Brande, for example, found no categorical difference or chemical

---

[67] Taylor, S.C. Metropolis Water Bill, 1851, Evidence, Q 12200.

[68] Ibid., Q 12211.

[69] S.C. Metropolis Water Bills, 1852, Evidence, Q 552. Hassall admitted this (S.C. Metropolis Water Bill, 1851, Evidence, QQ 4057, 4062).

[70] S.C. Metropolis Water Bills, 1852, Evidence, QQ 550-51.

distinction between living and dead organic matter in water. Asked if water with animalculae were not more unwholesome than water with an equal quantity of dead organic matter, Brande replied, "upon my word, I think it is unimportant, as a matter of health."[71] Similarly the engineer Nathaniel Beardmore regarded Hassall's studies as retrograde, because Hassall had looked at the wrong things: "he would have done great benefit to science by his facts, if he had given us chemical analyses of the various waters experimented upon, so that the constituent elements of the Fungi, Algae, Diatomaceae, etc., should be determined, and we should then have known the chemical nature of the inhabitants of our waters."[72] Knowledge of the chemical constitution of microorganisms could confirm their harmlessness. In a later water controversy in Liverpool the chemist Thomas Spencer argued that many "animalcules" were really innocuous siliceous plants, and that being made of silica meant that "after death, their decayed bodies have little tendency to offensive decomposition, and hence . . . they produce no deteriorating effect on the water."[73]

The microscopists recognized that medicine was turning increasingly to chemical explanations of impure water and that excepting the possibility that some microorganisms were pathogenic, the significance

---

[71] S.C. Metropolis Water Bill, 1851, Evidence, QQ 3841-42.

[72] [N. Beardmore], "Water supply," Westminster Review 54(1851): 190.

[73] T. Spencer, Liverpool Corporation Waterworks. Report on the quality of the Rivington water, the reservoirs, and the physical characteristics of the district (London: Nichols, 1857), pp. 30-32.

of microscopic analysis depended on the unwarranted assumption that microscopy discriminated varieties of chemical impurities better than chemistry did. The negligible bases of microscopical analysis emerged when the microscopists were closely questioned about the significance of their studies. To the General Board of Health Hassall asserted: "all living matter in water used to drink, since it is in no way necessary to it, and is not present in the purer waters, is to be regarded as so much contamination and impurity, is therefore more or less injurious, and is consequently to be avoided." Sensing that this explanation was unsatisfactory, Hassall returned to the only uncontrovertable alternative, that some microorganisms might be the direct causes of disease.[74] Redfern ran into similar trouble during his grilling by the 1852 select committee. Asked if the organisms in Thames water were harmful, he replied that it was difficult to say owing to the complexity of disease causation. As this answer was unsatisfactory, Redfern stumbled on, offering equivocal answers which betray the fundamental weakness of the microscopists' case. First he insisted that water containing microorganisms "cannot be otherwise than injurious. I am sure there are very few persons who would take up the last half tumbler of many river waters and drink it." The committee responded by suggesting that lots of articles of food carried microbes and that we inhaled them constantly. Finally Redfern found refuge in the same stand Hassall had taken. He warned that some microorganisms might

[74] GBH MWS App. III, p. 57.

"take up residence" in the intestine.[75]

Microscopical studies done later in the decade finally confirmed the irrelevance of microscopical water analysis. In 1854 Hassall carried out microscopical studies of the cholera epidemic for the General Board of Health's Committee on Scientific Inquiries. He examined waters especially associated with the cholera, including the water of the Broad Street pump. He found no organisms not characteristic of non-choleragenic water. Hassall reverted to worrying about the absorption of bad air by the water, and to regarding the "living productions" merely as indicators of organic contamination.[76]

The hope that the species of organisms might provide useful information about the nature of the pollution was also undermined during the 1854 cholera epidemic. In a General Board of Health study of the cholera at Sandgate James Brittan and Robert Etheridge of the Bristol Microscopical Society classified aquatic organisms according to the organic impurity of their habitat. They concluded: "This water yields a large amount of infusorial life, . . . not only as regards percentage of life, but variety, and all of them are such as would be detected in waters yielding a considerable amount of decomposing substances, which are necessary for their existence." But the same

---

[75] S.C. Metropolis Water Bills, 1852, Evidence, QQ 11368-89.

[76] Hassall, "Report on the microscopical examination of different waters (principally those used in the Metropolis) during the cholera epidemic of 1854," in G.B.H. Medical Council, Report of the Committee for Scientific Inquiries in Relation to the Cholera Epidemic of 1854, P.P., 1854-55, 21, [1980.], App. 8, pp. 235-41, 217.

conclusion -- the existence of decomposing substances -- could be made either on the basis of inspection or by chemical analysis. In asserting that organisms showed the water contained "organic matter in a state of disintegration and decomposition" and that the "characters of the organisms found, their known peculiarities as to habitat, food, etc." indicated the amount of organic matter in the water, Brittan and Etheridge were using biology to measure chemical variables. They admitted that it was organic matter in general that was a "predisposing agent to disease" and that if this were dissolved it would be undetectable microscopically.[77] Thus the microscopical analysis of water, even in its most advanced form where species were correlated with types or degrees of water contamination, remained an inadequate way of approximating chemical impurities. The chief contribution of a microscopical study of the London water supply done by Hassall in 1857 was to show the ineffective filtration practiced by the water companies. Hassall also detected their use of water taken on the incoming tide, finding the organisms of brackish water in the companies' supplies.[78]

---

[77] J. Brittan and R. Etheridge, in Report of T.E. Blackwell, C.E., to the President of the G.B.H., on the Drainage and Water Supply of Sandgate in connexion with the outbreak of cholera in that town, P.P., 1854-55, 45, (82.), pp. 20, 22.

[78] Hassall, "Report on the microscopical examination," pp. 235-36, 252; Hassall, Report to the Rt. Hon. Wm. Cowper, M.P., President of the G.B.H., on the Microscopical Examination of the Metropolitan Water Supply; under the Provisions of the Metropolis Water Act, P.P., 1857, 13, [2137.], pp. 12-14.

IV. The Institutionalization of Self-Purification

Between 1850 and 1867 the concept of the self-purification of rivers evolved from observation to dogma. It became a principal defense of pollution. Although by 1867 there were few data showing the extent of the process and no significant investigations of the mechanisms by which it worked, there was nevertheless an impressive collection of plausible scientific arguments to explain how such a process must work.

For writers in the first half of the century self-purification had been that tangible factor that distinguished rivers from standing water. In rivers pollution apparently disappeared, while a polluted pond or cistern became increasingly offensive. They attributed this purifying capability of rivers to aeration and motion: water "left stationary, or deprived of access to air" soon acquired "taint."[79] This taint was the essence of insalubrity, and it was believed to represent either the absorption of bad air or the putrefaction of dissolved organic matter.[80] For this reason there was widespread opposition among reformers to reservoirs and household cisterns.[81]

But while the distinction between rivers and standing water might be recognized, and the importance of air and motion noted, it was

---

[79] "Domestic chemistry," p. 55.

[80] GBH MWS Report, pp. 31-35.

[81] Hassall believed allowing water to stand deprived it of "life" (he was speaking figuratively); while R. Angus Smith favored reservoirs only because in them dissolved organic matter became organisms removable by filtration (GBH MWS App. III, p. 113; Hassall, Microscopical Examination, p. 27).

difficult to characterize those processes that kept rivers pure. The Westminster reviewer who defended the Thames supply in 1830 spoke of the effect of motion being to "dissipate, decompose, or precipitate offensive" organic matter, and then explained: "We use these three words, because what the chemical process is perhaps remains to be ascertained, but the offensive matter disappears."[82] The reviewer also recognized that the atmosphere was involved in the process and thought that it helped to dissipate volatile and odoriferous matter and caused some dissolved matters to precipitate.[83] By the early 1850s the activity of dissolved oxygen was being frequently pointed to as the active agent in self-purification and the explanation of the association of purification with motion and exposure to air. In their 1851 report Hofmann, Graham, and Miller were satisfied that oxidation, together with excessive dilution, led to the disappearance of sewage poured into the Thames.[84] With typical acuity Hassall recognized the kinetics of oxidative self-purification, observing that water absorbed atmospheric oxygen at a rate proportional to the degree to which the water was deficient in dissolved oxygen. Like the scavenging activities of microorganisms, this was "a beautiful manifestation of contrivance."

> The oxygen of the water is constantly varying in amount, . . . it being continually used up in the oxidation of the various substances present, and in

---

[82] "Thames water question," Westminster Review 12(1830): 35.

[83] Ibid., p. 36.

[84] Graham, Miller, Hofmann, "Chemical report," p. 385.

> the respiration of the animal productions contained in the water, and were it not for the law, that in exact proportion with this reduction in the amount of oxygen is the power of absorption of the water augmented, these living productions would be in constant danger of perishing.[85]

In an age when the biological effects of pollution were widely believed to result from organic and inorganic poisons that the pollution contained, Hassall's recognition of the effects of the oxygen demand of pollution in causing the suffocation of animals was remarkably acute.[86]

In the 1850-52 London water controversy the company representatives included oxidative self-purification as one of their proofs of the salubrity of the existing supply, but the questions of the reliability of the process and its mechanisms did not emerge as significant issues in the controversy. Taylor argued that "all water contains a quantity of air, and the sulphuretted hydrogen and ammonia are all oxydized, . . . and the more that water is shaken and agitated, . . . the more quickly does this sewage matter disappear."[87] The process took a few hours and required only reasonable aeration.[88] Thomas Hawksley confidently upheld the safety of New River water on the grounds that the Hertford sewage that had contaminated it quickly disappeared, while J.T. Cooper maintained that cities had been built on

---

[85] Hassall as Lancet Analytical Sanitary Commission, "Records of the results," p. 219.

[86] H.B.N. Hynes, The Biology of Polluted Waters (Liverpool: Liverpool U.P., 1960), p. 65.

[87] S.C. Metropolis Water Bills, 1852, Evidence, Q 448.

[88] Ibid., QQ 516, 10255.

rivers precisely because such a process existed, enabling rivers to receive urban wastes yet remain pure a short distance downstream.[89]

By the late 1850s however, the hygienic meaningfulness of self-purification had become a subject of controversy. The issue was not the existence of some process involving oxygen which converted organic pollutants into harmless materials, but the reliability of such a process and its significance in terms of distinct conceptions of the nature of the harmful substances in polluted water. The coming of this controversy was recognized in a prescient Builder editorial of 1852:

> The very process of fermentation whereby, as argued, such rivers as the Thames purify themselves of town feces in their course, is pointed out as an influence poisonous and pernicious to health (if not in fact, as we suspect, fermentative of pestilent decomposition and putrefaction in the blood itself).[90]

The Builder was recognizing the increasing importance of the zymotic process in pathology and etiology, and the editorial reflects the Victorian ambivalence to natural purification that was discussed in Chapter 3. Clearly self-purification involved decomposition. Judged by Taylor's H_2S pathology the process was safe because its products were quickly oxidized or dissipated. But to those concerned with the zymotic process itself, such natural purifications were of doubtful salubrity at best.

The first major outbreak of this controversy occurred in 1857 in

---

[89] Ibid., QQ 3522 (Hawksley); 677-83 (Cooper).

[90] "Editorial," The Builder 10(1852): 294.

regard to Bazalgette's great metropolitan main drainage project, and it was concerned primarily with the effect of pollution on the salubrity of the urban atmosphere rather than its effect on water supplies. Late in 1856 the Metropolitan Board of Works had decided that Bazalgette's plan was the cheapest and best way of dealing with greater London's sewage. The Minister of Works had final say, however, and commissioned engineers Douglas Galton, James Simpson, and Thomas Blackwell to evaluate the project. These three issued a huge and highly critical report which Bazalgette and his colleagues G.P. Bidder and Thomas Hawksley answered with a production of equal size, and there followed an exchange of assertions and denials which lasted until the great stink of 1858 made some resolution imperative (and Bazalgette's was the only completed scheme available). A variety of highly technical questions were at issue between the rival groups of engineers, including the size of the sewers relative to the amount of water they would need to carry, the healthfulness of the towns near the outfalls, the potential of the incoming tide to return sewage to central London, and the ability of the river to purify itself.

In the referees' report the latter question was dealt with by the chemists A.W. Hofmann and Henry Witt. In 1856 Hofmann, in a report on the quality of the London water supply, had eloquently argued that the issue of potable water quality had to be considered from the perspective of Liebig's zymotic theory. Hofmann applied the same criteria to the non-potable water of the lower Thames, concerning himself not with sensible qualities, but with the amount and character of the decomposition that went on in the water. Since a comprehensive study of the course of the decompositive changes going on in the Thames was out of the question, Hofmann turned his attention to the amount of putrefactive change occurring in the river. He found that organic matter levels did not increase substantially despite the pollution the river received in its course through London. To Hofmann this was not proof of purification -- of the disappearance of organic matter -- as much as it was an indication of the excessive amount of putrefaction that must be occurring in the metropolitan Thames and the atmosphere above it. He believed agitation of the river water by tides and steamboats increased the rate of putrefaction while the mudbanks stored putrescible materials which gradually dissolved and helped to maintain the concentration of putrefaction in the Thames. Therefore, despite the relatively small amount of organic matter it contained, the London Thames was a dangerous, rapidly putrefying river, and the construction of Bazalgette's outfalls would make it even worse.[91]

For Bazalgette, Bidder, and Hawksley, on the other hand, this same decompositive process was purification. Sewage was highly unstable and purified rapidly, emitting dangerous ammonia, even as it

---

[91] A.W. Hofmann and Henry Witt, "Chemical investigations relating to the metropolitan main drainage," in Copies of a Letter . . . addressed to Captain Douglas Galton, R.E., James Simpson, C.E., and Thomas E. Blackwell, C.E., directing them to consider Plans for the Main Drainage of the Metropolis . . . and their Report . . . , P.P., 1857 sess. 2, 36, (233.).

flowed through the sewers. Purification continued in the river, and they concluded:

> so soon as the refuse of life has become thoroughly blended with water, chemical changes of the most important character become gradually effected by oxygenation [sic] (without putrefaction) of some, if not all, of the combustible elements, and the consequent decomposition and rearrangement of other elements into inorganic forms nowise injurious to the health of animated beings.[92]

The engineers admitted that in principle this process had limits, but in practice sewage diluted with ten to twelve volumes of river water could be assumed to purify safely. They defended their conclusion by pointing to other English rivers -- the Mersey, Aire, Don, and Soar -- where they had "remarked with wonder the rapidity of the oxydising process by which Nature causes the disgusting feculae of towns to become resolved into pure and inoffensive substances."[93] Using the data of Hofmann and Witt, Bazalgette, Bidder, and Hawksley argued a contrary conclusion. Since London's sewage did not increase the organic content of the Thames, and since there had been no long-term increase in organic matter in the river since the Hofmann, Graham, Miller study of 1851, it followed that the Thames was in a condition of equilibrium, that it was purifying organic matter as quickly as it received it.[94] They observed that the "influence of sewage on the physical and hygienic condition of the

---

[92] Report to the Metropolitan Board of Works by Messrs. Hawksley, Bidder, and Bazalgette, P.P., 1857-58, 47, (419.), p. 40.

[93] Ibid., p. 40.

[94] Ibid., pp. 40-41.

Thames" had been "mistaken . . . and exaggerated."[95]

Both parties recognized that the data on organic matter in the Thames showed the conversion of organic to inorganic matter. But here agreement ended. Bazalgette, Bidder, and Hawksley relied on the senses to indicate dangerous impurities -- specifically H_2S and NH_3 -- and therefore trusted the apparent purification of the river. Hofmann worried about obscure and imperceptible modes of putrefaction -- the "peculiar condition" in which nitrogenous organic matter became the active agent of zymotic disease. Putrefaction was a necessary stage in the conversion of organic to inorganic matter, a great deal of conversion was going on in the London Thames, and regardless of the apparent qualities of the water, the river might at any time become the site of an especially morbid form of putrefaction.[96]

Bazalgette, Bidder, and Hawksley also included a collection of subsidiary arguments in their rebuttal. They tried to show that much of the organic matter in the Thames was not from the sewers but from the sea, brought in on the tide.[97] They consulted Henry Letheby, medical officer of health for the City of London, in spite of his expressed skepticism of the efficacy of self-purification in estuarine waters.[98] Letheby took particular exception to the conclusion the

---

[95] Ibid., p. 41.

[96] "Referees' Report," p. 28.

[97] Report to the M.B.W., pp. 43-44. See also Wm. Odling and A. Dupre, "The composition of the Thames," 29th Report of the B.A.A.S. (Aberdeen, 1859) (London: J. Murray, 1860), Sections, pp. 75-76.

[98] Report to the M.B.W., p. 39. In fact Letheby actually supported their view. He was worried about the production of H_2S by the reduc-

government referees had based on float experiments, that objects dumped into the river at the proposed outfalls would oscillate to and fro with the tides, repeatedly returning upstream to the metropolis before finally reaching the sea. This conclusion assumed that sewage was like an "unchangeable log of wood," Letheby observed, when in fact it disappeared.[99] Letheby consulted Dr. Robert Barnes, senior physician on the Dreadnought hospital ship which was moored in the lower Thames, and Barnes too affirmed that "sewage as sewage is not permanent; it is rapidly converted into living and inorganic forms."[100]

Upon receiving this rebuttal, Galton and Simpson (Blackwell was in America) issued their own rejoinder to it. Substantial purification did not occur in sewers, and conclusions based on the self-purification of inland streams were inapplicable to estuarine conditions. The metropolis sewage might make little quantitative contribution to the river's load of organic matter, but its contribution was of the worst variety -- high in phosphorous and nitrogen. The referees ridiculed the Bazalgette, Hawksley, Bidder contention that the elements comprising organic matter were themselves harmless -- the worst poisons of organic chemistry were composed of harmless elements.[101]

---

tion of marine sulfates by putrefying sewage, a problem that would become worse the further down river the outfalls were located. The referees wanted to move the outfalls downstream further from the metropolitan area, so Letheby's doubts about the self-purification of estuarine waters supported the M.B.W. plan.

[99] Ibid., p. 74.

[100] Ibid., p. 80.

Bazalgette, Bidder, and Hawksley denied this rejoinder, and if the great stink had not intervened, these parties might have battled inconsequentially for years.

This self-purification debate set a pattern for subsequent debates. The opposing parties might hold similar views regarding the perceptible or chemically measurable disappearance of organic contaminants from rivers, but they held quite different estimations of the hygienic significance of this process. The unanswered question of the identity of the morbid poisons of zymotic diseases inspired an analytical and/or purificatory nihilism among many sanitarians, and in the 1860s this skepticism became the chief argument of those combatting the dogma of self-purification.

V. Snow, Hofmann, and the Beginnings of Analytical Nihilism

Both those sanitarians concerned with establishment of universal sewage recycling and those working to reform the quality, quantity, cost, and management of water supplies based their arguments in part on the belief that meaningful self-purification did not exist. John Snow's theory of the "mode of communication of cholera," announced in 1849, revised and re-published in 1855, greatly expanded the range of arguments reformers could use to condemn polluted water. Snow

---

[101] _Copy of a letter to the Rt. Hon. Lord John Manners, M.P., 1st Commissioner of H.M. Works, from the government referees for the main drainage of the Metropolis, in answer to the Report made by Messrs. Bazalgette, Hawksley, and Bidder to the M.B.W. upon the Report of the Referees_, _P.P._, 1857-58, 48, (403.), pp. 10-11.

rejected the idea that the sensible filth in the Thames caused cholera. The pathological changes of the intestine that occurred in cholera convinced him that the disease entered the body orally, and this conclusion led to his concern with polluted water supplies and to his epidemiological studies of the Broad Street pump and the south London water fields. Snow suggested that the cholera poison might be some marginal form of life which was transferred from the bowels of sufferers into the water drunk by future victims and used this suggestion to meet difficulties in the water-borne etiology of cholera. A biological poison would require an incubation period, and this explained why the disease did not immediately ensue when the victim drank the contaminated water. A biological poison was necessarily particulate, and therefore not all who drank specifically contaminated water could be expected to develop the disease.[102]

But although he compared the cholera poison to a living organism, Snow refused to accept the fungal theory of cholera causation of Budd and the Bristol microscopists. Snow admitted the cholera poison resembled a living thing, but steadfastly refused to say it was one. Instead, he invoked Liebig's chemistry to avoid the issue, arguing that the contact process of chemical change characterized physiologi-

[102] John Snow, "On the mode of communication of cholera," in Wade Hampton Frost, ed., Snow on Cholera -- A Reprint of two papers by John Snow together with a biographical memoir by B.W. Richardson and an introduction by Wade Hampton Frost (New York: Commonwealth Fund, 1936), p. 113; "Water supply in relation to health and disease," J. Public Health and Sanitary Review 1(1855): 130-40.

cal processes and that there was therefore no basis for distinguishing vital from non-vital organic phenomena.[103] With this mysterious contagious particle Snow effectively obscured the relationship between pollution and disease. Snow had created the idea of a primal malignant entity which occupied the border between the living and the non-living and might at any time occupy polluted water, enter the body when that water was drunk, and undermine the normal condition of the intestine. This entity was distinct from "ordinary decomposing animal matter," and Snow was sure that these "morbid poisons [could not] be detected by analysis."[104]

The value of Snow's thesis for the public relations of water reform was enormous. Briefly, where heretofore the reformers had demonstrated the danger of polluted water by pointing to the perceptible impurities -- microorganisms, for example -- as proof of insalubrity, Snow's theory allowed them to exploit the public's imagination. The hypothesis that the zymotic morbid poison of cholera (and by implication of other diseases) was an infinitely small and intense-

---

[103] J. Snow, "On continuous molecular changes more particularly in their relation to epidemic diseases," in Frost, Snow on Cholera, pp. 150-52, 156-57. P.E. Brown suggests that Snow's unwillingness to be pinned down on this issue arose from his desire to distinguish his views from William Budd's clear advocacy of a microbial poison. I believe Brown has not given sufficient attention to the pervasiveness of Liebigian chemistry (Brown, "Autumn loiterer, pp. 519, 523).

[104] Select Committee on the Public Health Bill and the Nuisances Removal Bill, Evidence, P.P., 1854-55, 13, (244.), QQ 119-20, 122-23, 173. Snow was defending trades such as slaughtering, in which putrefying matter was produced. He pointed out that such trades did not contribute to epidemics.

ly virulent particle was sufficiently flexible to accomodate a great breadth of epidemiological data, and the particles were so loosely defined that the reformers could use them as the basis for promulgating all manner of disquieting notions about undetectable and unremovable contamination. In the 1860s these particles came to be known as germs.

After the mid 1850s the inability of chemical analysis to show whether water was safe, and the uncertainty whether any processes of purification could make polluted water safe increasingly dominated discussions of water quality. As Liebig's zymotic theory became popular among sanitarians, Snow's portrayal of the cholera poison as an ill-defined and unmeasurable zymotic entity was amplified, though most stayed away from his uni-factoral commitment to specifically polluted water supplies as the means by which that disease spread, and from the quasi-biological nature of his contagion. The Board of Health's Committee for Scientific Inquiries, which studied the 1854 cholera epidemic, conceived of the cholera poison as a ferment and wished for a study of the course of decomposition during the epidemic.[105] In an 1857 paper, Frederick Crace-Calvert, a Manchester manufacturing chemist, pointed out that the River Medlock could be feasibly purified with lime to such a degree that its water would be analytically superior to many domestic supplies, yet still advised against drinking it because some "nitrogenous residue" might not be

---

[105] G.B.H. Committee for Scientific Inquiries Report, pp. 48, 37.

removed.[106]

The replacement of standards of sensibility by analytical and purificatory nihilism is apparent in studies of the London water supply undertaken for the General Board of Health in 1856 to determine whether changes in water handling procedures required by the 1852 Metropolis Water Act had significantly improved the supply. To answer the question three of the Board's engineers studied the condition of the upper Thames, while Hofmann and Lyndsay Blyth, lecturer in natural philosophy at St. Mary's hospital, analyzed the waters the companies supplied. The engineers, William Ranger, Henry Austin, and Alfred Dickens, viewed the increasing discharge of raw sewage by upstream towns as a serious threat to the health of London. They had no faith in chemistry's ability to show whether purification occurred. The soluble (so they thought) cholera poison might enter the river at any time, and "to what extent the danger of such pollution may be obviated by the atmosphere, or by other causes, in so great a length of flow . . . we do not suppose that any body is capable of determining."[107] In their opinion the incompetence of chemical analysis to discover dangerous pollution was illustrated by the fact that the Southwark and Vauxhall company's water, which had been associated with

---

[106] F. Crace-Calvert, "On the purification of polluted streams," JRSA 4(1856-57): 506.

[107] Wm. Ranger, H. Austin, & A. Dickens, "Examination of the Thames . . . ," in Reports to the Rt. Hon. Wm. Cowper, M.P., president of the G.B.H., on the Metropolis Water Supply; under the provision of the Metropolis Act, P.P., 1856, 52, [2137.], p. 91.

the most concentrated cholera mortality during the epidemic of 1848-49, contained the smallest amounts of organic matter of the waters examined by the 1851 commission (Hofmann, Graham, and Miller). They concluded: "chemical analysis does not at present convey an exact understanding of the danger to health which a particular water may occasion."[108] Thames water would not be safe until Chadwick's sanitary schemes were completed and every town had a sewage irrigation farm.

Hofmann and Blyth were equally skeptical about the abilities of chemistry to measure the wholesomeness of water. They regarded the measurement of organic matter in water as "one of the most delicate operations in analytical chemistry," and wrote: "still greater is the difficulty of examining into the nature of the organic matter." They admitted: "very little is known of the nature of the ill-defined substances which constitute the organic matter in water." They questioned the significance, and even the existence of Mulder's crenic and apocrenic acids. Potability, they asserted, had to be considered with regard to the zymotic theory. In a statement frequently quoted during the next two decades they wrote:

> it is now generally admitted, that the substances which constitute the organic matter in water act by no means in consequence of being poisonous themselves, but by undergoing those great processes of transformation, called decay and putrefaction, to which all animal and vegetable matter is subject when no longer under the control of vitality. . . . These putrefactive processes either give rise to the formation of

---

[108] Ibid., pp. 92-93, 95.

> poisonous bodies, or -- and this is far more probable -- they act simply as ferments, exciting similar processes of decomposition in the substances composing the living animal organism.[109]

Since these putrefactive processes were especially evident in nitrogenous materials, Hofmann and Blyth asserted that "the deleterious character of organic matter in water is proportionate to the amount of nitrogen it contains."[110] Unfortunately there were no reliable techniques for measuring the minute concentrations of organic nitrogen typically present in potable waters. Hofmann and Blyth suggested a technique could be developed in which the total combined nitrogen in the evaporative residue of a sample was first determined, and then the nitrogen existing as ammonia, nitrites, or nitrates subtracted from this total. The difference would be "the quantity of still putrescible nitrogenous substance, [which was] . . . the necessary data for a comparison of the organic matter present in different waters."[111] Blyth fell ill, however, and no such technique was developed until 1867 when Edward Frankland took up the problem. Still, the views of Hofmann and Blyth were influential, and in arguing that water analysis could only be useful when it could measure things it was now unable to measure, Hofmann and Blyth greatly reinforced the growing conviction that there was no way to prove that water was safe.[112] By the mid 1860s nihilism had triumphed. It had

---

[109] Hofmann and Blyth, "Report on the chemical quality of the supply of water to the metropolis," pp. 4-5.

[110] Ibid., p. 5.

[111] Ibid., p. 6.

[112] While organic nitrogen remained unmeasurable until 1868, by the

acquired philosophical justification from the skeptical chemist Sir Benjamin Brodie and in the hands of Robert Rawlinson, J.T. Way, and later Edward Frankland became the central premise of the opposition to the use of once-polluted water.

---

early 1860s many chemists had begun to emphasize the results of the potassium permanganate process, which supposedly gave a measurement of the putrescibility of a water sample. A known amount of potassium permanganate was added to the sample and the amount of oxygen it lost (presumably in oxidizing organic matter) and, for some chemists, the rate of oxygen loss, was assumed to be proportional to the putrescibility. The original permanganate test had been developed in 1850 by the Danish chemist G.B. Forchhammer, and by 1865 there were several versions of it in use. Some measured the oxygen absorbed from a cold, others from a boiling, solution of $KMnO_4$. Usually permanganate continued to oxidize materials in the water indefinitely, and there were differences of opinion among analysts regarding the period at which the oxygen lost was a meaningful indication of salubrity. Worse, there were fundamental weaknesses in the technique. Permanganate would oxidize incompletely oxidized inorganic compounds such as nitrites in the water. Despite assertions that the matter oxidized by permanganate was the harmful matter in water, the process did not inspire enough confidence among sanitarians, or even among chemists, to slow the growth of nihilisitic attitudes about water quality to any substantial degree. See W.A. Miller, "Observations," pp. 120-24; Dr. Woods in CN 6(1862): 306-7; CN 11(1865): 283-85; Lancet, 1865, ii, pp. 165, 605; Edward Frankland, "On the water supply of the Metropolis," Proc. Royal Inst. 5(1866-69): 114-16; [B.H. Paul], "Water analysis for sanitary purposes," BMJ, 1869, i, pp. 428, 497; Thomas Spencer in disc. of E. Byrne, "Experiments on the removal of organic and inorganic substances in water," MPICE 27(1867-68): 6-9; C.W. Heaton, "The future water supply of London," Quarterly J. of Science 6(1869): 240.

VI. THE TRIUMPH OF NIHILISM, 1860-68

> Sanitary pursuits produce a kind of intoxication which raises the intellect of a genuine theorist above the vulgar rules of induction.
> -- Chemical News*

The mid 1860s were critical years in the development of ideas about water purity in much the same sense as were the early 1850s. After an interlude of relative quiet during the early 1860s, the London water controversy exploded again, in part as a result of the east London cholera epidemic of 1866, in part due to attempts to put into law recommendations of the Royal Commission on the Sewage of Towns (1857-64) and the first Royal Commission on Rivers Pollution (established 1865). During this period self-purification emerged as as political issue. Promoters of sewage farming and those working for passage of anti-pollution laws found it prudent to ignore the concept or to deny the effectiveness of self-purification, while for polluters, self-purification became an increasingly important justification.

Both the epidemic of 1866 and the investigations of the commissions underscored the fact that on water quality matters scientists were flying blind -- epidemiology showed that polluted water caused disease but the identity of the disease-causing factors, and their behavior and the means of their removal were unknown. Ignorance of

---

* CN 18(1868): 213.

what purification was, coupled with uncertainty about what chemical analyses were measuring, led to a situation in which conclusions about the safety of a water sample were largely independent of analytical results; instead they depended on the analyst's choice of morbid poison. Alfred Swaine Taylor defined the harmful substances in water as measurable inorganic chemicals and found the water of the River Lee safe; even when nihilists Robert Rawlinson and J.T. Way affirmed their utter ignorance of the ultimate causes of disease they were in effect characterizing morbid poisons since they were arguing that water-supply policy be based on the worst case -- the assumption that water-borne morbid poisons were undetectable and unremovable.

This chapter examines the politicization of self-purification and the development of analytical and purificatory nihilism both as a philosophical principle and a political stratagem. It also considers two cases in which systems of water analysis were embedded in essentially circular forms of reasoning, cases in which analysis faithfully confirmed the social-policy assumptions of the analyst: Taylor's 1868 proof that his analysis discriminated safe from virulent water; the adaptation of medical theory by Rawlinson and Way in 1866 and again in 1868 to demonstrate the harmfulness of polluted water regardless of the findings of epidemiology or chemical analysis. A third case is the subject of the next chapter: Edward Frankland's development, between 1866 and 1868, of a wonderfully complicated system of water analysis, which included substantial improvements in the chemical techniques of analysis, but in fact discriminated between safe and virulent water only after that decision had been made on other grounds.

I. Self-Purification and Analytical Nihilism in the early 1860s

The tumultuous scientific polemic that had characterized discussions of water quality throughout the 1850s was absent during the early 1860s. While widespread doubts continued regarding the safety of river and shallow well water, the effect of putrefying rivers on the health of the surrounding area, and the abilities of chemists to quantify these dangers, the absence of major water-borne epidemics during these years prevented the transformation of these concerns into serious campaigns for reform. Instead, another water-related issue occupied the attention of scientists, politicians, and the public: faith in the brave new world of profitable sewage recycling peaked in the early 1860s. A great many sanitary scientists were involved in the sewage recycling movement, and it provided a new context and new reasons for opposing the doctrine of self-purification.

First, some "reactionary" engineers, notably Thomas Hawksley and Joseph Bazalgette, had argued that sewage purified rapidly as it flowed through sewer pipes, this being just one instance of the incredibly rapid purification sewage underwent. All scientists involved in the recycling movement agreed that such a process would deprive sewage of much of its manurial worth, for Hawksley and Bazalgette were thinking of the rapid conversion of nitrogenous organic matter into valuable, but volatile, ammonia. Therefore both the Brady select committee on the sewage of towns (1862) and the Montagu committee on the sewage of the metropolis (1864), each devoted to the promotion of sewage farming, took ample testimony from chemists and sanitarians who were prepared to maintain that self-purification substantially did not exist, or that the conditions within sewers were so different from those in rivers that no comparison was warranted. Little in the way of an increased understanding of the mechanisms of self-purification emerged from these hearings, however, since the experts responded to difficult questions either with ad hoc equivocations or with brazen and unverified assertions. J.T. Way, for example, offered a plausible but entirely hypothetical argument to account for the assumed differences in the decompositive environments in sewers and rivers, citing the distinct CO_2:oxygen ratios in the atmosphere of each environment. The agricultural chemist Augustus Voeckler insisted that sewage remained unchanged as it flowed along in rivers and would qualify his opinion only slightly in the face of persistent allusions of the observed self-purification of rivers. Such extemporaneous scientific thinking was all too typical of expert witnessing in water issues, but at least Way, Voeckler, and their colleagues had the excuse that no detailed studies of the self-purification of rivers had yet been done.[1]

---

[1] S.C. on the Sewage of Towns, Evidence, P.P., 1862, 14, (160.), QQ 254-66 (Smith); 656-62 (Hofmann); 844 (Way); 1261-79 (Voeckler); Select Committee on the Sewage of Towns (Metropolis), Evidence, P.P., 1864, 14, (487.), QQ 832 (G.W. Hemans); 3095 (S. Gael). Also see J. Mechi in CN 1(1860): 175; and Thomas Hawksley, C.E., "The utilisation of town sewage," The Builder, 1861, i, p. 166.

The recycling advocates also objected to the doctrine of self-purification in general, because it was continually being invoked by local authorities as a reason why the expensive and untested recycling schemes of the enthusiasts and the government were unnecessary. The Royal Commission on the Sewage of Towns (whose bias is revealed in its full title as: "the commission appointed to inquire into the best mode of distributing the sewage of towns and applying it to beneficial and profitable uses") maintained that whether or not purification occurred, sewage pollution was disgusting and should therefore be stopped.[2] More interesting is the strategy of the subsequent Royal Commission on the Prevention of the Pollution of Rivers (RPPC) established in 1865. Two of the three RPPC members (R. Rawlinson and J.T. Way) were veterans of the Sewage of Towns commission, and the RPPC regarded its chief mission as promulgating the recycling technology developed by its predecessor. The existence of a process of self-purification of rivers was clearly antithetical to the reforms the commission was promoting, and to an astonishing degree the commission, which was after all supposed to be investigating the problem of river pollution, ignored the issue of stream self-purification.

The commission held extensive hearings between 1865 and 1867 on the pollution of the Thames, the Lea, and the rivers of industrialized Yorkshire. The issue of self-purification arose rarely -- only when the commission was examining witnesses whose opinions confirmed

---

[2] See Chapter IV, note 45.

theirs, or when obstreperous defenders of pollution forced consideration of the issue. In their first report they admitted that "as the sewage travels down with the flowing water of the river, a process of oxidation goes on which tends to purification"; but then pointed out that this was inconsequential since "in the general opinion of medical men" it was the quality of the organic matter, not its quantity, that was of concern.[3] Usually their questions about river pollution reflect the assumption that whatever was poured into the river stayed in it. At Enfield on the Lea, where treated sewage ran several miles through a ditch before entering the river, the commission queried: "At all events whatever there is in your [sewage effluent] water when it passes away from your works goes into the river Lee?"[4] When farmer William Delano maintained that the Lea was after all not so badly polluted, since aquatic weeds absorbed the impurities, the commission responded to Delano's "opinion" by terminating his testimony: they wanted no details about the efficacy of self-purification.[5] The only sort of self-purification the commissioners allowed was a purely mechanical sort -- a deposition of suspended impurities -- and therefore the observation of self-purification

---

[3] RPPC/1865/1st Rept. (River Thames), P.P., 1866, 33, [C.-3634.], p. 18. Also see RPPC/1865/2nd Report (River Lee), P.P., 1867, 33, [C.-3835.], p. xi; where they admitted that the effluent from the Luton sewage works (lime precipitation process) "seems to become purer," after a few miles of flow in the river.

[4] RPPC/1865/2nd Rept., Vol. II, Evidence, P.P., 1867, 33, [C.-3835-I.], Q 3935.

[5] Ibid., QQ 4139-40.

proved that the impurities in question had been only suspended. Rawlinson inquired of J.T. Child, a Halifax dyer: "Are you aware that the great mass of the colouring matter is not chemically combined with the water, but is mechanically suspended in it?" Child answered diplomatically: "It is mechanically suspended to some extent, I think, but not entirely." Rawlinson persisted: "Is not that proved in a grand scale in your rivers, because as they flow downwards by exposures to the sun and the air, and the earth forming the bed of the river, the water loses its colouring matter as it goes to the sea? Showing that it is not a chemical but a mechanical combination?" Christopher Sedgwick, chairman of the local board of Skipton, was one of the few with some definite conception of the capabilities of self-purification. When he maintained that Skipton sewage mixed with four volumes of water would be purified in a river falling 14 feet/mile over 10 to 12 miles, Sedgwick was told that he was lucky to have plenty of water to "dilute" the sewage.[6] The commission's philosophy was summarized by Way in a question to R.H. Sedgwick: "It might be said," Way observed, "that the evil is at the beginning of the streams that feed a river, and that the evil grows worse the lower you come down?"[7]

The possibility that sensible, and sometimes even chemical

---

[6] RPPC/1865/<u>3rd Rept., Vol. II, Evidence</u>, <u>P.P.</u>, 1867, 33, [C.-3850-I.], QQ 11876-78; 14645-48, 14651. For other examples of the commissioners' attitude toward self-purification see <u>2nd Rept., Evidence</u>, Q 4443; <u>3rd Rept., Evidence</u>, QQ 2778-79, 3423-27*, 8272-73, 8433.

[7] RPPC/1865/<u>3rd Rept., Evidence</u>, Q 14707.

indicators of water quality were useless and could be downright dangerous received considerable, if sporadic attention during the early 1860s. In what became an overtaxed simile an 1864 Lancet editorial called for water supplies to be like Caesar's wife -- above suspicion -- since chemical analysis could not detect harmful contamination.[8] Sanitarians crusading to end the use of frequently polluted shallow wells pointed out that waters with the best appearance were usually the most polluted. The "sprightliness" of pump waters came from decomposition products of the cesspool matter with which they were usually contaminated.[9] River water might be equally deceiving. Robert Rawlinson noted that the Thames was best when muddiest, worst when clear.[10]

Besides these anxieties that sensible standards were untrustworthy guides to water quality, there was some direct evidence supporting purificatory nihilism. Putrescibility was commonly regarded as an indication of impurity, and filtration did not effectively remove dissolved putrescible matter from water. Boiling or the addition of antiseptic chemicals might keep water from putrefying, but these techniques were not practical on a large scale. There existed "no efficient artificial method . . . to purify, for drinking and culinary

---

[8] "Editorial," Lancet, 16 April 1864, pp. 142-43.

[9] Lankester, "London pump waters," BMJ, 21 June 1862, pp. 673-74; Henry Letheby, "Report on the sanitary condition of the City of London for the quarter ending 22 June 1861 . . . ," CN 4(1861): 76-77.

[10] S.C. Sewage of Towns (Metropolis), Evidence, QQ 4202-3.

purposes, water which has once been infected by town sewage. . . . it is always liable to putrify [sic] again."[11] As Lord Robert Montagu put it, in introducing his River Waters Protection Bill to the House of Commons, "let them not lay the flattering unction to their souls that filtration offered any adequate remedy."[12]

Despite these repeated warnings, many, particularly water and sewage works engineers and the municipal officials who employed them, continued to evaluate their work by sensible standards. The first Rivers Pollution Commission encountered this tendency frequently during its hearings. The East London Water company's engineer, Thomas Greaves, admitted of the company's reservoirs that "no doubt the deposit in the bottom is not so sweet and clean as the water in the upper part, but they [reservoirs] are generally bright, sometimes very bright." Greaves was eventually forced to acknowledge that brightness was "not altogether" a sign of safety.[13] Likewise, James Muir, the New River company's engineer, when asked if Tottenham effluent entered the Lea "in a pure state, responded: "Yes, in a tolerably clear state."[14] W.T. Pledge, chairman of the Luton board of health, was astonished to learn that the lime sewage process his board employed only removed suspended matter -- "brightness" had always been his criterion for judging the success of the treatment.[15]

---

[11] Ibid., p. vi.

[12] Hansard's Parliamentary Debates, 177(8 March 1865), pp. 1313-14.

[13] RPPC/1865/2nd Rept., Evidence, QQ 932-46.

[14] Ibid., Q 1082.

[15] Ibid., QQ 1380-83.

Late in 1865 the status of analytical and purificatory nihilism changed dramatically. Hitherto nihilism had been pragmatic, an expression of concern by water chemists who recognized the gap between the methodology of water analysis and knowledge of the identity and behavior of morbid poisons in water, and thus distrusted purification because they could not determine whether it was effective. Yet even the sanitarians who had the greatest doubts about reliability of purification and the abilities of analysis recognized the need for water and did their best to evaluate water quality in each case they investigated. In 1865 nihilism became programmatic. Sir Benjamin Brodie, Bt., (the younger), professor of chemistry at Oxford, added to the empirically based nihilism of the water analysts an a priori nihilism reminiscent of Hume. Brodie had been one of the dons who testified before the first Rivers Pollution Commission at its Oxford hearings in November 1865. The analytical and purificatory nihilism he introduced on that occasion and clarified a year and a half later before the Royal Commission on Water Supply was drawn from two distinct lines of argument.

Brodie's first argument was based on chemistry rather than philosophy. Considering the problem of the oxidation of organic matter from the perspective of pure chemistry, Brodie pointed out that most organic substances were difficult things to oxidize; oxidative self-purification was therefore unlikely.

> With regard to oxydation, we know that to destroy organic matter the most powerful oxydising agents are required, we must boil it with nitric and chloric acid and the most perfect chemical agents. To think

> to get rid of the organic matter by exposure to the air for a short time is absurd.[16]

Thus the contention that "the oxydizing power acting on sewage in mixture with water after a distance of any length is sufficient to remove its noxious quality" was in Brodie's view "simply impossible."[17]

Complementing Brodie's chemical deductions was his devastating and intransigent skepticism. Brodie candidly admitted apparent purification: "it is . . . singular how very little the impurities cast in [it] affect the river; the Cherwell by my house is not in a very impure condition, though it is only a few hundred yards below the outlet of the sewers," and he suggested that the deposition of sewage at the outfall might account for this.[18] But in Brodie's opinion these changes in appearance had absolutely no relation to the question of the hygienic quality of the water. No matter how much nastiness got removed, and no matter how it got removed, the poisons of sewage were unknown, and there might remain a "quantity . . . quite sufficient to be injurious to health."[19] Brodie's nihilism emerged even more strongly in his testimony to the 1868 Water Supply Commission when the Duke of Richmond pressed him to say when sewage-polluted water would become safe to drink. Brodie repeatedly maintained that the question was unanswerable. As an illustration, he suggested the case

---

[16] RPPC/1865/<u>1st Rept., Evidence</u>, Q 1497.

[17] Ibid., QQ 1493-94.

[18] Ibid., Q 1497.

[19] Ibid., QQ 1495-96, 1500-2.

of a mixture of 1% sewage in a glass of water with air bubbling through it, conditions loosely imitating aeration in a stream. Then Brodie lectured the Duke:

> The question is at what time would that tumbler of water become in such a state that anyone of us would be willing to drink it off, how many days, weeks, months, years would elapse? That I understand to be the problem, and I am sure I cannot solve it; but I can only say that when you have once put sewage into the water I should be rather reluctant to drink it.[20]

Anyone with a modicum of sense could answer such a question, Brodie affirmed, and water quality was a determination to be made by experience, not by science.

Brodie's doubts (which he shared with his Oxford colleagues, despite their admiration of the biological processes that nature provided for cleansing streams) inspired the analytical and purificatory nihilism practiced in the last years of the 1860s and throughout the 1870s, and which dominated the issue of water quality until the bacteriological analyses of Koch were made available in the mid 1880s. For Brodie skepticism was a luxury he could afford, for he was not actively involved in advising the public on water quality. Others, however, -- and Edward Frankland in particular -- developed from Brodie's philosophical nihilism an array of arguments, analytical procedures, and modes of representing results and of communicating with the public that allowed them to maintain their active role in effecting sanitary improvement, while at the same time admitting that they

[20] Royal Commission on Water Supply, Report, P.P., 1868-69, 33, [4169.], Evidence, Q 6989.

were working in total ignorance of the noxious elements in polluted water.

II. Defending Chemistry's Honor: A.S. Taylor in 1868

The changes that had occurred in the understanding of water quality between the early 1850s and the late 1860s are nowhere more clearly reflected than in the testimony of Alfred Swaine Taylor to the 1868 Select Committee on the River Lea Conservancy Bill. In the face of the increasingly nihilistic attitudes of many of his colleagues, and the nihilism even of the M.P.s who questioned him, Taylor retained intact the perspective he had held in the early 1850s. Self-purification worked reliably and was an hygienically meaningful process. Water conveyed disease only by means of poisons that could be easily discovered by chemical analysis.

The River Lea Conservancy Bill was a response to the East London cholera epidemic of 1866 and to the subsequent investigation of it by the first Rivers Pollution Commission (Rawlinson, Way, and Harrison). The commissioners recommended that all sewage not first purified by sewage irrigation be kept out of the Lea, and it called for the establishment of a powerful conservancy board whose duties would include enforcement of this prohibition.[21] The 1868 bill that embodied these recommendations was opposed by many -- by upstream towns such as Luton and Hertford which would be required to improve their

---

[21] RPPC/1865/<u>2nd Rept.</u>, pp. xxiv-xxvi.

sewage works; by the East London and New River water companies, concerned with protecting the reputation of the existing supply.

Taylor appeared on behalf of Luton (and indirectly of Hertford). His analyses (for organic matter by ignition and mineral matter, with nitrates and ammonia examined qualitatively) of Luton's sewage effluent and of Lea water from below the Luton outfall showed that these were not as pure as most potable waters, but Taylor nevertheless regarded them as quite satisfactory since "there was no trace in either sample of the compounds which render sewer waters noxious or poisonous, namely, the sulphide of ammonium and sulphuretted hydrogen."[22]

Taylor was the ideal analyst to defend Luton. He believed that the proof of purification was the absence of those sensibly noxious and chemically measurable gases of putrefaction, towards the removal of which the lime sewage treatment processes employed at Luton (and Hertford) were principally directed. He stated that his 37 years analyzing water allowed him to say definitively whether water was safe or harmful. Taylor's rejection of nihilism was quite as axiomatic as Brodie's assertion of it: "I deny the existence of that which cannot be discovered."[23] Unlike many analysts, Taylor had a clear conception of the chemical distinction between safe and unsafe water:

> If I find a water containing upwards of 40 or 50

[22] A.S. Taylor, "Report on analysis of two samples of water marked Luton no. 1 and no. 2," in S.C. River Lea Conservancy Bill, Evidence, P.P., 1867-68, 11, (306.), pp. 172-73.

[23] Ibid., Q 2400.

> grains in an imperial gallon, of solid contents, and four or five grains of organic matter, especially if there are nitrates with it, I do not care to inquire what the influence is upon the population; I say at once it is not wholesome water.[24]

In those rare occasions when analytically pure water still appeared to be causing disease Taylor claimed that a bio-assay technique -- ingestion of the water by animals -- surely would provide a reliable basis for evaluating the water.[25]

Since gases of putrefaction existed neither in Luton's effluent nor in the river below the outfall, Taylor concluded that Londoners need not be concerned about the dilute Luton sewage that contaminated their water. The proof of self-purification, he repeatedly asserted, was that after intermixture with water and a few miles' flow in a river, the chemist could find no traces of sewage in river water.[26] Taylor's faith was not simply a function of his positivism. He believed that rapid purification was characteristic of nature. Urea, for example, changed rapidly into ammonia. Aquatic organisms assimilated pollutants. Where substantial organic matter remained in a river some distance below a sewage outfall, Taylor assumed that it was not a remnant of sewage itself, but instead represented sewage which had been biologically recycled, "decomposed and recomposed," possibly more than once.[27] If a river were not overloaded Taylor

---

[24] Ibid., Q 2405.

[25] Ibid., QQ 2401-2.

[26] Ibid., QQ 2413, 2416.

[27] Ibid., QQ 2422-25.

believed it was true that "when nature is left to itself . . . the counteracting chemical resources . . . bring matters very much back to their original condition."[28]

So long as Taylor was questioned by those sharing his confidence in chemistry and providence he ran into no trouble. Under examination by Sir George Bowyer, however, the weaknesses in Taylor's position were uncovered. Taylor admitted that air-borne morbid poisons, such as the typhus poison, were chemically undetectable.[29] He would not admit the same situation might hold true for water-borne poisons, but again fell back on bio-assay techniques, arguing that if apparently disease-causing water did not contain chemically detectable poisons, its virulence could still be determined by feeding it to animals.[30] In practice neither Taylor, nor to his knowledge any of his colleagues, had ever met with this difficulty -- he claimed to have found a chemical cause of disease in every disease-causing water he had been given to analyze. Dysentery among the troops in Aden, for example, had been due to oxides of magnesia in the water they drank.[31] Yet Bowyer persisted, and the assumptions upon which Taylor's reasoning was based finally emerged:

> I have been consulted in cases where they [waters] have unfortunately produced illness, and I have found on analyzing them a sufficient cause for the illness. In other cases they [waters] have been said to produce

---

[28] Ibid., Q 2426.

[29] Ibid., Q 2513.

[30] Ibid., Q 2515.

[31] Ibid., Q 2516.

> illness, but on analyzing them I have found them perfectly wholesome, and have told the parties that the illness must be referred to something else.[32]

Bowyer pressed on, suggesting that it was really impossible to say how minute a portion of sewage might be lethal. Taylor maintained that on finding sewage matter in water he condemned that water; but the imputation of Bowyer's questions -- that polluted water might contain noxious sewage materials undetectable by chemistry -- effectively remained unanswered.[33]

As Bowyer recognized, the foundations of Taylor's water analysis were crumbling. His assertion that harmful materials in water were chemically detectable was based on the assumption that they were, and was in no sense a conclusion induced from Taylor's accumulated experience analyzing water. If epidemiological evidence were given priority over chemical analysis Bowyer was right: there indeed existed noxious materials in sewage that Taylor's techniques could not detect.

III. The Nihilism of Rawlinson and Way

The civil engineer Robert Rawlinson and the chemist John Thomas Way also testified before the Select Committee on the Lea Conservancy Bill in 1868. Rawlinson, chief of the sanitary engineers the Local Government Act Office employed to examine plans for local sanitary

---

[32] Ibid., Q 2518.

[33] Ibid., QQ 2527-30.

improvements, and Way, formerly chemist to the Royal Agricultural Society, constituted two thirds of the first (appointed 1865) Royal Commission on Rivers Pollution. Both men were advocates of sewage farming, opposed the doctrine of inevitable self-purification, and both had adopted the views Brodie had put forth in his testimony to the Rivers Commission in late 1865 -- that one could never know if polluted water had become pure. Their testimony before the 1868 Lea committee and before an earlier select committee considering the commission's recommendations (the Select Committee on the Thames Navigation Bill of June 1866) marks the emergence of Brodie's nihilism as a central premise in the political argument that polluted water supplies ought to be abandoned.

Like Brodie, Rawlinson and Way argued that since morbid materials were unknown, public welfare was best served by avoiding the use of once-polluted water. They exploited the pathologists' confusion. If medical men could not identify morbid poisons, how, they wondered, could chemists be expected to measure them. Central in their conception of the morbid substances in water was the idea of

> some indefinable matters, most probably of an animal character, which are capable of setting up a kind of fermentation in the blood, and producing disease; those matters may be infinitesimally small, but they are still capable of doing great mischief.[34]

This notion of undefined "peculiar conditions" of putrefaction as the cause of zymotic disease -- essentially Hofmann's 1856 notion --

---

[34] Ibid., QQ 1942-43.

allowed Way to explain why sewage only occasionally produced acute effects; at other times, he maintained, sewaged water simply predisposed a person to disease.[35]

Unlike Taylor, both Way and Rawlinson acknowledged that epidemiology was the proper science for determining whether water was harmful. Yet their epidemiology was of a strange sort. By applying an extremely flexible set of interpretive principles to the epidemiological data that described the health of a population drinking once-polluted water, Rawlinson and Way were able to condemn the water no matter what the data showed. The concept of predisposition was doubly useful. It explained why not all who drank infected water during an epidemic period were struck by disease -- not all were predisposed; and it explained why sewage-polluted water did not always cause acute disease -- sometimes it was putrefying only in its non-acute, predisposing mode, harmful but not virulent.[36] In this manner a considerable variety of epidemiological data could be subsumed. That normally Lea water had no effect on the health of Londoners proved only that normal putrefaction was occurring. On the other hand, the excessive death rate of the metropolis as a whole might be ascribed in part to the cumulative effects of drinking polluted Thames and Lea water.[37] Way's statement that water-borne diseases could only be understood inductively by epidemiology in fact

---

[35] Ibid., Q 1957.

[36] Ibid., Q 1994.

[37] Ibid., QQ 1919-23; 1927-28.

hid a mental process as full of specious reasoning as Taylor's circular justification of water analysis.[38]

Rawlinson and Way also used the poor understanding of the nature of morbid poisons to conclude that methods of purification were inadequate. Rawlinson believed filtration was a mechanical process which removed only the solids in water and concluded therefore that it had no effect on quasi-soluble morbid poisons.[39] Way regarded filtration as to some degree a chemical (oxidative) process but he did not credit it with reliable purification.[40] Further, Way regarded chemical sewage treatment as mere deodorization, claiming that "matter of a most objectionable character" remained in the effluent.[41] Even sewage irrigation, which both championed, could not be relied upon to produce potable effluents.[42]

---

[38] Ibid., Q 2056.

[39] Ibid., QQ 2714-15. For Rawlinson, solubility was not incompatible with germiness. Like many of his contemporaries, Rawlinson used the germ simply as a name for specific morbid poisons of unknown characteristics. All that could be known of germs came from the behavior of disease -- these entities had no existence independent of the pathological process. Also see Q 2743; and Chapter 3 above.

[40] S.C. Thames Navigation Bill, Evidence, P.P., 1866, 12, (391.), QQ 3167-69.

[41] S.C. Lea Conservancy, Evidence, Q 1850.

[42] S.C. Thames Navigation, Evidence, Q 2684. The stance of those who were at the same time purificatory nihilists and advocates of sewage irrigation is interesting. Usually, as with Way, Rawlinson, and Frankland, they affirmed that irrigation could not reliably produce potable effluents but at the same time acquiesced in the domestic use of water from rivers that received irrigation-treated sewage. Moreover, they promoted irrigation projects even when they knew the effluents from these would end up in streams used for drinking water. See RPPC/1865/2nd Rept., pp. xiii-xiv.

Complementing this distrust of purification was a distrust of chemical analysis. Way felt that chemistry could sometimes offer sanitarians useful information, but acknowledged -- on occasion with what seems a brazen complacency for an analytical chemist -- its utter inability to detect morbid poisons. Just as Taylor had been certain that his colleagues could detect morbid poisons, so Way was equally confident that scientists agreed they could not.[43]

This systematic skepticism -- about the identity of the causative agents of disease, about the meaning of epidemiological evidence, about the trustworthiness of purification and analysis -- left the concept of river self-purification, which had been so central in Taylor's case, without foundation. Way and Rawlinson concluded that self-purification probably didn't always happen and that no one could tell whether or not it had happened. Rawlinson told the Lea committee that self-purification might well exist but it could never be proved to have removed noxious matter.

> it [the self-purification of rivers] is a thing that you can only follow in your imagination, and . . . if all the chemists in Europe told me that after the sewage had gone into the water that there would be no injurious effects from it I simply would not believe it. I am now daily drinking Thames water, knowing, . . . all the abominations that go in . . . under certain conditions this effete matter may go on to corruption, and if it passes into my system will do me serious injury.[44]

---

[43] Ibid., Q 5270; S.C. Lea Conservancy, Evidence, QQ 1939, 1959-62, 2001, 2576.

[44] S.C. Thames Navigation, Evidence, QQ 2725, 66-69, 119-23, 128-29. Rawlinson supported this conclusion by pointing out that after a flow of thirty miles the Bridgewater canal usually appeared pure but sometimes still stank, showing that self-purification was not always completed in that distance. Rawlinson's example is

Way adopted the same detached complacency toward the issue as Brodie had. Asked how far down river Luton's sewage would have to travel before it was purified, Way blithely pleaded ignorance. The barrister Sargood, who represented Luton, was taken aback by Way's persistent refusal to state an opinion. Way admitted he could find nothing "positively dangerous to health" in the Lea below Luton, but this made no difference.[45] In contrast with Brodie, Way also admitted the existence of oxidative self-purification. He accepted Liebig's conception of decomposition as a rapid process in which decomposing matters induced further decomposition.[46] "I have no doubt that a great amount of oxidation and destruction of these organic matters goes on; it must be so," he acknowledged.[47] "In the abstract they [sewage poisons] are all removed by natural causes."[48] Yet self-purification remained for Way only an ideal process and even Sargood's persistent questioning elicited no number of miles of flow which could be trusted to effect the destruction of noxious sewage matters. Moreover, Way rejected the burden of having to prove that polluted rivers remained unsafe. In his view, so long as rivers

---

interesting in that it relies on an olfactory standard of purity. Elsewhere in his testimony Rawlinson referred to the subtle ferments of Hofmann and suggested these might not be removed, but here, perhaps for the benefit of a lay committee, he chose a more graphic example. Rawlinson, therefore, appears to have adopted a particular version of the filth-disease relationship for a particular rhetorical occasion.

[45] S.C. Lea Conservancy, Evidence, QQ 1880-1900.

[46] Ibid., QQ 1979-81, 1989.

[47] Ibid., QQ 1987-91.

[48] Ibid., Q 2034.

had received sewage it was demonstrated that they had been made unsafe and it was the removal of the noxious matters that required to be proved.[49]

> I agree that it [self-purification] is so in abstract; but I think if it is urged as a reason why that which is known to be in the water has been taken out, we should call upon other people to show that it has been taken out. I do not believe in the power being adequate to the object.[50]

Inability to detect morbid poisons and an ignorance of what agencies removed them meant to Way that the only proven means for protecting water supplies was to prevent sewage from ever getting into them.

> We know when it [sewage] goes in; it is a difficult thing to say and we do not know when it is taken out or how it is taken out. But one thing we can do, we can prevent its going in, but we can never ensure its removal.[51]

In the spring of 1868 -- even before the select committee on the Lea bill met -- the first rivers pollution commission had been dissolved: Harrison and Rawlinson had decided they could no longer work together.[52] A replacement commission was quickly established, one whose investigations would be far more thorough and whose conclusions would be far more influential than those of Rawlinson's commission. The second commission inherited its predecessor's strong anti-pollution stance, but vastly extended the scientific

---

49 Ibid., QQ 1995-96; 2035.

50 Ibid., Q 2036.

51 Ibid., Q 2004.

52 The Builder, 25 January 1868, p. 59; Public Record Office, HO/74/3, pp. 409, 412, 417, 426.

foundations of that advocacy by developing political and technological applications of analytical nihilism. Edward Frankland was its chemist, and it is Frankland who was responsible for the further development of the perspectives of Way, Rawlinson, and Brodie.

VII. THE RADICALIZATION OF A WATER SCIENTIST: EDWARD FRANKLAND AND THE LONDON WATER CONTROVERSY, 1866-69

> A great deal of nonsense was talked about organic matter in water.
>
> -- Thomas Hawksley*

The return of cholera to London in July 1866 after an absence of 12 years came as a shock to public health officials and ended the complacency with an improved metropolitan water supply that had characterized the early 1860s. One of the most active investigators of the 1866 epidemic was William Farr, the statistician of the Office of the Registrar General of Births, Deaths, and Marriages, and in that capacity one of the leading epidemiologists of the mid-nineteenth century. Associated with Farr in studying the epidemic was the chemist Edward Frankland, who a year earlier had succeeded A.W. Hofmann as director of the Royal College of Chemistry and, in a minor function, as water analyst to the Registrar General.

The impotence of sanitary science in the face of the cholera -- the inability of water analysis to detect cholera-bearing water or to suggest any reliable means of purifying water -- led Frankland to the conclusion that rivers were unacceptable as a source of urban water supplies because they could never be shown to be safe. Instead, urban water supplies must come from uninhabited watersheds

* Thomas Hawksley, C.E., in disc. of E. Byrne, "Experiments on the removal of organic and inorganic substances in water," MPICE 27(1867-68): 17.

or from deep wells.

From the late 1860s until the mid 1880s Edward Frankland dominated British water science. He was a leader both in the scientific effort to learn more about what pollution and purification meant and what analysis showed and in the political battles to obtain better water supplies for the British public and to pass stronger laws preventing the pollution of rivers. He served as official water analyst for London from 1866 until his death in 1899, as chemical consultant for the Royal Commission on Water Supply (1867-69), and was the most active member of the second Rivers Pollution Commission (1868-74). Moreover, Frankland was one of the rulers of British science, a member of those bodies which decided what would be published, and of that elite which determined professional advancement.[1] As the professor at the Royal College of Chemistry he was holder of the most prestigious chemistry professorship in Victorian Britain.

Yet Frankland's centrality is by no means simply a function of his powerful position within the scientific establishment. He led in all ways. He experimented while others opined. Frankland collected more evidence, dealt with it more systematically, and wrote and thought more clearly than any of his colleagues. But Frankland invariably took extreme views on issues of water quality. Reviewing

---

[1] On this aspect of Frankland's career see Ruth Barton, "The X-Club: Science, Religion, and Social Change in Victorian England," Diss. U. of Pennsylvania, 1976, in passim.

the vast array of water-related problems industrialized Britain faced, Frankland beheld an inefficient management that endangered both health and prosperity. The long-term solutions he recommended to meet these problems -- acquisition of unpolluted water supplies and a revolutionary technique of sewage treatment to foster health, use of by-products and enforcement of mutually advantageous effluent standards to promote prosperity -- were challenged from all sides. It is this combination, of Frankland's stature within the scientific community, of the excellence of his science, and of the extreme views he advocated, that make Frankland the central figure in water quality issues during this period. What this means is that the views of physicians, engineers, and other chemists on water matters at this time must be evaluated as responses to Frankland, for it is clear that statements were often made principally to support or to oppose Frankland.

This chapter examines the origins of Frankland's views on water quality and the structure of his system of analysis. The next considers Frankland's disproof of the process of oxidative self-purification, and the credibility of Frankland's ideas in regard to the context of distrust considered in Chapter Two.

I. The Epiphany of Edward Frankland

Edward Frankland also testified at the June 1866 hearings on the Thames Navigation Bill. Frankland represented Thames basin paper manufacturers who contended that their pollutions did not adversely affect the potability of Thames water. In their defense Frankland argued that all types of pollution were not equally harmful. The vegetable extracts from paper mills were of no concern, for Frankland, like Hofmann, believed putrefying animal matter caused disease. He noted that the Thames was most dangerous in summer: although self-purification was most rapid during this season, the relatively deoxygenated condition of the warmer water made putrefaction more likely.[2] The key to Frankland's testimony, however, was the argument that under normal circumstances, the oxidative self-purification of the Thames would destroy the paper mill effluents. He refused to say how many miles of flow this would take, but he was certain it happened. Counsel for the bill quoted to Frankland Brodie's remarks to the rivers commission from the previous November -- that oxidative self-purification of rivers virtually did not exist -- and asked Frankland if he agreed with the statement. Frankland didn't:

> We find that water which has got into this putrid condition, or which has been mixed with sewage, if it is allowed to run a certain number of miles, depending on the volume of water with which the sewage is mixed, and the temperature at the time, if it is

---

[2] Select Committee on the Thames Navigation Bill, Evidence, P.P., 1866, 12, (391.), QQ 5138-39, 5107-8, 5180.

> allowed to run a sufficient number of miles, this organic matter almost entirely disperses, and is burnt up, in fact, with the water, so that although it may be a great question as to how far the mixture will have to run, I do not agree with the answer, that in no distance would it be possible that the sewage matter could be removed in this way.[3]

Frankland's support of the paper manufacturers was especially significant. In June 1865 he had been appointed by the Registrar General as official analyst of the London water supply. The paper mill effluents were therefore deemed safe by the very one entrusted with ensuring the safety of the water into which these effluents were poured.

During his first year in office Frankland stuck closely to the techniques, and more importantly, to the perspectives of his predecessors: Hofmann, who had held the post for the first few months of 1865; and R. Dundas Thomson, who had served from 1857 to 1864. If anything Frankland was even more conservative than Thomson and Hofmann, for in his own way each had attempted to convince public and government of the folly of relying on a polluted water supply. Hofmann had issued the manifesto of analytical nihilism in his 1856 statement that the obscure nitrogenous matters and modes of putrefaction responsible for water-borne diseases were analytically undetectable.[4] Thomson had tried to impress upon the public the in-

---

[3] Ibid., QQ 5170-71. See also QQ 5164, 5112.

[4] A.W. Hofmann and Lyndsay Blyth, "Report on the chemical quality of the supply of water to the metropolis," in Reports to the Rt. Hon. William Cowper, M.P., President of the General Board of Health, on the supply of water to the metropolis under the provisions of the Metropolis Water Act, P.P., 1856, 52, [2137.], pp. 4-6.

feriority of the London supply by adopting a format for his analytical results which compared London water with distilled water.[5] Yet both had tempered their concern with complacency; neither had crusaded for better water with the zeal that Chadwick and Hassall had shown or that Frankland would show.

Frankland's first annual report (March 1866) reflects an understanding of water quality indistinguishable from that of most of his fellow analysts. He used Hofmann's improved version of the ignition process for organic matter and complemented it with determination of the oxygen absorbed from potassium permanganate. He believed "the most pernicious kinds [of organic matter] were those which were most easily oxidised" -- those undergoing putrefaction. In his March report, as he would in his June testimony, Frankland commented on the self-purification of rivers and pointed out that although organic matter levels were highest in winter (there were fewer oxidizing agencies such as vegetation during this season) the water was safest in winter because putrefaction was least likely to occur.[6]

During the next two years Frankland's views about water quality and water analysis changed radically. He came to believe pernicious

---

[5] RPPC/1868/6th Rept., "The domestic water supply of Great Britain," P.P., 1874, 33, [C.-1112.], pp. 250-52; G.R. Burnell, "On the present condition of the water supply of London," JRSA 9(1860-61): 171-73.

[6] Edward Frankland, "Water supply of the metropolis during the year 1865-1866," J. Chem. Society 19(1866): 240-42; idem, "Observations on the London waters," CN 12(1865): 302; "Analysis of metropolitan waters in January, 1866," CN 13(1866): 72.

matters were by no means easily oxidised and that self-purification -- to the minimal extent such a process existed -- was greater in the summer. He carried out a "complete revolution" in the methodology and interpretive principles of water analysis. The source of Frankland's epiphany was the cholera epidemic of 1866.[7]

Cholera reached London in July 1866 and the epidemic peaked in early August.[8] The geography of the epidemic suggested that the disease was carried by the water the East London company distributed from its reservoirs at Old Ford on the lower Lea. As William Farr's water analyst Frankland was actively involved in the investigation of the epidemic. Farr had meticulously studied the courses of earlier cholera epidemics in 1848-49 and 1854 and the position of water analyst to the Registrar General had been created to obtain the data necessary to determine the epidemiological relationships between the mortality of London's citizens and the quality of London's water.[9]

The birth of Frankland's nihilism can be discovered in his reports to Farr during the late summer and autumn of 1866. In his

---

[7] Edward Frankland, "On the proposed water supply of the metropolis," Proc. Royal Institution 5(1866-69): 347.

[8] William Farr, Report on the Cholera Epidemic of 1866 in England, Supplement to the 29th Annual Report of the Registrar General of Births, Deaths, and Marriages in England, P.P., 1867-68, 37, [4072.], Chart 4 after p. xc. On this epidemic also see William Luckin, "The final catastrophe: cholera in London, 1866," Medical History 21(1977): 32-42.

[9] "Reports on the examination of Thames water," JRSA 31(1882-83): 74-76, 87-90, esp. p. 88.

letter to Farr of August 4, during the height of the epidemic, Frankland was still finding East London water analytically satisfactory; indeed, it contained less organic matter than usual and gave generally better analytical numbers than it had in July. Frankland concluded:

> Chemical analysis . . . does not reveal any exceptional degree of pollution in this water. It must be borne in mind, however, that chemical investigation is utterly unable to detect the presence of choleric poison amongst the organic impurities in water, and there can be no doubt that this poison may be present in quantity fatal to the consumer, though far too minute to be detected by the most delicate chemical research.
>
> It is thus that the occurrence of cases of cholera, and of choleric diarrhoea, upon the banks of any of the streams, from which the water supply of London is derived, may at any moment diffuse this poison over large areas of the metropolis.[10]

Chemistry, Frankland had quickly decided, offered no help to those hoping to detect the cholera poison before it produced disease.

Soon Frankland arrived at the same conclusion in regard to technologies of water purification. In his August 4th letter Frankland had recommended a massive program of filtration through animal charcoal which he believed would remove from water all organic matter. He also recommended treating the water with potassium permanganate, widely sold as Condy's fluid, a disinfectant. Since potassium permanganate would preferentially oxidize any putrescent matter in the water, it is apparent that at this time Frankland still

---

[10] Frankland to the Registrar General, "The purification of water," CN 14(1866): 71; also in Farr, The Cholera of 1866 in England, Appendices, pp. 116-17.

regarded morbid poisons as peculiar conditions of putrefaction.[11] By the end of August Frankland had decided that these techniques, though helpful, were not completely reliable, and even boiling could not be relied upon to purify choleragenic water. Boiled water (as tea) had disseminated lethal cholera to an east London family.[12]

By early November Frankland had demonstrated experimentally the undetectability of the morbid poison of cholera. He had taken the "rice-water" evacuations of a cholera victim, diluted them with 500 parts distilled water, filtered the mixture through filter paper, and analyzed the still-turbid mixture. The diluted, filtered evacuations absorbed .04 parts/100,000 oxygen from permanganate; normal Thames water took .07 parts. Filtration through animal charcoal helped somewhat, but Frankland pointed out that 1 part cholera evacuations diluted with 1000 parts ordinary river water would be analytically undetectable.[13]

As early as September 1866 Frankland had addressed the possibility that the cholera poison might be the germ of an organism. The ability of the poison to remain virulent after short boiling was characteristic of germs, he thought: boiled infusions were quickly repopulated.[14] In November he again referred to the possibility of

---

[11] Frankland to the R.G., "The purification of water," p. 71.

[12] Farr, The Cholera Epidemic of 1866 in England, App., pp. 131, 125.

[13] Ibid., p. 144; and Report, p. lxix. Farr credits Chauveau with this experiment.

[14] Ibid., App., p. 125, footnote. On the vagueness of the early germ theory see J.K. Crellin, "Airborne particles and the germ theory: 1860-80," Annals of Science 22(1965-66): 49-60; and idem,

cholera germs.[15] But Frankland's conception of germs at this stage was not yet distinct from the notion that the virulence of a water was in some manner associated with its putrefaction. Even as he pointed to germs as a possible cause of cholera, he continued to advocate potassium permanganate as a disinfectant, a suggestion that the Chemical News (Crookes, its editor, had been one of the perpetrators of the 1865 germ theory of the cattle plague) rightly regarded as illogical: potassium permanganate was "successfully resisted by living organisms. . . . without apparent inconvenience."[16] In late November Frankland stressed the importance of his discovery that dissolved gas ratios provided the best test of putrescibility and that the drop in oxygen solubility with increasing temperature was usually the immediate cause of putrefaction of water, an observation whose importance was linked to the assumption that putrefaction was intimately associated with the development of virulence in the water.[17]

"The dawn of the germ theory: particles, infection, and biology," in F.N.L. Poynter, ed., Medicine and Science in the 1860s (London: Wellcome, 1968), pp. 57-76.

[15] Farr, The Cholera Epidemic of 1866 in England, App., p. 144.

[16] "The cholera poison," CN 14(1866): 109. Two weeks earlier Crookes had reprinted an article from the Medical Times and Gazette, "Condy's fluid and carbolic acid," in which the immunity of living organisms to Condy's fluid was pointed out: "the amoeba, paramecium, colpods, and other disgusting broods are not in the least affected by water too reddened [Condy's is made of $KMnO_4$ which is pink] to be drinkable." Both Condy's fluid and carbolic were disinfectants intended to deal with a Liebigian morbid poison of transitory matter (CN 14(1866): 84).

[17] Farr, The Cholera Epidemic of 1866 in England, App., pp. 145-46; "On the effect of temperature on organic matter in water," CN 14 (1866): 275.

By March 1867 Frankland had already partially worked out a new system of water analysis, a set of procedures which in their mature form decidedly loosened the theoretical ties that bound the morbid poisons of zymotic diseases to putrefying animal matters. In a lecture at the Royal Institution (where he was one of the professors of chemistry) Frankland descanted on the complete unreliability of the incineration and potassium permanganate processes for determining the organic matter in water -- both of which he had employed until a month previously. Replacing these were new techniques for determining the organic carbon and nitrogen in water. Frankland had adopted Hofmann's 1856 suggestion for finding the quantity of organic nitrogen by subtracting separately determined ammonia, nitrites, and nitrates from the total nitrogen obtained by a combustion analysis of the evaporative residue of a sample of water.[18] Despite his partial commitment to some undefined version of the germ theory, Frankland still regarded the existence of putrescible nitrogenous organic matter as the primary index of potable water quality. Organic nitrogen probably represented animal contamination, Frankland thought, but its source, animal or vegetable, was hygienically irrelevant.[19]

---

[18] Hofmann and Blyth, "Report on the chemical quality," p. 6. Between February 1867 and March 1868 Frankland was determining organic nitrogen in this indirect manner and it was not listed on his analytical tables. By early 1868 he and Henry Armstrong had developed means for its direct estimation (but for a necessary correction for ammonia), and it was thereafter listed explicitly.

[19] Edward Frankland, "On the water supply of the metropolis," Proc. Royal Institution 5(1866-69): 113-17.

In this March lecture Frankland introduced the concept of "previous sewage contamination," a "convenient expression" for a number obtained from the total amount of inorganic nitrogen compounds a water contained, and which purported to represent the minimum amount of sewage or "analogous" animal organic matter a river had received and successfully oxidized into ammonia, nitrites, and nitrates. "Previous sewage contamination" (PSC) had considerable rhetorical value as a device for reminding Londoners that they drank purified, diluted sewage, and this in part explains Frankland's use of the title. At the same time the increasing importance that PSC came to have in Frankland's ideas about water quality reflects the growth of his belief that nitrogenous organic matter alone was not an adequate measure of water quality. Resistant zymotic germs, introduced when water was polluted by sewage, might remain virulent even after all the sewage had been oxidized, transformed from present sewage contamination into "previous sewage contamination."[20]

New, better methods of water analysis had not changed Frankland's views about purification or the detectability of morbid poisons. He still maintained that "no absolutely reliable protection" from cholera existed. Charcoal filtration helped, but it was not entirely adequate, nor could a short period of boiling be trusted to purify -- germs could withstand such treatment.[21]

---

[20] Ibid., pp. 117-19. See also Christopher Hamlin, "Edward Frankland's early career as London's water analyst, 1865-1876: the public relations of 'previous sewage contamination'," BHM 56(1982): 56-76.

[21] Frankland, "On the water supply of the metropolis," p. 125.

In a July paper in the Quarterly Journal of Science Frankland clarified his conception of germs. Cholera germs were principally air-borne, he believed, but they grew best in a medium of sewage. Therefore, sewage-polluted drinking water was especially dangerous during an epidemic period. The sewage matter might support multiplication of cholera germs outside the body and disease would result when the victim drank the resulting swarm of sewage-nurtured germs, or the victim might take in the sewage separately from the germ -- the sewage would support multiplication of the germ in the victim's gut and the disease would result. In Frankland's view

> the fact . . . remains clear and incontrovertible that amongst the materials which, during the prevalence of the disease, produce cholera when taken internally, sewage is the most effective; and this property sewage preserves, although in a diminished degree, even when largely diluted.[22]

At this stage in the evolution of Frankland's ideas about germs the quantity of nitrogenous organic matter was still important, since germs depended on this pabulum of organic sewage matter for their multiplication. For this reason, too, "previous sewage contamination" remained only a "convenient expression." It might reveal the quantity of sewage water had received, but since PSC reflected only organic matter which had already undergone oxidative purification, it in no sense measured the virulent capacity of water. Both in his March lecture and in his July article Frankland used previous sewage

[22] Edward Frankland, "The water supply of London and the cholera," Q. J. Science 4(1867): 314-15.

contamination to remind the public they drank sewage, yet in neither did he try to derive from the concept the notion that the water was anything other than purified:

> it must be consolatory to the drinker of Thames water to know that the whole of the fecal matter is so completely oxidized before it reaches the water cisterns of London as to defy the detection of any trace in its noxious or unoxidized condition.[23]

Yet in the July article Frankland maintained skepticism toward techniques of purification. Even long boiling might not kill germs. Still, he continued to recommend boiling and filtration through animal charcoal.[24]

By November 1867 Frankland had weaned his cholera germs of their sewagy nutriment. Earlier in the year he and William Odling had been appointed chemical analysts for the Royal Commission on Water Supply. Frankland, who rarely missed an opportunity to express his views on water quality, presented the commission with a lengthy and digressive appendix to the joint report he and Odling made on the quality of lake and stream waters from Cumberland and north Wales. Here he wrote:

> In view of the opinions now very generally entertained, with regard to the propagation of certain forms of disease by spores or germs contained in excrementitious matters, the search for inorganic nitrogen compounds or previous sewage contamination . . . is second only in importance to that for actual sewage contamination.[25]

---

[23] Ibid., p. 322; Frankland, "On the water supply," p. 118.

[24] Frankland, "The water supply and the cholera," p. 328.

[25] Report of the Royal Commission on Water Supply, P.P., 1868-69, 33, [4169.], Appendix D, p. 20. [Cited as RCWS, Report.]

Frankland was arguing here that germs came with sewage, and that since "dilution with large volumes of water does not, as in the case of unorganized substances in solution, render organized germs innocuous," the mere fact that once-polluted water contained no traces of the putrescible organic nitrogen of sewage was inconsequential. Instead, the presence of the oxidation products of this putrescible nitrogen -- ammonia, nitrites, and nitrates -- constituted "a record of the sewage or other analogous contamination to which the water has been subject," and indicated the possible presence of resistant germs of disease.[26] Thus "previous sewage contamination," hitherto only a rhetorical device to remind Londoners of the sewage they drank, became suddenly one of the principal bases on which Frankland evaluated water.

In acknowledging the possibility that diseases were caused by resistant water-borne germs, Frankland was by no means asserting that Hofmann's zymotic pathology had been a mistake. Resistant germs did not replace peculiar putrefaction, they merely stood as one more reason why sewage-polluted water ought never to be used for domestic purposes. Immediately following his discussion of germs, Frankland quoted the 1856 assertion of Hofmann and Blyth, that zymotic diseases happened when special forms of putrefaction were communicated from nitrogenous organic matter to the tissues of the human body.[27]

---

[26] Ibid., pp. 20-21.
[27] Ibid., p. 21.

Frankland retained this ambivalence throughout the 1870s. His conception of germs gradually lost its tenuousness and germs eventually became the primary, but never the only hazard in the use of sewage-polluted water. Moreover, the increased importance he was placing on previous sewage contamination could be justified no matter which form morbid poisons took and in fact was rendered more important by the ambiguity of the nature of morbid poisons. In a January 1868 Chemical Society lecture introducing the new system of water analysis that Frankland and his student H.E. Armstrong had developed, Frankland made this clear. Previous sewage contamination was significant for two reasons. First, although it showed "the salutary change from organic to inorganic has been effected at the time the sample of water was collected for analysis, it by no means follows that it will be equally complete under future altered conditions as regards temperature, exposure to air or vegetation, and comparative volume of pure water."[28] After considering the possibility that oxidation might not always be completed, Frankland returned, in the strongest terms yet, to the problem of resistant, water-borne germs:

> Now excrementitious matters certainly, sometimes, if not always, contain the germs or ova of organised

---

[28] Frankland and H.E. Armstrong, "On the analysis of potable waters," J. Chem. Soc. 21(1868): 107. In the winter of 1868 such warnings were particularly apt; owing to the increased precision of the recently developed technique for the estimation of organic nitrogen, Frankland had discovered 'actual sewage contamination,' -- unoxidized organic nitrogen -- in some London waters (RCWS, Evidence, QQ 6387-89, but see RCWS, Appendix AJ, p. 87).

> beings; and as many of these can doubtless retain their vitality for a long time in water, it follows that they can resist the oxidizing influences which destroy the excrementitious matters associated with them. Hence great previous sewage contamination in a water means great risk of the presence of these germs, which, on account of their sparseness and minute size, utterly elude the most delicate determinations of chemical analysis.[29]

By early 1868, therefore, Frankland had worked out a philosophy and set of procedures for water quality evaluation that remained fundamentally intact until the late 1880s. With this philosophy Frankland perplexed the water supply commissioners in February 1868; it justified the recommendations he made as chemical member of the second rivers pollution commission (to which he was appointed in April 1868); it was the basis for Frankland's continual criticism of the London water supply, and for what many of his colleagues regarded as the unphilosophical formats with which he presented analytical results to the public.

It is clear that between the summer of 1866 and the spring of 1868 Frankland's understanding of the relationship between zymotic disease and polluted water underwent a considerable evolution. But what selective forces directed this evolution?

The transformation of Frankland's ideas from putrefying organic nitrogen to resistant germs cannot be explained as a function of the advance of medical science. Ideas about germs as causes of disease were certainly available during these years. Evidence consistent

---

[29] Frankland and Armstrong, "On the analysis of potable waters," p. 107.

with the contention that diseases were caused by particulate, probably living, water-borne entities was also available. Yet nothing in these ideas or in this evidence could render Frankland's description of water-borne germs anything more than plausible. Frankland's germs were a product of imagination, not a discovery of microscopy.

When Frankland spoke of "opinions now very generally entertained" (and it may be noted that to entertain an opinion means to consider it rather than to accept it), he was referring to an array of vague notions that there existed ultimate, undiscoverable, malignant particles, quarks or point-atoms of disease, whose behavior was in some manner to account for the specificity of diseases and for the multiplication of morbid matter that took place during the course of a zymotic disease. As discussed in Chapter Three, three of the scientists who had studied the cattle plague -- Lionel Smith Beale, Robert Angus Smith, and William Crookes -- had come up with germ theories. Yet Beale's ideas of morbid bioplasm -- "diseased," not "disease" germs -- was so different from the "bacterial" explanation of disease that in his 1883 taxonomy of disease theories Benjamin Ward Richardson made a major distinction between Beale's germ theory and the "parasitic theory" of the Glaswegian physician John Dougall, who held that distinct species of bacteria caused distinct forms of disease.[30] By the mid 1870s Frankland was citing the work

---

[30] B.W. Richardson, The Field of Disease. A Book of Preventive Medicine (London: MacMillan, 1883), pp. 829-40; and L.S. Beale, "On the nature of the morbid poison -- contagium of cattle plague," in

of Chauveau in support of his germ theory, but Chauveau's work consisted of dilution studies of vaccine matter and showed only that this matter was particulate. Other researchers were similarly imprecise in their descriptions of the nature of germs and in their ability to connect specific organisms with the disease process. Pasteur found some microbes resisted oxidation under certain conditions. Hassall and R. Dundas Thomson had observed an excessive number of vibrios in the evacuations of victims of the 1854 cholera epidemic.[31] Probably the studies most helpful to Frankland were those John Burdon Sanderson was carrying out for John Simon's Privy Council Medical Department on the "Intimate Pathology of Contagion" and the "Origin and Distribution of Microzymes in Water," studies which Sanderson had begun in 1867 but which were not published until 1870 and 1871.[32] However important these investigations may have been in the development of the germ theory, the fact remains that by early 1868 Frankland had extracted a very clear and distinct conception of water-borne morbid poisons from the tangle of

---

Royal Commission on the Cattle Plague, Third Report, P.P., 1865, 22, [3656.], Appendices, p. 145.

[31] Farr, The Cholera of 1866 in England, pp. xliv-xlv.

[32] John Burdon Sanderson, "Introductory report . . . on the intimate pathology of contagion," in 12th Annual Report of the Medical Officer of the Privy Council, for 1869, P.P., 1870, 38, [c.-208.], Appendix 11, pp. 229-56; and idem, "Further report of researches concerning the intimate pathology of contagion. The origin and distribution of microzymes in water, the circumstances which determine their existence in the tissues and liquids of the living body," in 13th Annual Report of the Medical Officer of the Privy Council, for 1870, P.P., 1871, 31, [C.-349.], Appendix 5, pp. 48-69.

ideas associating filth with disease that comprised pathological and etiological theory in the 1860s and 1870s.

Throughout the late 1860s Frankland was advocating the acquisition of an unpolluted water supply for London, and it is likely that the germ theory at which he had arrived by early 1868 was justified primarily by its utility in the achievement of this goal. Frankland had not discovered resistant, water-borne germs during this period, but he had taken hypotheses about the nature of morbid poisons which were both consistent with his observations of the 1866 cholera and at the same time pointed most strongly to the folly of relying on natural and artificial techniques of purification to make the Thames and Lea safe and the untrustworthiness of all analytical confirmations of that purification. As had Chadwick, Frankland was subjecting hypotheses to tests of political utility.

Early in 1866, even before cholera had made the issue critical, the question of how London should be supplied with water had reappeared as an important political issue, regaining much the same status it had held in the early 1850s. Several well-developed schemes for supplies which avoided the polluted Thames and Lea were presented. J.F. Bateman, who had supplied Manchester with good water from the Pennines, proposed an aqueduct that would supply London with water from the head of the Severn watershed in the Welsh hills. At about the same time G.W. Hemans and Richard Hassard developed a scheme for obtaining London's water from the Cumberland lake district. Consideration of these schemes, both of them apparently economically and technically viable, generated a technical and medical controversy of major proportions in which all water analysts of stature were involved. The central issue was whether a new water supply was necessary, so as in 1850-52 the very existence of the London water companies was at stake. Statements made by water analysts during the 1866-69 period must be considered with regard to the overriding importance of this political issue.[33]

Frankland made no secret of his opposition to the use of Thames and Lea water as London's supply. He began his March 1867 Royal Institution lecture by describing the causes of the 1866 cholera and concluded that whatever mistakes the technicians in charge of the East London company's reservoirs had made, the epidemic had been fundamentally a consequence of reliance on polluted water.[34] This led him directly to a consideration of the alternatives which he endorsed in general as practical solutions to the problem. In his July 1867 article Frankland followed a similar strategy. He regarded the Cumberland and Welsh schemes as "truly worthy of this age of engineering triumphs," and contrasted the Thames with the wonderfully pure water Glasgow obtained from Loch Katrine.[35] As had Chadwick, Frankland vividly described the hazards of hard water in contrast with the softness of the alternatives: "The process of ablution

---

33 Lancet, 24 February 1866, p. 213; 3 March 1866, p. 235; The Builder, 6 May 1865, p. 313.

34 Frankland, "On the water supply of the metropolis," pp. 109-11.

35 Idem, "On the water supply and the cholera," p. 319.

. . . in hard water is essentially one of dyeing the skin with the white insoluble greasy and curdy salts of the fatty acids contained in soap."[36]

What advantages did Frankland find in the conclusions about morbid poisons and germs he had adopted by early 1868? Frankland's conception of germs involved several sets of choices about the nature of morbid poisons. As Figure 7.1 illustrates, each time he was faced with alternative characteristics of water-borne morbid poisons he chose the characteristic which held out the more serious consequences for the use of once-polluted water, yielding finally a conception of morbid poison against which there could be no defense. Frankland's choices during this year and a half can be understood as deductions of maximum political utility from the observations he had made during the 1866 epidemic (Figure 7.1). What these choices meant was that "cholera, typhoid fever, and other diseases" were carried by germs in sewage-polluted water; germs which could not be detected nor removed by filtration, boiling, or most disinfectants; which were unaffected by the dilution of sewage; which could withstand the loss of the sewage medium in which they had entered the water; and which were remarkably resistant to the oxidation that had destroyed that sewage. There remained no means by which sanitary engineers could defend the public from Frankland's germs once these had entered the water.

---

[36] Ibid., pp. 317-19, 321, 323.

Figure 7.1

Character of Morbid Poison

Choices	Comments
1. A. Chemically detectable B. Chemically undetectable	Underlined choices made on an empirical basis, summer and autumn, 1866
2. A. Susceptible to existing techniques of purification B. Not susceptible to existing techniques of purification	
3. A. Soluble organic matter (can be purified by dilution) B. Germ (cannot be purified by dilution)	Deduction from 1 & 2, autumn, 1866
4. A. Germ coexistent with and effectively dependent on organic matter B. Germ independent of organic matter	Switch from A. to B. between July 1867 and November 1867
5. A. Germ resistant to oxidation B. Germ susceptible to oxidation	Choice A. implied in November 1867; made explicit by January 1868

Judged in terms of the history of the germ theory, Frankland's views seem remarkably progressive. From confusion he had drawn an idea which turned out to be fundamentally correct. Yet Frankland's progressivism in this instance is less a function of scientific excellence than it is of a scientist's commitment to the public welfare. It is also a testament to Frankland's immense faith in the ability of Victorian engineering to provide the water supplies that were necessary to improve the public's health. Frankland's germ theory, and as we shall see, the whole of his revolutionary system of water quality evaluation supported, and one is tempted to say was constructed to support, those water supply reforms he thought necessary.

II. Water Quality 1866-1869

Many of Frankland's colleagues were similarly involved in wrestling with the problems of how to purify water reliably and how to demonstrate that it had been purified during these years. Certainly there were ample opportunities for discussing these issues. The 1866 East London cholera ignited a great deal of debate. It was investigated by four departments of government: the Registrar General, the Board of Trade, the Privy Council Medical Office, and the first rivers pollution commission; and by the Association of Medical Officers of Health and Lancet Analytical Sanitary Commission outside of the government. Testimony on the epidemic was also taken by the 1867 Select Committee on East London Water Bills. The adequacy of existing supplies for London and the advantages and practicality of alternative supplies were considered during the extensive hearings of the 1866-1869 Royal Commission on Water Supply. Controversial scientific papers -- on filtration at the Institution of Civil Engineers; on water analysis at the Chemical Society -- provoked lengthy and bitter discussions.[37]

---

[37] Farr, The Cholera of 1866 in England; Report by Capt. Tyler to the Board of Trade on the Quantity and Quality of the water supplied by the East London Waterworks company . . ., P.P., 1867, 58, (339.); J. Netten Radcliffe, "Cholera in London especially in the eastern districts," in 9th Annual Report of the Medical Officer of the Privy Council, P.P., 1867, 37, [3949.], Appendix 7, pp. 264-331; RPPC/1865/

Besides providing an excellent picture of water quality evaluation at this period, these investigations are also important in that they provided a special rhetorical environment for water science, which by fostering plausible speculation acted as a forcing bed for scientific ideas. As is the case with Frankland's resistant germs, concepts were often expressed in strong and vivid language and extreme statements were made in an effort to persuade a scientifically illiterate audience that water must or must not be safe. From commissioners or cross-examining barristers scientists frequently faced penetrating questions which called for opinion rather than proved fact. Germs were talked of a good deal, perhaps more so than at any time prior to the actual discovery of disease germs in the 1880s, but there was little unity in the usage of the term.

Frankland's conclusions from the 1866 cholera epidemic -- that chemistry could neither discover the poisons of cholera nor suggest how they might be reliably removed -- were by no means unique. During and immediately after the epidemic several of Frankland's colleagues embraced nihilism and the germ theory, and these included some who subsequently became staunch defenders of nature's purifying ability and of the power of chemical analysis to prove water safe. Particu-

---

2nd Report, (River Lee), P.P., 1867, 33, [C.-3835.]; The Lancet Analytical Sanitary Commission, "On the epidemic of cholera in the east end of London," Lancet, 1866, ii, pp. 157-60, 217-19, 273-76; E. Byrne, "Experiments on the removal of organic and inorganic substances in water," MPICE 27(1867-68): 1-54; Frankland and Armstrong, "On the analysis of potable waters," pp. 77-108; Disc. in CN 17(1868): 45-47, 60, 72, 79-81, 127-28.

larly interesting are the cases of William Crookes and Henry Letheby.

Despite having been one of the developers of the germ theory of the cattle plague, Crookes had no patience with Frankland's portrayal of cholera as a disease caused by undetectable and unremovable water-borne particles. Crookes' germs were air-borne; they were in sewer gas and they infected cisterns. Chemical News editorials during the winter of 1867 denied the east London epidemic had been water borne: "Never more than now has water-supply as a cause of epidemic disease been in less favour."[38] Basing his conclusions in part on the 1851 Hofmann, Graham, Miller study, Crookes maintained "there is no evidence, physical or chemical, to show that this [organic matter in London water] is otherwise than harmless."[39] By April 1867, however, Crookes recognized that the East London company's polluted supply and poor handling techniques had caused the epidemic, yet was relieved to find the fault had been the engineers' rather than the chemists'.[40] By the spring of 1869 Crookes had done a complete turnabout. In editorials he defended Frankland's brand of water analysis from the attacks of Henry Letheby, the water companies' analyst. Previous sewage contamination was no obscurantist rhetorical mechanism, "the phrase simply states an indubita-

---

[38] "London water supply: past, present, future," CN 15(1867): 49.

[39] Ibid., pp. 37-38.

[40] "London water," CN 15(1867): 190.

ble fact."[41] He recognized that Frankland and Brodie were not denying that self-purification existed, but simply challenging the assumption that it must always be reliable. The burden of proof, Crookes agreed, lay with Letheby and the companies.

> We know as an absolute fact that tons upon tons of human excrement are thrown into the waters of the Thames and Lea before we drink them, and we have a right to look for absolute demonstration of the complete destruction of all this filth, and not only to the filth itself -- lifeless organic matter; -- we must also be convinced that it is impossible, at all times and under all circumstances, for living matter -- the low forms of life and their germs with which the processes of disease and putrefaction appear to be so closely connected -- to retain their vitality under the conditions assigned by Dr. Letheby, before we can be content to accept his statement, or can learn to look with indifference on previous sewage contamination.[42]

Noxious matters were probably "far beyond the reach of the test-tube and the balance," and were likely resistant to oxidation: "it has yet to be demonstrated that the dissolved oxygen of the water has any greater destroying effect upon them [germs] than it does upon the fishes which inhabit the same streams."[43] Subsequently, however, Frankland and Crookes became antagonists on water quality issues. By the early 1880s Crookes had become one of the water companies' analysts and one of the most abusive critics of Frankland's analytical techniques and interpretive philosophy.

---

[41] "Dr. Letheby on the methods of water analysis and on 'previous sewage contamination'," CN 19(1869): 231.

[42] Ibid., pp. 231-32.

[43] Ibid.

Henry Letheby, medical officer for the City of London and professor of chemistry at the London Hospital, underwent an opposite intellectual evolution during this period. Letheby had succeeded John Simon as the City's medical officer in 1856 and had continued Simon's activist policies.[44] By the early 1860s Letheby had recognized that specifically polluted water was a route of zymotic disease transmission and was campaigning to close the City's cesspool-polluted shallow wells.[45] During the 1866 cholera epidemic Letheby adopted the same skeptical attitude toward purification as had Frankland. Better filtration had improved the quality of London's water, but even the best filtration would "never be sufficient to insure such a purity of water as the complete removal of those subtle agents of disease which even the most refined appliances of chemistry have failed to discover." Absence of measurable organic matter did not mean absence of "minute germs of disease."[46] Worse, if living germs were responsible for cholera -- and "unquestionably" they were -- oxidative purification, whether self-purification in rivers or artificial oxidation by Condy's fluid, could not be counted on to destroy these presumably resistant organisms: "the analogies of physiology are against such a supposition, and they warn us not

---

[44] Royston Lambert, <u>Sir John Simon and English Social Administration</u> (London: McGibbon and Kee, 1963), pp. 241-42.

[45] Henry Letheby, "Report on the sanitary condition of the City of London for the quarter ending 22 June 1861," <u>CN</u> 4(1861): 76-77; idem, "City pumps," <u>CN</u> 4(1861): 260-62.

[46] "Composition and quality of the metropolitan waters in July 1866," <u>CN</u> 14(1866): 83.

to receive it even as a possible fact."[47] The chemist "would be putting forth very dangerous propositions," Letheby observed, "if by relying on his science alone he ventured to dogmatise on so difficult a subject."[48]

In promulgating these sanitary principles Letheby had issued what was for July 1866 an extremely radical doctrine. In his complete distrust of chemical analysis and his lack of faith, based on a germ theory, in oxidative purification, Letheby had articulated exactly those principles which would form the basis of nihilism in the 1870s, and which Frankland, who would be the leading nihilist during that decade, would not entirely accept until the spring of 1868. Still, Letheby's views at this time were far less radical than they at first seem. Unlike Frankland, Letheby believed boiling or strong disinfectants would destroy germs. Far more significantly, Letheby's germs entered the water after it had left the companies' filters; purification, therefore, was the responsibility of consumers, not the companies.

By the end of 1866 Letheby had already begun to retreat from the radical positions he had taken during the epidemic. He contended that there was no evidence showing that the organic matters in the companies' waters were anything other than "the harmless products of vegetable and infusorial growths."[49] Not only had the

---

[47] Ibid.

[48] Ibid.

[49] Letheby, "Report on the quality and quantity of the water supplied to the metropolis in the year 1866, as compared with the year

water of the East London company not been the cause of the cholera, but the drinkers of that water had exhibited a "singular" exemption from the cholera that surrounded them.[50] In a February 1867 report he affirmed that analysis accurately reflected the wholesomeness of river water.[51]

Still, although he made clear his satisfaction with the London water supply, Letheby remained stuck with his conception of resistant germs, and even if East London water had not been the vehicle of the 1866 cholera, it might at any time become the vehicle of cholera if Letheby's description of germs were accurate. In June 1867 testimony before the Select Committee on the East London Water Bills, Letheby backed away from his earlier description of germs. Letheby argued that since the nature of morbid poisons was unknown, assuming all things equal there was only a small chance that morbid poisons were in fact the resistant germs he had depicted the previous summer.

> it does not appear to me that it [failure of self-purification] is very probable, because it is only assuming one condition of thing of which we know nothing, and that is, that the agent of these special zymotic diseases . . . is a living thing. If those diseases are brought about by the action of putrefying matter on the human body, then I say that putrefying matter cannot exist in running water for any distance; but if you assume that it is a living germ, then I say

---

1851," in S.C. East London Water Bills, Report, P.P., 1867, 11, (399.), App. 3, pp. 315-16.

[50] Ibid.

[51] Lancet, 9 February 1867, p. 185.

> that it is possible that it may be in the water without being discovered.[52]

In Letheby's view there was about one chance in four (acceptable odds, apparently) that the morbid poison conformed to his 1866 description. And though willing to admit in principle that such germs might exist, Letheby ignored this possibility in practice and employed a conception of morbid poison -- putrefying matter -- which justified his methods of water analysis and his confidence in self-purification. Like Frankland, Letheby dealt with unanswered questions of disease etiology by treating as true the alternative most consistent with the social goals that held his loyalty. Frankland's choice maximized public safety; Letheby's supported the water companies and, as we shall see, justified fundamental principles on which local sanitarians worked.

This queer combination of uncertainty in principle and certainty in practice characterizes Letheby's testimony to the 1867 select committee. Having admitted that he did not know what were the harmful substances in polluted water, Letheby was quite pleased to note that he had found none of them in the East London company's water. He pointed out that "We have no positive test for the discovery of the proportion of organic matter," and then maintained that organic matter could be determined by multiplying the oxygen absorbed from potassium permanganate by eight: "I . . . rely upon

---

[52] S.C. East London Water Bills, App. 11, Evidence, Q 702. Also see Q 639; and RCWS, Evidence, Q 3898.

my experience in making it [oxygen absorbed] evidence of the absolute proportion."[53] He admitted, "we do not . . . understand the immediate agents of disease," but declared that self-purification happened since "nothing of sewage can be discovered by chemical means."[54] He believed that cholera could be spread when the excreta of those with the disease contaminated drinking water, but concluded, on the basis of analyses, that the East London company's water had not been responsible for the cholera in 1866 and that Snow had probably been mistaken about the water-borne nature of the 1854 cholera.[55] He refused to say how long cholera evacuations remained virulent, yet asserted that sewage diluted with twenty volumes of water and after flowing nine miles in a river was harmless when drunk.[56] He was outraged by the sewage-pollution of the Thames, but confident that self-purification made its water safe.[57]

Yet simply to list Letheby's contradictions or to cast him as sanitary villain is to do him an injustice. To be sure, Letheby took the water companies' money for defending their interests.[58]

---

[53] S.C. East London Water Bills, Evidence, QQ 2344, 2369; App. 11, Evidence, QQ 586, 589.

[54] S.C. East London Water Bills, Evidence, Q 2369; App. 11, Evidence, QQ 639, 693-94, 696; RCWS, Evidence, QQ 3894-96.

[55] S.C. East London Water Bills, App. 11, Evidence, QQ 643-49, 679, 713, 747, 758-66; RCWS, Evidence, QQ 3901-6, 3910.

[56] S.C. East London Water Bills, App. 11, Evidence, QQ 715, 732-33.

[57] Ibid., Evidence, QQ 701, 635, 632-34; RCWS, Evidence, QQ 3879, 3891.

[58] RCWS, Evidence, QQ 3861, 3909.

In fact, the principles upon which Letheby and his fellow metropolitan medical officers of health based their activities were consistent with the interests of the water companies. Letheby believed that local insanitary conditions -- cesspools, sewer gas, unclean cisterns, bad wells, untrapped drains, and yards full of filth -- caused epidemics generally and had been responsible for the 1866 cholera. As the Lancet sympathetically pointed out, such a preoccupation was a fundamental premise of the medical officers' profession. They fought for local sanitary reform by promising that with it would come freedom from disease. To the extent that the epidemic was a result of the external factor of water supply their authority and the immediacy of the reforms they advocated was diminished. Thus Letheby's defense of the company was equally a defense of the principles to which he and his colleagues were committed. Not surprisingly Chadwick too rejected the conclusion that the epidemic had resulted from impure water supplied by the East London company.[59] Contradictions between theory and practice developed because Letheby accepted the obligation of the analytical chemist to make the best possible approximation of the virulent potential of a water, whatever the inadequacies of the science which defined virulence. Letheby's support for the companies came from his recognition that they supplied the best water practically

[59] Lancet, 8 September 1866, pp. 273-76, 293-94; Chadwick in disc. of Byrne, "Experiments on the removal of organic substances," pp. 32-35.

available. Despite the sewage it contained, Thames water was far better than that taken from local shallow wells. For the remainder of his life (he died in 1876), Letheby continued to uphold, on empirical grounds, the reliability of traditional water analysis, the adequacy of the metropolitan water supply, and the efficacy of the self-purification of rivers.

The struggles of Frankland, Crookes, and Letheby -- to reconcile the still unresolved issue of the identity and nature of morbid poisons with ideas of the beneficence of nature, of the directions and demands of sanitary progress; to fulfill one's duties as a protector of the public while justifying one's own scientific practice -- characterize the evidence sanitary scientists gave to the Royal Commission on Water Supply. The commission -- a duke, a London alderman, the head of the Thames Conservancy, two civil engineers, and a geologist -- was ill-equipped to deal with the complicated chemical and engineering evidence it heard. Its job was to examine schemes for supplying London with unpolluted soft water and to judge the fitness of the existing supply.

The parade of experts who testified before the commission represent a different generation of water scientists from those who argued about the quality of the metropolitan supply in 1850-52. Such stalwart upholders of the water companies as Brande and Taylor had gone, and their constituency was represented by Letheby. Chadwick was not involved, though William Miller, Robert Angus Smith, and Thomas Hawksley remained active and influential. Some issues had changed. The absence of microscopical debate during the 1867-69 hearings is striking; surely organisms still inhabited the water, yet the attention of sanitarians was focused on obscurer forms of organic matter. Hardness remained an important issue and the experts still could not agree whether hard water was good or bad. The question of how polluted water was related to disease was more unsettled than ever. Snow's ideas, the various zymotic hypotheses, the undefinable germs -- none of these resolved anything, they merely increased the options sanitarians had to explain disease. The commission was told the east London cholera had been caused by everything from miasmatic emanations of unspecified filth to water-borne germs derived from the excreta of cholera victims. Because there were more options for explaining water-borne disease there were more opinions of the operations "most proper to be followed in the [water] analyses, . . . and the value and meaning of the results obtained."[60]

Compared to the 1850-52 hearings, the hearings of 1867-69 are characterized by a greater diversity of issues, a broader range of interests represented, and, most importantly, by a pervasive uncertainty regarding the meaning and measure of pure water. As Letheby's struggles with the concept of disease germs illustrate, the 1866 cholera epidemic had shaken the foundations upon which water evaluation had been based and had confirmed suspicions of the

---

60 RCWS, Report, p. lxxiv.

analytical and purificatory nihilists. This uncertainty is evident even in the testimony of those committed either to support or to oppose existing supplies.

Some sanitarians -- John Simon, William Farr, and Edmund Parkes, for example -- advocated techniques of purification but found themselves unable to say exactly what purification was. John Simon convincingly argued that chemists were unable to detect morbid poisons. "Very competent persons" believed water polluted by the excreta of cholera victims "would contain demonstrable microscopic germs, each carrying an infectious power." Thus Simon, like Frankland, strongly denied the efficacy of purification by dilution. But while he used these germs to explain how terribly serious was the need for unpolluted water, Simon refrained from the conclusion usually associated with them -- that none of the usual methods of purification could be relied upon to remove them. Instead, sewage irrigation, followed by a 15 to 20 mile flow in a river would serve to remove germs. Simon recognized a great danger in the use of polluted water, but like so many of his colleagues was unable to say how serious the problem was.[61]

William Farr, Frankland's boss at the Registrar General's Office, was unsure what to make of his analyst's results. He was glad "to learn that as far as chemistry can determine the original sewage contamination had been converted into comparatively innocuous

---

[61] RCWS, Evidence, QQ 2838, 2815-19.

inorganic compounds before delivery to customers," but worried that "eggs of worms or the seeds of disease in some way would often remain unchanged" by the processes of purification that converted sewage into innocuous compounds.[62] Like Simon, Farr trusted sewage irrigation would make sewage drinkable.[63]

Edmund Parkes, the army's professor of hygiene, exposed the commissioners to the utter confusion that permeated the understanding of the relationship between filth and disease. Parkes doubted plain sewage -- sewage not containing the excreta of cholera patients -- could produce cholera when ingested, but he did think sewaged water together with certain atmospheric conditions would produce "a tendency to develope it [cholera] with great rapidity."[64] He would not say that boiling would free water of "seeds of disease," but noted it had little effect on the amount of organic matter contained in the water.[65] Purification was essentially a process of oxidation and could be accomplished by filtration through soil or by long flow in a river.[66]

It is clear that Parkes was struggling to organize a huge body of concepts and observations. Some of the phenomena of cholera -- the periodicity of epidemics, the unexplained way epidemics traveled

---

[62] Ibid., QQ 2852, 2861.
[63] Ibid., Q 2886.
[64] Ibid., QQ 3121-22.
[65] Ibid., QQ 3169-70.
[66] Ibid., QQ 3177, 3151-57, 3180-88.

across vast and empty areas -- suggested the involvement of special atmospheric conditions. Moreover, each of the multitude of versions of the identity of the cholera poison bore its particular complement of consequences for purification. Although he broached the possibility that the cholera poison might be a living solid, Parkes tended to treat it as non-living soluble organic matter when he considered its removal. Thus boiling, because it removed little organic matter, was of doubtful utility, while oxidative purification in earth or rivers would probably be more effacacious. Ultimately, Parkes concluded, the only way to tell whether water was safe was to see what happened to those who drank it.[67]

A quite different uncertainty pervaded the testimony of the water analysts the commission heard. William Odling, Benjamin Brodie, and William Miller repudiated the notion that there existed some single set of chemical determinations which could show the quality of a water. At the same time they managed to convey the message that chemists were the proper people to judge the goodness of water.

All three denied that inorganic nitrogen compounds represented previous sewage contamination, therein undermining the only bit of certainty Frankland had felt water analysis could provide. In Odling's view the amount of nitrates did "not give any ground on which to estimate the . . . proportion of sewage."[68]

---

[67] Ibid., Q 3170.

[68] Ibid., Q 6459.

> a proportion of the nitrates which the sewage itself does furnish, is in one case destroyed and in the other case is not; and so far as the history of the water is concerned, in the one case when the nitrates are destroyed, that water may show but a very small amount of previous sewage contamination, whereas it might have had a much larger amount than the other.[69]

Each of the three chemists pointed out that nitrates in water had not necessarily come from sewage, and that the amount present in a sample bore no relationship to the amount that had been present. And having undercut the slim basis on which Frankland's interpretations of water quality rested, these chemists offered nothing to replace it.

Each was sure a process of oxidative self-purification existed; each was equally unwilling to say when that process made polluted water safe. Odling thought self-purification occurred "to a very considerable extent," but was uncertain whether that extent was sufficient and worried about the passage through the water companies' filters of dissolved sewage materials.[70] Miller arrived at the profound conclusion that previously contaminated water was safe "in the majority of instances," but that there might be "cases in which danger is produced."[71] Sewage-contaminated water which had flowed 4 or 5 miles in a river was "very much better for human consumption," but Miller would not assert that all sewage would have been destroyed, nor was there any way to tell if it had

---

[69] Ibid., QQ 6462, 6473-75, 7004-5, 7074-82.

[70] Ibid., Q 6458, QQ 6479-80.

[71] Ibid., Q 7082.

been.[72]

Among the chemists testifying before the commission, Sir Benjamin Brodie was both the greatest doubter and the most content with uncertainty. Like Simon, he maintained that the quality of a water could only be discovered epidemiologically: did the population drinking the water suffer excessively from zymotic disease?[73] By contrast with the post hoc certainties of epidemiology, water analysts were chasing a poison which was being continually diluted, being destroyed at an unknown rate, and which they would not recognize if they found.[74] Brodie cast doubts on every component of water evaluation. Rivers seemed to purify spontaneously, yet there was no way of telling when such purification was complete.[75] Sewage disappeared more quickly percolating through soil than flowing in rivers, but again there was no way to detect completion.[76] There might be such things as disease germs; they might pass through filters.[77] The new analytical processes developed by Frankland and

---

[72] Ibid., QQ 7062, 7071. Coming from Miller such nihilistic views are particularly interesting for they suggest Miller's consideration of the hypothesis that morbid poisons were independent of chemically measurable organic matter. In 1859 Miller had examined the decrease of dissolved oxygen in the Thames as it flowed through central London. He asserted before the R.C. Water Supply that this diminution was "direct proof of the effect of oxygen in destroying those organic contaminations which are thrown into the river," but would not say that this process was proof of purification (Q 7088).

[73] Ibid., QQ 6991, 7013.

[74] Ibid., Q 7011.

[75] Ibid., QQ 6983-91.

[76] Ibid., Q 6992.

J.A. Wanklyn were "imperfect" and improperly verified.[78] Frankland's PSC was invalid because nitrates might come from sources other than sewage.[79] Some techniques of purification might improve water quality, none were completely trustworthy. Brodie admitted epidemic cholera might be spread by specifically polluted water in the manner John Snow had suggested, but there was no proof the London supply had caused such epidemics in the past, and only epidemiologists could tell if it had.[80]

Where the commissioners were offered certainty, that certainty came either from strange quarters or was linked with dubious and transparent assumptions. Robert Angus Smith, the last chemist to testify, claimed he could detect sewage, but the unorthodoxy of Smith's system of water analysis and his inability to describe it coherently to the commissioners diminished the effectiveness of his testimony. Smith retained the view he had held in 1848. Because not all types of organic matter were equally harmful, the animalcular population of a water rather than its chemical composition

---

[77] Ibid., QQ 6993-94.

[78] Ibid., QQ 6997-7003. Brodie desired blind tests in which analysts were asked to distinguish waters, some of which were known to be safe, others known to cause disease. Such tests were not done until 1880 when J.W. Mallet of the University of Virginia and the U.S. National Board of Health undertook an exhaustive experimental study of the reliability of existing methods of water analysis. Mallet found none of the processes terribly good. The most easily available copy of Mallet's report is in <u>Chemical News</u> 46(1882): 63-66, 72-74, 101-2, 108-11.

[79] RCWS, <u>Evidence</u>, QQ 7004-5.

[80] Ibid., QQ 7039-44.

afforded the most sensitive indication of the type of organic matter present. Smith's central procedure was an examination of the "vital forms" produced when a water was allowed to stand, and he believed that the "amount of locomotion and the evident vitality and flexibility of the forms were indices of the impurity or of the condition of the organic matter."[81] Smith didn't associate particular species with degrees of harmfulness; instead the number of organisms, their size, and their disgustingness were measures of water quality.[82] He believed germs had been responsible for the 1866 cholera and that if "carefully nursed" they would be detectable microscopically.[83] Supplementing this examination of the organisms a water engendered was a seven part classification system for organic matter based on such variables as putrescibility, animal or vegetable origin, and recency.[84]

---

[81] Ibid., Q 7209.

[82] Ibid., QQ 7236, 7242.

[83] Ibid., Q 7211.

[84] Smith's ideas about water analysis appear in their clearest form in his paper on "The examination of water for organic matter," which was based on a water analysis manual he had written for the Indian health authorities in 1865. This paper appeared in slightly different versions in the Proceedings of the Manchester Literary and Philosophical Society (3rd series, 4(1871): 37-88); and in Chemical News (19(1869): 278-82, 304-6; 20(1869): 26-30, 112-15). It is clear that Smith was aware of the difficulty of the problem of relating disease to decomposition and that he recognized, for example, the ambiguous nature of putrefaction as a process which was both malignant and purifying. Smith was also remarkable in his awareness of Pasteur's conception of fermentation and putrefaction and in his attempt to fit Pasteur's conception of germs into the mainstream of British thinking on water quality issues. Throughout his career he maintained his belief that aquatic organisms revealed much about the sources of water pollution, and Smith was one

By contrast with Smith's ineffable wandering answers about the nature of germs, the significance of putrefaction, oxidation, and nitrates, some witnesses had quite definite ideas about what morbid poisons were like and what processes reliably removed them, ideas with consequences which invariably supported the interests these witnesses represented.

On 20 June 1867 James Alfred Wanklyn told the commission about the new process of water analysis which he would describe to the Chemical Society that evening.[85] Wanklyn believed this "albuminoid ammonia" process would revolutionize water analysis and gave the commission the impression of water analysts pushing one another aside in their eagerness to adopt the process and pay homage to its inventor. Wanklyn had no patience with analytical nihilism. It was quite possible that inferior processes of water analysis might be incapable of detecting water-borne morbid poisons, but the arrival of Wanklyn's process would put an end to such problems. Wanklyn knew all about germs. They were "pretty stable" and only broken up by a caustic solution of potassium permanganate, strangely enough the exact means by which albumin was converted to ammonia in

---

of the first British sanitarians to recognize the value of Koch's gelatine culturing technique. Yet Smith was continually unable to make ideas into techniques and his views on water quality were almost completely ignored by his colleagues. It is not even clear that Smith himself followed the procedures he had set forth in the late 1860s.

[85] J.A. Wanklyn, E.T. Chapman, and Miles H. Smith, "Water analysis: determination of the nitrogenous organic matter," J. Chem. Soc. 20(1867): 445-54.

Wanklyn's analytical process.[86]

To enhance further the excellence of his analytical process Wanklyn denied the existence of self-purification. When such analysts as Letheby had maintained that rivers invariably were purified because no traces of sewage remained after a few miles of flow, they had been reckoning without the albuminoid ammonia process. Wanklyn argued that "albuminoid matter in sewage is extremely persistent" and claimed he could detect it several miles below a sewage outfall.[87]

The water companies were ably, if not always formally represented before the commission. Letheby and Thomas Hawksley pointed to the enormous amount of dissolved oxygen in the Thames and affirmed sewage oxidized rapidly and harmlessly. Letheby claimed sewage mixed with twenty volumes of water would be safe to drink after a flow of 12 miles and later lowered the requirement to 7 miles.[88] Hawksley was equally optimistic:

---

[86] Wanklyn noted: "I have no doubt whatever that it is dangerous, for this albuminoid matter may be organic germs of all kinds, and in fact, the very worst things that there could be in a water. . . . What are supposed to be the worst things that can be present in a water are germs that may set up a kind of vegetation within us, and if there are any of these germs in a water the albumin in the germs will be converted into ammonia by this process of mine, so that there is an actual measure of the possible amount of germ" (RCWS, Evidence, Q 5451). Though he continued to insist germs were made of albumin, Wanklyn successfully avoided answering the question of how one was to know when albumin was a germ, maintaining that his process detected germs and in a general manner equating albumin with germiness.

[87] RCWS, Evidence, QQ 5482-85.

[88] Ibid., QQ 3894-97.

> when any of these sewage matters are put into running water which contains a vast amount of oxygen, as much as 40 or 50 gallons of oxygen in 1000 gallons of water, it immediately attacks them and they become gradually decomposed and resolved into their elements, or are re-associated in perfectly pure states, nothing offensive remaining, and that has been going on ever since the creation of the world, the very Atlantic itself would ere this have been a cesspool if it were not for that process.[89]

In the 1867-69 inquiries, just as in the metropolitan sewage controversy of 1858, views of the self-purification of rivers were linked to views of the nature of morbid poisons. The water companies' supporters -- Letheby and Hawksley; local medical officers such as Thomas Orton; water company engineers such as James Simpson; and even the Thames Conservancy's engineer Stephen Leach -- all regarded traditional water analysis as a reliable index of safety and emphasized local insanitary conditions in explaining outbreaks of zymotic disease. The harmfulness of water was associated with its putrefaction, and impure water was often perceptibly impure. Harmful matters in water were "exceedingly decomposable in the presence of oxygen."[90]

The commissioners were not sanitary scientists, and evaluating the masses of often contradictory testimony was a task for which they were unsuited. They contrasted the seriousness of the issue --

> Water is a necessity of life; the consumers in a place like London have no power to choose their own source, but are at the mercy of the parties undertaking the

---

[89] Ibid., Q 5075.

[90] Ibid.

> supply; the health, even the life, of the inhabitants is in the hands of these parties; and it is therefore a matter of paramount public interest that the manner in which they exercise this immense power should be jealously watched, and efficiently controlled.[91]

-- with the disunity among the scientists investigating it:

> We have found that opinions are divided about it [water quality evaluation], but . . . the elements which enter into its determination are of very subtle character, and by no means admit of the satisfactory kind of treatment we are in the habit of expecting from the modern advanced state of physical science.[92]

Unable to make sense of the tangle of contradictory testimony, the commission to a large degree ignored science in favor of "other considerations." Since the science of the present proved unhelpful the commission relied upon the science of the past and decided that what ought to be was exactly what was. The 1851 chemical commission of Hofmann, Graham, and Miller (which had suffered from no uncertainty) had decided Thames water was satisfactory and self-purification reliable; despite cholera epidemics in 1854 and 1866, the 1867-69 commission confirmed its predecessor's verdict. From the masses of evidence it had collected the commission extracted testimony supporting these conclusions, thereby giving scientific opinion "due weight." Nature was beneficent: fish and weeds, prodigious surpluses of dissolved oxygen and the inherent instability of noxious matters, combined to effect self-

---

91 Ibid., Commission report, p. lix.

92 Ibid.

purification.[93] The commission was happy to find "the remarkable power of oxidation possessed by running water admitted more or less by all chemists."[94]

III. Nihilistic Water Analysis: The Structure of Frankland's System of Water Analysis

Surely the most vexing witness the commission heard was Frankland. As the commission's chemist he had in hand an enormous amount of pertinent chemical information, yet consistently Frankland upheld conclusions which seemed to the commissioners -- as we shall see, quite rightly -- independent of that evidence.

By the beginning of 1868 Frankland's revolution was over. During the preceding year and a half he had developed new analytical techniques, new ways of interpreting results, new analytical formats, and new mechanisms for communicating analytical results to the public. Various aspects of this synthesis were made public in a lecture to the Chemical Society on 15 January 1868; in his testimony before the water supply commission on 27 February; and in a Royal Institution lecture on 3 April. The evolution of Frankland's ideas about water quality has already been traced, but what were the components of his revolutionary system and how were they applied?

---

93 Ibid., p. lxxiv.

94 Ibid., pp. xciii, cii.

First, Frankland had put the measurement of the organic matter in water on a far sounder basis. With his student H.E. Armstrong, Frankland had adapted the combustion procedure commonly used for the ultimate analysis of organic materials to deal with the minute concentrations of organic contaminants typically present in water. In this procedure the sample of water was evaporated with a mild reducing agent (H_2SO_3) which would destroy nitrites, nitrates, carbonic acid, and carbonates, leaving a residue in which the carbon and nitrogen could be assumed to have come from organic matter.[95] This residue was then inserted into a combustion tube, combusted in vacuo (lead chromate supplied the oxygen), and the amounts of organic carbon and nitrogen were calculated from the gases generated. Simple in principle, the Frankland/Armstrong combustion process was in fact extremely tricky to run, took two days to complete, and required equipment and skills beyond what many analysts possessed.[96]

As will be discussed in the next chapter, this combustion process was exceedingly controversial. By contrast the remaining processes included in Frankland's system were reasonably simple and

---

[95] Ammonia, which was usually present in very small quantities, still had to be determined separately and the result subtracted from the combustion results in order to obtain the correct amount of organic nitrogen.

[96] Frankland and Armstrong, "On the analysis of potable waters," pp. 88-98. The invention and early days of the combustion process are discussed in several surviving letters: Frankland to Armstrong, 8 January 1867, 21 September 1867, 20 April 1868, undated [late 1868 or early 1869], in Royal Society of London, Misc. Mss, vol. 10, #90-93; and J.J. Day to Armstrong, 23 November 1867, 14 December 1867, 4 April 1868, 6 May 1868, 21 June 1868, 27 March 1869 in Imperial College, Armstrong Papers, C241-C246. Day was an assistant in the

well-accepted. He measured total solids -- the weight of the sample after evaporation, nitrates and nitrites by Crum's process (agitation with mercury, which converts these to NO gas), ammonia by the Nessler test, and hardness.[97]

But the revolutionary character of Frankland's analytical system lay less with the processes he performed than with the principles he developed to interpret the results they gave. His combustion process was not simply a means for better approximating the amount of organic matter a water contained, and Frankland rejected the assumption of most of his colleagues that in general water was impure in proportion to the amount of organic matter it contained. For Frankland the proportion of the two organic elements to one another was more important than the actual quantity of organic matter present. He believed each type of contamination would produce a characteristic carbon:nitrogen ratio, and in general the worst contaminations -- sewage for example -- contained the lowest ratios. Thus when the geologist Joseph Prestwich, a member of the water supply commission, triumphantly pointed out to Frankland that lake water, which Frankland advocated, contained a higher concentration of organic nitrogen than the Thames, Frankland cooly noted that the ratio of organic elements in lake water ranged from 95:1 to 13:1 while Thames waters by contrast gave ratios of 8:1 and 6:1.[98] In

---

water laboratory.

[97] Frankland and Armstrong, "On the analysis of potable waters," pp. 87, 101-3.

[98] To avoid confusion I have normalized these ratios, giving the

practice, however, Frankland only considered these ratios in questionable cases; in regard to London water, known to be contaminated with sewage, he habitually spoke of the mere presence of organic nitrogen as actual sewage contamination, ignoring entirely the accompanying concentration of carbon.[99]

The history of a water, a knowledge of the sources of its pollution, was for Frankland the fundamental principle by which to interpret analytical results. The organic elements were important because they provided information about a water's history. Similarly, nitrates, nitrites, and ammonia were important because they showed water had once been polluted -- by sewage or one of its analogues, Frankland insisted -- and not because they showed purification had occurred; that putrescible matter had been oxidized.[100]

But if a concern for history justified Frankland's preoccupation with the oxidation products of organic nitrogen, an eye to rhetorical effectiveness justified "previous sewage contamination," one of the titles under which these data appeared. By expressing

---

carbon first. Frankland was inconsistent in his ordering of them. Compare RCWS, Evidence, Q 6289, with Frankland, Water Analysis for Sanitary Purposes, with hints for the interpretation of results (Philadelphia: Presley Blakiston, 1880), pp. 91-94.

[99] RCWS, Evidence, Q 6405.

[100] It was not Frankland, but R. Angus Smith, who, in a rare moment of lucidity, explained the significance of these nitrates most clearly: "They are disliked because of their previous bad company. They have been in contact with sewage when they can be shown to have been formed recently. If, however, we are sure that the company is to be kept out we do not require to be afraid of such reformed characters unless they become personally offensive" (RCWS, Evidence, Q 7247).

these products in terms of the amount of average London sewage that would contain an equivalent amount of them Frankland was able to turn small numbers into large ones and to give his readers a vivid notion of what proportion of the water they drank had once been sewage. The words Frankland used to describe the concept to a Royal Institution audience reflect its rhetorical purpose:

> a half-pint glass of it [Royal Institution well water] contains nearly a quarter of a pint which has previously been in the condition of average London sewage, besides a dessert spoonful of actual or unoxidized sewage.[101]

Previous sewage contamination was not Frankland's only rhetorical trick. He also referred to the nitrates, etc., as "the skeleton of sewage," implying that they were the discoverable remains of sewage. Germs were another mechanism. In early 1868 Frankland's conception of germs was still exceedingly tenuous. He frequently stated his ignorance of the nature of water-borne morbid poisons, but germs -- germs with definite sizes and capabilities no less -- continually emerged as a model of what morbid poisons might be like, a model which supported the conclusion that once-polluted water could never again be used safely. Life, for example, might be exactly the quality a morbid poison required to maintain its virulency against the destructive forces of oxygen during the long flow between the sewage outfalls of upstream towns and the intakes of the London water supply. Thus an egg -- a germ -- might survive such

---

101 Frankland, "On the proposed water supply," p. 361.

treatment while the same organic matter which constituted the egg would have been destroyed had it been in an unorganized form.

> I will say that if you were to break an egg and beat up the contents and mix them with Thames water at Oxford, the organic matter so introduced into the Thames might under favourable circumstances and probably would be entirely destroyed and converted into mineral matter before it reached Teddington; but if you were to throw an egg in without being broken, it would be carried down by the stream and would reach Teddington with its vitality undestroyed.[102]

Likewise, the reason the public could find no security in filtration was that germs might be so small as to pass through filters: "I should not be prepared to say that after any amount of filtration we should be guaranteed from the presence of those minute germs, which being smaller in some cases, as they have been proved to be, than blood globules, would pass through the pores of the chalk, for instance, like human beings pass through the streets of London."[103]

These examples illustrate the use Frankland was making of the germ theory. It was his rhetorical workhorse; an argument incontrovertible because hypothetical; a concept sufficiently vague as to be adaptable to a wide range of rhetorical requirements, yet concrete enough to convey a vivid image -- of a poison resisting the elements, of a poison able to slip between the pores of a filter. Since disease germs had not been discovered Frankland could quite justifiably maintain that they were not in all sewage or that they were where sewage was not. They might be what made sewer gases

---

[102] RCWS, Evidence, Q 6372.

[103] Ibid., QQ 6401, 6244.

harmful, even if the sewage itself never infected water: "these gases from sewers in all probability contain also floating in them organic germs, and it is those, which, when conveyed into the water in this way [i.e., sewer gases absorbed by water], are productive of mischief afterwards."[104]

Another aspect of Frankland's rhetorical facility was his ability to dodge difficult questions by misconstruing them and responding with irrelevent answers. Thus, asked if germs must have at some time been in sewage, he responded that not all sewage contained germs.[105] Similarly, to Prestwich's question about the supposed harmlessness of the organic nitrogen in lake water -- "would not the organic nitrogenous substances in any of those waters be liable to putrefaction in the one case [lakes] in the same way as in the other [rivers]" -- Frankland tendered an answer which referred to putrefaction as one of the steps by which organic matter was purified -- became inorganic -- and offered a lengthy digression on the failure of sewage to disappear during flow in a river, never touching on the salubrity of the putrefaction of organically contaminated lake waters.[106]

The commissioners had considerable trouble understanding the significance of previous sewage contamination, chiefly because they could not comprehend the premise upon which it was based: that no

---

[104] Ibid., Q 6270.

[105] Ibid., Q 6403.

[106] Ibid., Q 6291.

techniques of purification could be trusted and that chemical analysis could never show whether harmful materials existed in water. If previous sewage contamination represented sewage that was no more, why then worry about it, they wondered. Their incomprehension is not wholly Frankland's fault. He made his position clear early in his testimony:

> I consider that water contaminated with sewage contains that which is noxious to human health. There is no process practicable upon a large scale by which that noxious material can be removed from water once so contaminated, and therefore I am of opinion that water which has once become contaminated by sewage or manure matter is thenceforth unsuitable for domestic use.[107]

This was straightforward enough, but the commissioners assumed that despite this strong statement, Frankland must, like any other chemical analyst, rely on his analyses to determine whether or not water was safe. They were repeatedly disabused of this assumption. Frankland blithely agreed that the nitrates he measured in London water and which he called previous sewage contamination were harmless. Nitrates were only important because they showed sewage had formerly been present and there were no reliable processes for purifying sewage-polluted water. The Duke of Richmond, Frankland's questioner, took this response to mean that the water was good -- Frankland, a chemist, had found nothing harmful in the water; it must, therefore, be safe.

---

[107] Ibid., Q 6222.

> But although they [nitrates] are there in that state, you state, do you not, that there is nothing in them [nitrates] that could be injurious to health, and therefore the water is a wholesome water to drink?[108]

"I did not intend my statement to go so far as that," Frankland replied. While nitrates themselves were safe, they might indicate the presence of something dangerous. "Other portions of that sewage" might have escaped filtration and might still be in the water "in too minute a quantity to be capable of detection by chemical analysis."[109]

Frankland acknowledged that purification could improve water. He eventually developed a scheme of standards for classifying potable waters in which previous sewage contamination was a central component. Purification processes that lowered PSC levels were therefore assumed to be making water safer.[110] Similarly, Frankland enthusiastically supported measures to make Thames and Lea water safer. He saw considerable room for improvement in filtration since some of the companies' filters did a better job of purifying the same water than did those of other companies. Sewage treatment plants for upstream towns would help too. But while all techniques of purification might improve the water, none would change Frankland's opinion that a polluted water supply was inherently unsafe.[111]

---

[108] Ibid., Q 6238.

[109] Ibid.

[110] RPPC/1868/<u>6th Report</u>, p. 17

In fact, what so exasperated the commissioners was the underlying philosophical structure of Frankland's water analysis. In Frankland's hands, Brodie's wholesale rejection of the question of when polluted water became safe became the very stuff and basis of water analysis. Frankland had constructed a system of water analysis on the premise that the analyst could never discover whether water contained anything harmful. In this construction Frankland was advocating a water analysis wholly distinct from the practice of his colleagues, and even from the practice of those who admitted morbid poisons were beyond the reach of analytical chemistry. Hitherto chemists had recognized that despite their inability to isolate the material causes of water-borne diseases they could approximate these by measuring substances -- organic matter -- which were presumably required for water to be harmful. Frankland rejected even this limited assumption. Water with a bad history was to be allowed no repentence, no matter how much its analytical appearance was improved by processes of purification. Most importantly, however, Frankland had liberated the concept of impure water from all analytical constraints. Oxidative self-purification might occur, organic matter might disappear entirely, yet within Frankland's system the water might still be condemned as dangerously impure.

By diagramming the procedures Frankland used to interpret his

[111] RCWS, Evidence, QQ 6235-36, 6246, 6376, 6390-92, 6381, 6396, 6417-18.

analytical results it is possible to expose the assumptions which underlay the studies on water quality issues he carried out in the 1870s and which will be discussed in the next chapter. Moreover, a diagram of Frankland's complicated, apparently circular, system of decision making can serve as a frame of reference for discussion of less complicated, but similarly circular systems for veryifying or denying purification which other analysts and bacteriologists employed during the 1870s, 1880s, and 1890s.

Frankland believed that the sewage contamination of the Thames and Lea was "self-evident from the circumstance that sewage flows into the Thames and Lea."[112] Figure 7.2 shows how Frankland's interpretive procedures confirmed this conclusion. First, Frankland considered the organic nitrogen concentration. Organic nitrogen was assumed to represent "actual sewage contamination" and its presence therefore provided analytical justification for condemnation.[113] But such "unoxydised sewage" was only rarely present in the London water supply which was known to be polluted with sewage. Therefore Frankland turned to a test "second only in importance to [the search for] actual sewage contamination": previous sewage contamination.[114] If nitrates, etc., were present in substantial quantity water might be condemned on grounds we have considered previously -- germs contained in the original sewage might have

---

[112] Ibid., Q 6223.

[113] Ibid., QQ 6405-8.

[114] RCWS, Appendix D, p. 20.

Figure 7.2 Frankland's Explanation of Analytical Results for Waters Presumed Bad (Composite diagram)

Test	Result	Verdict	Explanation
1. Organic nitrogen	present	water bad	Represents actual sewage contamination
	absent	go to test #2, PSC	
2. Previous Sewage Contamination (nitrates, nitrites, ammonia)	present	water bad	Organic nitrogen may not always be entirely oxidized, or resistant and undetectable germs may still be present
	absent	water bad	PSC is a minimum; vegetation may have removed all nitrates, etc., but germs may still be present, and vegetation cannot be assumed to be always effective

resisted oxidation or oxidation sometimes might fail even to destroy putrescible organic matter. Suppose, however, that Frankland found little or no previous sewage contamination. Water could still be condemned because PSC represented the minimum amount of sewage with which the water had been contaminated. Vegetation and other natural agencies removed the products of sewage oxidation from water. The polluted Lea water supplied by the East London company showed a previous sewage contamination of zero sometimes during the summer months, yet Frankland continued to condemn it -- a fact that the companies' analysts, Odling, F.A. Abel, and Letheby, found ridiculous. Thus, whether or not polluted water contained any actual or previous sewage Frankland condemned it.[115]

The most controversial component of Frankland's interpretive system was the expression of nitrates, etc., as previous sewage contamination, yet this assumption was the foundation of Frankland's historically oriented water analysis. For nitrates, nitrites, and ammonia to be hygienically significant the assumption had to be made that the various organic matters which produced them were approximately equivalent in their potential to harm health. Frankland justified this assumption in two ways. First he tried to expand the set of potentially harmful substances to coincide with the set of materials generating nitrates, etc. The vagueness of his germ theory helped him here. Second, he maintained that innocuous organic nitrogenous matters -- vegetable matters, in particular -- never yielded these oxidation products to water when they decayed. Frankland argued this assertion empirically. Water from areas uncontaminated by animal matter never contained nitrates; wherever nitrates were found in water, they could be traced to animal contamination, usually either sewage or manured fields.

Frankland exploited the multifactoralism characteristic of pathology and etiology during this period to defend PSC. PSC was never defined unambiguously. Within a single report Frankland defined it as representing "sewage and other analogous contamination"

---

[115] Ibid., Appendix AK, pp. 78-79.

and "contact with decomposing matters."[116] Sometimes the term appeared as "previous sewage or manure contamination." Since both animal manure and human sewage produced nitrates, etc., Frankland had to assume that manure was as serious a threat to health as sewage. To grant manure any lesser status would have undermined the significance of PSC.[117] Pathological rationale for this assumption was provided by the zymotic explanation of how pollutants affected health articulated by Hofmann and Blyth in 1856, and which Frankland was still quoting as late as 1873.[118] Unlike John Snow, Frankland was not interested here in narrowing the understanding of what made water virulent, or in establishing the safety of normal putrefaction, for such an admission would diminish the utility of the concept.

Certainly the oddest of the hypothetical morbid matters that Frankland came up with to justify the validity of PSC was the decaying chalk-dwelling microbes of Antoine Béchamp. During his testimony to the water supply commission, Frankland was repeatedly pressed to defend the contention that the nitrates in water which had filtered through hundreds of feet of chalk had the same hygienic significance as nitrates contained in sewage-polluted river water. The problem arose with regard to the water supplied to parts of London by the Kent company, which obtained its supply from wells

---

116 Ibid., Appendix D, p. 20.

117 RCWS, Evidence, Q 6227.

118 RPPC/1868/6th Report, p. 5.

sunk deep in the chalk formation of southeastern England. The Kent's water nearly always contained more PSC than any of the river-derived supplies, a fact which continually plagued Frankland and which probably led to his abandonment of PSC in his London reports in 1876.

In his 1868 testimony Frankland argued that the high nitrate levels in the Kent's water were significant because the company's wells were recharged by drainage from manured land. He admitted that water which had percolated through 250 feet of chalk would be preferable to water which had not been so thoroughly filtered, but at the same time maintained that not even "filtration through a considerable stratum of chalk could be relied upon to free the water perfectly from such germs."[119] To this Frankland added another reason why this chalk-filtered water might be dangerous: a living, nitrifying ferment might exist in the chalk. The idea came from Béchamp, who had claimed chalk strata functioned as a butyric ferment and had attributed this function to organisms within the chalk. These chalk dwellers might be quite as dangerous as sewage:

> If the chalk . . . contains a vast number of living organisms, it is conceivable that those organisms in their decay may give rise to quantities of nitrates and nitrites and of ammonia, and that it may be that some of those matters found in the chalk water are derived from this source; but I would not on that account think that water was preferable because it had been in contact with putrefying animalculae rather than

---

[119] RCWS, Evidence, QQ 6229-30, 6233-35, 6240, 6244.

in contact with putrefying sewage.[120]

This statement suggests the extremes to which Frankland was going in the spring of 1868 in an effort to establish a solid theoretical framework with which to defend his new analytical categories. By the end of 1868 he had decided that the chalk filtration was an adequate purification; in subsequent years he was much taxed with explaining why the high PSC of the Kent's water was not significant.[121]

Frankland complemented extension of the definition of virulent water-borne substances with a restriction on which nitrogenous organic substances would produce nitrates. He admitted in principle that vegetable decay could contaminate water with nitrates, nitrites, and ammonia. In practice, however, this did not happen:

> vegetable organic matter has never been observed to yield these products [nitrates, nitrites, and ammonia]. I have sought for them in vain in scores of samples of water heavily contaminated with vegetable matter and taken from both [sic] springs, rivers, and surface drains.[122]

In subsequent years Frankland's critics frequently argued that nitrates, etc., did sometimes come from vegetable decay, and in a few cases they even cited examples in which this was supposed to be occurring. Frankland carefully checked each of these counter instances and as late as 1876 he was still arguing that nitrates,

---

[120] Ibid., QQ 6334-35. Also see Lancet, 1866, ii, p. 677.

[121] RCWS, Appendix AX(2), p. 105.

[122] Ibid., Appendix AJ, p. 89; Evidence, QQ 6329-31, 6227-28, 6296, 6318.

etc., in water were never products of vegetable decay.[123]

But what of unpolluted waters that for some reason gave a poor analytical showing? Frankland frequently found himself in the position of advocating waters from mountain lakes heavily contaminated with nitrogenous extracts of peat. By September 1868 he had also come to favor use of deep well waters, such as the Kent company supplied, which were heavily contaminated with "previous sewage." In such cases Frankland's interpretive system provided excuses for the poor showing of these waters (see Figure 7.3).

In water with negligible organic matter and minimal previous sewage contamination Frankland of course had no problem justifying a favorable verdict. In waters with substantial PSC, but still no actual sewage contamination, Frankland could argue that PSC in certain cases was simply not worrisome. The nitrates in the Kent water were unimportant, he was maintaining by late 1868, because it was unlikely that germs could percolate through great depths of chalk. Sometimes good water contained considerable amounts of organic nitrogen presumably derived from peat, as was the case with the water from the Welsh mountains and the Cumberland lakes. In such cases Frankland turned to the carbon:nitrogen ratio. Peaty waters normally contained far more organic carbon than organic nitrogen, and Frankland used this characteristic to distinguish

---

[123] Frankland to the Registrar General, 10 July 1869, in Papers relating to the Metropolitan Water Supply, P.P., 1872, 49, (99.), pp. 33-36; and Frankland, "On some points in the analysis of potable waters," J. Chem. Soc. 3rd series 1(1876): 829-33.

Figure 7.3 Frankland's Explanation of Analytical Results for Waters Presumed Good (Composite diagram)

Test	Result	Verdict	Explanation
1. Organic nitrogen	present	go to test #2; organic carbon	
	absent	go to test #3; PSC	
2. Organic carbon	high	good	C:N ratio is high for harmless vegetable organic matter.
	low	good	During oxidation of vegetable organic matter carbon oxidizes faster than nitrogen. During oxidation of animal organic matter nitrogen oxidizes faster than carbon. Therefore, ratios of organic elements of animal and vegetable extracts will become increasingly similar. Old sewage looks analytically like old peaty matter.
3. Previous sewage contamination (nitrates, nitrites, ammonia)	high	good	As with Kent company water sometimes PSC is still high after extensive filtration which can be assumed to have removed all disease germs.
	low or absent	good	

waters polluted by vegetable extracts from those polluted by sewage.[124] Unfortunately, as he gained experience analyzing water Frankland began to find cases in which this rule didn't hold. Sometimes wholly blameless waters produced carbon:nitrogen ratios giving them the analytical appearance of sewage. The explanation for this, Frankland believed, was to be found in the distinct kinetic phenomena of vegetable and animal decomposition. In peaty waters organic carbon was destroyed more rapidly than organic nitrogen, while in animal extracts, such as sewage or manure, organic nitrogen disappeared more rapidly than organic carbon. As time went on vegetable and animal extracts gave increasingly similar analytical returns. On this basis Frankland could argue that an unpolluted peaty water with the analytical characteristics of a sewage-polluted water was simply old.[125]

If Frankland's water analysis was simply a series of procedures for confirming conclusions made on other grounds, what, one may ask, justified its existence?

If we assume that Frankland intended water analysis simply to supplement other types of knowledge and not to provide the sole basis for deciding whether or not a water was safe, then there are no grounds for convicting him of circular reasoning. Different classes of water sources -- lakes, springs, rivers -- might quite

---

[124] RCWS, Evidence, Q 6291.

[125] RPPC/1868/6th Report, pp. 6-8.

legitimately be evaluated according to different sets of chemical standards. If on the other hand water analysis was to be the sole basis for decision, then Frankland's interpretive system was a deception, for analytical returns, as we have seen, were not the principal basis on which Frankland decided a water's quality. In regard to Frankland's views in 1868 it is impossible to answer this question. By 1873, perhaps as a result of experience accumulated during five years' use of his system, his position had become clear: chemical analyses were to be considered only in light of information about a water's source. River waters could not be compared with lake and spring waters.[126]

Further, the inherent inability of chemical analysis to provide an exclusive basis for determining water quality did not mean that analytical information was not useful. In Frankland's hands water analysis was less a basis for sanitary decision making than a tool for doing science. Possessing a reliable method of tracing minute changes in the concentrations of chemical species in river water, Frankland learned a great deal about the chemical dynamics of rivers -- about how water composition responded to seasonal changes, to floods, to droughts, and to changes in the composition of the

---

[126] A.H. Allen, "On some points in the analysis of waters, and the interpretation of results," The Analyst 1(1877): 61-65; G.W. Wigner, "Water analysis," The Analyst 1(1878): 208-15; E.A. Parkes, A Manual of Practical Hygiene, 6th ed., by F.S.B. Francois de Chaumont, 2 vols. (New York: Wm. Wood, 1883), Vol. 1, pp. 65-75, 99, 101-6; C.B. Fox, Sanitary Examinations of Water, Air, and Food. A Handbook for the Medical Officer of Health (Philadelphia: Lea & Blakiston, 1878), pp. 155-60, 170-71.

river's bed.[127] Always there remained the possibility of finally fulfilling the purposes for which Farr had created the post of water analyst to the Registrar General: someday the vast accumulation of analytical data on London's water might supply the basis for a correlation of disease with water purity. Finally, Frankland's analyses could show probable improvement or deterioration of the water supply. They could not, of course, show when purification was complete simply because the curve of purification was necessarily asymptotic -- no one knew where the axis of purity lay, nor even the function that described the curve approaching it. While Frankland never lost his conviction that London needed an unpolluted water supply, he continued to work for the improvement of the existing supply. In his monthly and annual reports during the 1870s and 1880s he frequently called upon the companies to improve their water-handling techniques and developed analytical procedures -- microscopical examination, turbidity determination -- designed to monitor the efficiencies of the companies' filters. Even in the spring of 1868, during the height of the agitation for a new water supply, Frankland pointed out that substantial improvements could be made in the existing supply.[128]

From Frankland's perspective, therefore, water analysis, if never conclusive, was worthwhile and provided useful information.

---

[127] Frankland, "On some points in the analysis of potable waters," pp. 843-46.

[128] RCWS, Evidence, QQ 6235-36, 6246, 6376.

But what consequences for the social relations among scientists and between scientists and society were generated by Frankland's analytical practices? The first half of this question will be considered in the next chapter; suffice to say here that Frankland's reasoning processes accentuated the context of distrust described in the second chapter. In regard to the second part of the question, however, William Odling's 1884 comments are enlightening. Odling, who at the time was working with William Crookes and Charles Meymott Tidy as an analyst for the water companies, considered Frankland's practices

> an abuse of chemistry, that a chemist . . . should state and summarise the results of his analyses in such a fashion as to make it appear that the unwholesomeness, which he really infers on other grounds, is strictly deducible from the results of his periodical chemical examinations.[129]

Odling's complaint is well-founded. It is easy to get the impression from Frankland's reports that chemical analyses are the sole basis for judging water quality.

Yet the fault was not wholly Frankland's. The public, members of the Royal Commission on Water Supply, and members of parliament wanted very badly to make this inference. When Brodie insisted that he could not tell when water-borne morbid poisons were finally destroyed he was met with a succession of pleas to give some opinion, some number of miles of flow in which this could be assured to have

---

129 William Odling, "On the chemistry of potable water," <u>CN</u> 50(1884): 206.

happened.[130] Experts were to be providers of certainty. We may recall the water supply commission's statement that water quality matters "by no means admit of the satisfactory kind of treatment we are in the habit of expecting from the modern advanced state of physical science."[131] They desired scientists to deal with the question in terms of absolute categories, to mark clearly the border between safe and unsafe. As we have seen, the leading analysts of the late 1860s, no matter which side they took on the London water controversy, accepted their inability to offer more than approximations of the behavior of morbid poisons in water.

But besides requiring of its scientists the ability to make absolute decisions, the public also assumed that when chemists justified their advice in the language of chemistry their statements were based on some set of chemical tests capable of distinguishing impure water. The water supply commission, for example, found it difficult to comprehend how Frankland could on the one hand assert the inability of chemistry to detect morbid poisons and on the other hand speak in chemical terms about greater or lesser degrees of purity. They assumed the existence of a definite relationship between chemists' advice and chemists' tests.

There are some glaring examples from the 1870s of how unfounded that assumption was. In 1879 Meymott Tidy related a case in which

---

[130] RCWS, <u>Evidence</u>, QQ 6988-91.

[131] RCWS, <u>Commission report</u>, p. lix.

a chemical witness had condemned as "horribly polluted" a water which that chemist's own analysis had shown to be quite pure. When this inconsistency was pointed out the witness admitted his astonishment but continued to maintain that the water was nevertheless polluted.[132] Similarly, in 1876 M.F. Anderson reported that his own analysis showed Uppingham water safe

> and yet I have every reason for believing it to be capable of producing typhoid fever, and that in this case typhoid germs were actually present in the water I examined.[133]

We have seen Frankland doing the same thing; in Frankland's case however, the excuses were built into his system of interpreting analytical results. Numerous modern examples might be brought forward to illustrate the persistence of the assumption that scientists' opinions are based on science.

Society asked its chemical analysts to perform illusions, and Frankland, along with most of his colleagues -- Brodie being the noteworthy exception -- obliged. All sets of standards were arbitrary because the nature of morbid poisons was unknown. Frankland's arbitrary standards differed from those of his colleagues in that Frankland's were based on a water's source, while those of Letheby and Taylor were based on arbitrary chemical standards such as the presence of hydrogen sulfide or the presence of nitrogenous organic

---

[132] C. Meymott Tidy, "Processes for determining the organic purity of potable waters," J. Chem. Soc. 35(1879): 65-66.

[133] "Uppingham water," BMJ, 1876, i, p. 308.

matter.

This combination of the public's demand for certainty from its analysts and the arbitrary basis upon which their advice necessarily was given enhanced the importance of scientific evidence in deciding issues of water quality and sewage treatment, at the same time poisoning any legitimacy this scientific enterprise possessed. By the 1870s statements about water quality were exclusively political. The participation of scientists in the politics of water supply and sewage treatment had led to a condition in which the validity of a conclusion was judged -- in most cases quite justifiably -- according to the circumstances of its origin rather than on the basis of its content or the evidence supporting it. In general this context had a negative effect on the development of water science in Britain. Scientists, linked by money, prestige, or institutions to various parties, became makers of excuses rather than seekers of truth. The next chapter examines how this context infected experimental investigations of the self-purification of rivers.

VIII. POLITICS AND CREDIBILITY: FRANKLAND AND SELF-PURIFICATION, 1868-1881

> Oxidation was the dust thrown . . . in the eyes of those demanding reform.
> -- Percy Frankland*

The cholera of 1866 had ignored sanitary progress. It had eluded the sewage treatment operations of upstream towns, resisted the self-purifying powers of the River Lea, paid no heed to the filtration clauses of the 1852 Metropolis Water Act. The 1869 report of the Water Supply Commission resolved nothing; cholera still baffled analytical chemists and sanitarians. The "if," "when," and "how" of self-purification remained as perplexing as ever, yet also more important than ever. Public health acts passed in 1872 and 1875, and a Rivers Pollution Prevention Act passed in 1876 broadened and deepened the sanitary obligations of municipalities. Even more effective as a spur toward confronting the issue of sewage purification were lawsuits against polluting municipalities, brought by downstream neighbors or by newly established riparian authorities such as the Thames Conservancy.[1]

---

* Percy Frankland, "The Upper Thames as a source of water supply," JRSA 32(1883-84): 566.

[1] For examples of towns threatened with legal action for polluting see: J.T. Bunce, History of the Corporation of Birmingham, 2 vols. (Birmingham: Cornish, 1878, 1885), Vol. 2, pp. 127-39, 169-73; William Hope in disc. of Society of Arts "Conference on steps to be taken to insure prompt and efficient measures for preventing the pollution of rivers, Dec. 1874," JRSA 23(1874-75): 91. Also see SR 2(1875): 27 (Harrowgate), 212, 247 (Tunstall), 283 (Darlington),

Science, both in terms of concepts and institutions, was little help. As regards concepts, there was no consensus about the principles of water and sewage purification. As for institutions, the preeminent forums for water and sewage science were courts of law and parliamentary select committees. Attempts were made during the 1870s to establish institutions for cooperative and non-partisan scientific study of the water and sewage problems of towns. Between 1869 and 1876 a British Association Committee on the Treatment and Utilisation of Sewage, funded largely by towns' contributions, experimentally studied sewage treatment processes.[2] Between 1874 and 1884 the Society of Arts held frequent conferences on water supply and sewage treatment, the idea being that these should serve as clearinghouses for practical information rather than opportunities for theoretical disputation by scientists and promoters.[3]

Unfortunately, the cooperative approach to municipal sanitary problems was not fruitful. The British Association Committee spent much of its time bickering about money, while the Society of Arts conferences became simply another forum for partisan debate.

---

368 (Burnley), 368, 430 (Thames towns), 430 (Hackney).

[2] "Reports of the committee on the treatment and utilisation of sewage," B.A.A.S. Reports; 39th meeting (1869): 59-87; 40th meeting (1870): 49-75; 41st meeting (1871): 166-89; 42nd meeting (1872): 135-75; 43rd meeting (1873): 413-51; 44th meeting (1874): 200-14; 45th meeting (1875): 65-81; 46th meeting (1876): 225-42.

[3] A Society of Municipal and Sanitary Engineers, again stressing a cooperative approach, was founded in 1873. More broadly based but with similar goals was the peripatetic Sanitary Institute of Great Britain, which held its first meeting in 1877.

Opposing interests were too real to disappear beneath the banner of sanitary cooperation. Victorian water science remained adversary, polemical, and litigious. Water policy, and more importantly, water science evolved in courts of law and Westminster committee rooms where expert witnesses -- scientists, engineers, physicians -- made their livings refuting one another. A close relationship developed between the scientific discourse associated with litigation and the ostensibly disinterested discourse of scientific societies. Often the papers and discussions at the Chemical Society, the Institution of Civil Engineers, or the Society of Arts are indistinguishable in substance from debates in explicitly litigious contexts. Sometimes, as with Charles Meymott Tidy's "River Water" (to be discussed in this chapter), the apparent purpose of a paper was to establish scientific respectability, the imprimatur of journal citation, for testimony the author would give in pollution cases.[4] It is noteworthy that the most important sewage treatment paper read at the Institution of Civil Engineers during the 1870s was by a barrister, C. Norman Bazalgette. To be sure, Bazalgette's specialty was sanitary law, and he was well acquainted with arguments pro and con various modes of sewage treatment.[5] Nevertheless, the acceptance by engineers, chemists, and physicians of a barrister as an authority

---

[4] Charles Meymott Tidy, "River Water," J. Chem. Soc. 37(1880): 267-327.

[5] He was the son of the famous sanitary engineer J.W. Bazalgette. C.N. Bazalgette, "The sewage question," MPICE 47(1876-77): 105-250.

on sewage treatment symbolizes the despair of those professions, and reveals how much rhetoric had replaced science in water quality matters.

Frankland was a central figure in this scientific enterprise. In the six uncompromisingly radical reports of the second (1868-73) Rivers Pollution Commission Frankland took extreme positions on the methods and capabilities of sewage treatment, the feasibility of a virtual ban on pollution, and the inherent unsuitability of rivers as sources of water supply. Throughout the 1870s and well into the 1880s he defended these positions frequently in courts of law and in testimony to parliamentary select committees. The commission's recommendations and effluent standards were the basis of unsuccessful anti-pollution bills in 1872, 1873, and 1875. A weak bill passed in 1876. Frankland continued his campaign for the abandonment of London's river-supplied water, using his monthly reports to the Registrar General as a vehicle for undermining public confidence in the existing supply.

Frankland's activism in water politics led his critics to ignore substantive aspects of his water science. His work was termed "political," and therefore invalid. The reception of "previous sewage contamination" (PSC) is one example. Recognizing only the blatant political implications of PSC, critics refused to consider seriously its utility as a measure of past contamination. By the end of the 1870s many water analysts were relying extensively on oxidation products of organic nitrogen as measures of previous contamination, yet Frankland was rarely credited.[6]

This chapter focuses on another example of the failure of water scientists to take Frankland seriously. In 1868-69 Frankland carried out river studies and laboratory experiments which convinced him that the oxidative self-purification of rivers was a fiction. Frankland's work was rudimentary, yet substantially more thorough than anything done previously. Notably, for more than a decade the defenders of river water made no effort to show where Frankland had gone wrong. They ignored his studies or denied his conclusions, but failed to bring forward similarly detailed counter evidence until 1880.

Here again, Frankland's credibility fell victim to his activism. There was no need to disprove his self-purification studies. Frankland was portrayed as one whose ambition and fascination with his own system of analysis had gotten the better of his judgment. His integrity was publicly questioned. A showy denial of oxidative self-purification was exactly the mumbo-jumbo to be expected as a complement to undetectable, unremovable germs.[7]

This chapter examines: 1) Frankland's disproof of oxidative

---

[6] See for example J.A. Wanklyn, "Further report on the drinking water in the eastern counties," BMJ, 1874, i, pp. 87-88.

[7] Royal Commission on Metropolitan Sewage Disposal [RCMSD], 1st Report, P.P., 41, 1884, [C.-3842.], Evidence, Q 10860, for example. See also Frankland's testimony before the select committee on the Cheltenham Corporation Water Bill (House of Lords Record Office, Minutes of Evidence, House of Commons, 1878, Vol. 5, 13 March 1878, pp. 246, 276, 283).

self-purification; 2) the chief factors responsible for the repudiation of Frankland's water science; 3) the eventual challenge to Frankland's river studies and experiments by Charles Meymott Tidy in the early 1880s; and 4) the failure of their debate to lead to an understanding of the biological self-purification of rivers.

I. Self-Purification Disproved

Between the summers of 1868 and 1869 Edward Frankland proved that the oxidative self-purification of rivers was a fiction -- a bold black sub-heading in the first report of the second Rivers Commission referred to "the alleged self-purification of rivers."[8]

Frankland's first anti-self-purification studies appeared in remarks appended to a November 1868 report he and William Odling did for the Royal Commission on Water Supply on the effect of sewage on the water of the upper Thames. Tracing changes in river composition was difficult, Frankland noted. One could not follow the same water down river. The Thames grew, receiving streams whose discharge and composition could be accurately characterized as well as unmeasurable quantities of uncharacterizable ground water. Having outlined reasons why self-purification studies were of doubtful validity, Frankland presented his results.

First, Frankland considered effects of sewage from individual

---

[8] RPPC/1868/1st Report (Mersey and Ribble Basins), Vol. 1, P.P., 40, 1870, [C.-37.], p. 18.

towns. Samples were taken immediately above a sewer outfall, two miles below it -- space enough for mixing, but presumably not for self-purification -- and five miles below. The results were inconclusive, he noted; changes were simply too small, too near the limits of his analytical techniques.[9] Frankland therefore adopted another means of determining the purifying capacity of the Thames: a comparison of the actual composition of the water with the composition that would result if all the sewage of the river-basin population were diluted with the amount of water in the river. The Thames basin above Lechlade -- fairly near the head of the river -- had a population of 58,015. The river at Lechlade contained 0.133 parts/100,000 organic carbon and 0.033/100,000 organic nitrogen. Since Lechlade was so near the head springs of the Thames, Frankland assumed that these figures could provide a ratio of population to pollution in a situation in which negligible self-purification had occurred. From this ratio one could estimate the mass of organic carbon and nitrogen supplied to the river by the total population of the Thames basin (above Hampton, where the water companies drew their supplies). Dividing this mass by the discharge of the river at Hampton would give the hypothetical concentration of pollutants there if no self-purification occurred. Frankland calculated that the 834,248 people in the Thames basin above Hampton should give Thames water a concentration of 0.239/100,000 organic carbon and

---

[9] RCWS, Report, P.P., 33, 1868-69, [4169.], Appendix AX(2), p. 107.

0.059 organic nitrogen. To check this estimate Frankland calculated the organic nitrogen that ought to be in the Thames from Liebig's data on the amount of "soluble organic nitrogen" excreted per person and obtained .062/100,000, close to the .059 obtained by assuming the Thames above Lechlade representative of the river as a whole. Thus, by two separate methods Frankland obtained similar estimates of the amount of sewage that entered the Thames above Hampton.

In fact the Thames at Hampton contained .260/100,000 organic carbon, about 9% above his estimate, but only .024/100,000 organic nitrogen, 59% under the estimate based on the Lechlade ratio, 61% under the estimate based on excreted nitrogen studies. Frankland suggested the increase in organic carbon was due to input of peaty or vegetable extracts, attributed the deficiency in organic nitrogen -- the chief mark of sewage -- to the "oxidation or precipitation of sewage matter." He did not comment on the concept of self-purification.[10]

These data, showing an apparent substantial reduction of a critical contaminant would appear to support self-purification, not to disprove it. It is a little surprising to find Frankland, barely

---

[10] The same numbers could be manipulated to give the impression of an enormous amount of purification. See ibid., p. 108, for Odling's calculation. See also [William Pole], "The water supply of London," Quarterly Review (American ed.) 127(1869): 242. Odling calculated reduction by including urea, which became inorganic primarily while sewage was still in sewers, and which accounted for 80% of excreted nitrogen. Frankland calculated reduction after the sewage left the sewers and therefore did not include urea in his calculation. Odling calculated a 95% reduction by including urea.

three months later, denying that self-purification existed. To be sure, his denial was only implied. He began his annual report on the London water supply for 1868 with the statement:

> It has been frequently asserted, but without proof, that the noxious organic matter of sewage, when discharged into a river of considerable magnitude, is entirely destroyed by oxidation after a flow of a few miles.[11]

Next, Frankland quoted Brodie's famous assertion that organic matter was too stable to be destroyed by anything other than strong oxidizers; that dissolved oxygen in rivers could therefore have only negligible effect on organic pollutants. This was, in fact, the statement Frankland had denied in June 1866.[12] Frankland went on to claim "my analyses of Thames water during the past year entirely confirm Sir. B. Brodie's opinion," following this immediately with a qualification: "they leave no doubt that, although oxidation does take place to some extent, a considerable proportion of the animal organic matter contained in the sewage of Oxford . . . etc., reaches Teddington [near Hampton and marking the division between the Thames estuary and the Thames as a river] in an unoxidized condition."[13]

Frankland was employing a particular concept of self-purification. He was not interested in the rate of disappearance or in the mechanisms by which it occurred, but in whether sewage matter had

---

[11] RCWS, Appendix AJ, p. 90.

[12] Select Committee on the Thames Navigation Bill, <u>Evidence</u>, <u>P.P.</u>, 12, 1866, (391.), QQ 5170-71.

[13] RCWS, Appendix AJ, p. 90.

completely disappeared. The data summarized in the annual report were on the composition of London water and showed only that organic matter was sometimes in the water, not how much had originally been there. Nor did the data on the upper Thames confirm Brodie. While these showed a "considerable" portion of organic matter remained, they also showed that a more considerable portion had disappeared.[14]

In the following months Frankland obtained evidence which more legitimately supported the conclusion he had made in the 1868 annual report. While touring Lancashire with the rivers commission, Frankland found three rivers apparently well suited for an investigation of self-purification: polluted rivers with relatively long stretches in which no new pollution or major tributaries were received. Between March and June 1869 Frankland took five sets of data on these rivers -- three sets on an 11 mile stretch of the Irwell, one each on a 13 mile stretch of the Mersey and a 13 mile stretch of the Darwen. In these five cases organic carbon was reduced in three, by 4.5%, 29.6%, and 20.8%; and increased in two, by 10.1% and 21%. Organic nitrogen was reduced in a different three, by 11.8%, 13.2%, and 17.9%; and increased in two, by 21.5% and 0.8%. Again Frankland qualified his remarks on the data with the observation that following a given body of water downstream was impossible.

[14] Odling later maintained that the studies proved self-purification (Odling in RCMSD, Evidence, QQ 16485-87). Insignificant improvements could also be taken as evidence of enormous self-purification (L. Flower in correspondence on Percy Frankland, "The bacterial purification of water," MPICE 127(1896-97): 139).

Still he felt sufficiently confident in his results to note that far from purifying themselves, relatively little oxidation took place in Letheby's prescribed distance of 12 miles.[15]

Frankland supplemented these observations with laboratory simulations of river flow. A 1:9 sewage-water mixture of known composition was agitated and "freely exposed to air and light every day by being syphoned in a slender stream from one vessel to another, falling each time through three feet of air."[16] Samples were taken for analysis at 96 hours and again after 192 hours, at which time organic carbon had only decreased 25%, organic nitrogen 33%. To Frankland these results suggested that a stream contaminated with 10% sewage, flowing 1 mile per hour, would still be considerably polluted after flowing 192 miles.

Finally, Frankland examined the rate at which dissolved oxygen disappeared, a test similar to the modern biochemical oxygen demand determination. Frankland divided a 5% sewage-water mixture of known composition into several stoppered bottles. These were kept at 17 C and exposed to diffuse light, one bottle being opened each day to determine the dissolved oxygen remaining in the water. After seven days 96% of the oxygen was gone, yet no complete oxidation (as measured by increased CO_2) had occurred. Frankland realized that preventing oxygen diffusion from atmosphere to water made the experiment an unrealistic representation of a river, but calculated

[15] RCWS, Evidence, QQ 3894-97.

[16] RPPC/1868/1st Rept., p. 20.

that even if oxidation had proceeded at its initial rate, only 62.3% of the sewage would have been completely oxidized in 192 hours.[17]

Frankland concluded that however one measured self-purification, it was clear that

> oxidation of organic matter in water proceeds with extreme slowness, even when the sewage is mixed with a large volume of unpolluted water, and . . . it is impossible to say how far such water must flow before the sewage matter becomes thoroughly oxidized. . . . there is no river in the United Kingdom long enough to effect the destruction of sewage by oxidation.[18]

Again Frankland quoted Brodie. Apparent self-purification was due, Frankland suggested, simply to the subsidence of solid matters.

Frankland's studies of river self-purification were primitive. Still, by contrast with what had come before, they were revolutionary simply because they were quantitative and experimental. It might be expected that Frankland's elevation of self-purification from casual observation to component of exact science would prompt others to refine these initial studies into a detailed understanding of the chemical dynamics of English rivers. The studies did serve as a model, but for the study of polluted Massachusetts rivers

---

[17] Ibid., pp. 20-21. These experiments appear to have been done quite early. J.J. Day speaks of them in a November 1867 letter (Imperial College, Armstrong Papers, C 241, J.J. Day to H.E. Armstrong, 23 November 1867). This is before Frankland was appointed to the Rivers Commission. It is therefore possible that the experiments were undertaken with reference to the controversy on the source of the London water supply.

[18] RPPC/1868/<u>1st Rept.</u>, p. 21.

by chemists and engineers of the Massachusetts State Board of Health. In Britain Frankland's studies were neither refined nor refuted until the early 1880s.[19] Instead they were rejected outright, or ignored.

Reasons for rejection may be found in inadequacies in the studies themselves. Frankland identified sources of complication inherent in self-purification studies but made only a half-hearted attempt to correct for these complications. His Thames studies were corrected for changes in river volume; for the northern rivers he assumed that volume had not changed in two cases and that it had doubled in the third. It was not clear that the few series of experiments on bottled sewage simulated critical conditions of streams. Frankland appeared to have hedged his bets by selecting for self-purification studies three of the worst rivers in Britain, rivers his colleagues would not have expected to present much evidence of self-purification. The Irwell was the worst of these, with

---

[19] With one exception (C.E. Austin, On the Cleansing of Rivers [London: R.J. Mitchell & Son, 1872], pp. 28-34). I will not discuss Austin's work because it was ignored by Frankland and his contemporaries. For Massachusetts: J.P. Kirkwood, "A special report on the pollution of rivers; an examination of the water-basins of the Blackstone, Charles, Taunton, Neponset, and Chicopee Rivers; with general observations on water-supply and sewerage," 7th Annual Report of the State Board of Health of Massachusetts, Jan. 1876 (Boston: Wright and Potter, 1876), pp. 23-174; idem, "Rivers pollution: an abstract of the Report to the State Board of Health of Massachusetts," Van Nostrand's Eclectic Engineering Magazine 16 (1877): 146-58; W.R. Nichols, "On the present condition of certain rivers of Massachusetts together with considerations touching on the water supply of towns. A report to the Massachusetts State Board of Health," 5th Annual Report of the State Board of Health of Massachusetts, Jan. 1874 (Boston: Wright and Potter, 1874), pp. 61-152.

ten times as much organic matter as the Thames at Hampton and all the inorganic rubbish industrial Manchester had to offer besides. Nor could the Irwell legitimately be termed a sewage-polluted river, since Manchester and most other northern towns were not yet water-closeted to a significant degree. Clearly, therefore, Frankland's rivers were not representative.[20]

Finally, the data showed purification was occurring. The conclusion that self-purification did not occur required a peculiar understanding of that concept. Concerned primarily with drinking water safety, Frankland saw self-purification as the reaching of an end, rather than as an ameliorative natural process. Thus, Frankland's colleagues could regard these studies not as examples of a new standard in water investigation, but as so much scientific rhetoric put forth to secure non-river water supplies.

More important, however, is the response to Frankland's water science in general. The disproof of self-purification was one small element of Frankland's water science, a network of methods of analysis, water quality standards, models of morbid poisons, beliefs about the abilities of technology, and rhetorical techniques for communicating with the public. For Frankland it was a system conceived to foster prosperity and health; for his critics it seemed designed only to serve the aggrandizement of Edward Frankland.

---

[20] C.M. Tidy, "River water," CN 43(1881): 113. [To be cited as Tidy, "River water II."] See also Tidy in disc. of Percy Frankland, "Upper Thames as a source of water supply," p. 447.

There was much that was controversial in Frankland's system, but criticism focused on central factors: the analytical procedures on which the rest of Frankland's conclusions were supposedly based, and on Frankland's integrity.[21] So long as Frankland's water science could be effectively challenged on such fundamental grounds there was no need to challenge such specifics as the disproof of self-purification. Roughly, during the early 1870s, Frankland's analytical procedures were his critics' main target. By about 1880 his analytical procedures had been largely vindicated, and Frankland's integrity became the focus of criticism.

II. Edward Frankland's Loyal Opposition

Almost everyone involved in water issues quibbled with some component of Frankland's system. The first (1870) report of the Rivers Commission is a catalog of the positions that prompted criticism.

Frankland cultivated the opposition of industry. Industry foolishly discarded as waste materials that could be profitably recycled. Harsh industrial effluents did not purify rivers as some had claimed. Industry had no social right to pollute, nor was pollution in its own interest. Stringent anti-pollution laws might pose

[21] For an example of the effectiveness of criticism of analytical methods in destroying a scientist's credibility see the cross-examination of William Dibdin in RCMSD, Evidence, QQ 15565-938.

a short-term hardship, but would strengthen industry in the long term.[22]

Frankland told chemists that his combustion process was the only reliable approach to estimating the organic matter in water. His proposed effluent standards would make the combustion process alone the legal means of measuring organic pollution.[23] Such legal sanction would have meant an end to water analysis for many chemists; whatever its merits, the combustion process required equipment few could afford, skills few possessed.

Chemical processes of sewage treatment were rejected categorically. The commission implied that such schemes were often fraudulent.[24] William Crookes was scientific advisor of the Native Guano Company, the specific target of the commission's scorn; Crookes' Chemical News opposed all aspects of Frankland's water science.

Sanitarians found the commission ambivalent toward local sanitary reform. While it considered sewage treatment necessary, the commission rejected the traditional justifications of sewage treat-

---

[22] RPPC/1868/1st Rept., pp. 9, 42, 158, 101. See also RPPC/1868/5th Rept. (Pollution arising from mining operations and metal manufacturers), P.P., 33, 1874, [C.-951.], p. 15. Compare these views of the second commission with the views of the first commission (RPPC/1865/3rd Rept. (Rivers Aire and Calder), P.P., 1867, 33, [3850.], pp. li, xiv). An able statement of the manufacturers' point of view is E.C. Potter, The Pollution of Rivers: by a Polluter (Manchester: Johnson and Rawson, 1875), p. 6. See also Select Committee on River Pollution, Lords Papers, 9, 1873, (132.), Evidence, QQ 180, 229, 250-57, 293, 342-47.

[23] RPPC/1868/1st Rept., pp. 112-17, 14.

[24] Ibid., pp. 55-58; RPPC/1868/2nd Rept. (The A.B.C. Process of Treating Sewage), P.P., 40, 1870, [C.-180.], in passim.

ment. Chadwick had maintained that sewerage and sewage treatment would lead to better health, and that sewage recycling would pay the cost of such improvement and was also necessary in order to conserve the nation's capital of fertilizer. The commission found towns on polluted rivers no less healthy than towns on unpolluted rivers.[25] It regarded sewage recycling as unprofitable, downplayed the preservation of plant nutrients.[26] Worse, according to the commission, no method of sewage treatment removed agents of disease reliably. No water that had received treated sewage was acceptable for domestic use.[27] Apparently, the only purpose of sewage treatment was to satisfy the sensibilities of downstream landowners.[28]

It is not surprising, therefore, to find Frankland's water science continually attacked during the 1870s. A sampling of criticism Frankland received during the 1870s shows two things: 1) a failure to consider seriously his philosophy of water quality evaluation;

---

[25] RPPC/1868/<u>1st Rept.</u>, p. 43.

[26] See James Blackburn to <u>Times</u>, "Sewage irrigation," <u>Times</u>, 27 Dec. 1871.

[27] RPPC/1868/<u>1st Rept.</u>, p. 112.

[28] It could be added that Frankland failed to court the support of anglers, the most politically influential group working for anti-river pollution legislation. Anglers were concerned chiefly with pollution from mines. Frankland was uncharacteristically lenient toward mine pollutions, believing that there was no technically effective means of avoiding them. The acidity and alkalinity standards in Frankland's proposed effluent regulations, though professedly included in the interests of the fish, would have allowed rivers to become too harsh for fish to survive. See <u>The Field</u>, 1865-1875; RPPC/1868/<u>5th Rept.</u>, pp. 2, 39, 49-50; S.C. River Pollution, 1873, <u>Evidence</u>, QQ 15, 136, 500-1.

and 2) an unconscious alliance among his chief opponents. Though motives for attacking Frankland varied, the arguments used against him remained the same.

Wanklyn's Objections, 1868 onward

The sharpest thorn in Frankland's side was James Alfred Wanklyn. His criticisms were the longest sustained, most bitter, yet most substantive. Between 1868 and about 1877 the question of whether Frankland's or Wanklyn's analytical process was the better dominated discussions of water quality evaluation in Britain. In Wanklyn's view the dispute concerned which process -- his "ammonia" or Frankland's combustion -- better measured the harmfulness of a water. As we saw, Frankland regarded such a quest as futile; analysis was to shed light on the history of a water. All the same, the debate was largely conducted on Wanklyn's traditional perception of the issues, it being assumed that the process that more accurately determined organic elements gave a better measure of harmfulness. Analysts waited for a victor, feeling that controverted questions such as self-purification would be automatically answered with the resolution of the Frankland/Wanklyn dispute.[29] The criticisms of Wanklyn are important: his continued invective damaged Frankland's credibility; his substantive objections to the combustion process

[29] Cornelius J. Fox, Sanitary Examinations of Water, Air, and Food. A Handbook for the Medical Officer of Health (Philadelphia: Lea & Blakiston, 1878), p. 58.

were used extensively by critics who had little knowledge of chemistry and whose reasons for attacking Frankland lay elsewhere.

In June 1867 Wanklyn described his new "ammonia" process to the Chemical Society.[30] Wanklyn thought albumin was the dangerous material in water because albuminoid substances putrefied rapidly. He believed a definite proportion of the albumin in water was converted to ammonia when the water was distilled with a caustic solution of potassium permanganate. The amount of this "albuminoid ammonia" was the main index of water quality.[31] Initial reaction was negative,[32] and barely half a year after its introduction Wanklyn's process appeared on the verge of being upstaged. In the lecture announcing his combustion process in January 1868, Frankland made short work of the Wanklyn process. Its results were discordant with those of the combustion process; it must, therefore, be

---

[30] J.A. Wanklyn, E.T. Chapman, and Miles H. Smith, "Water analysis: determination of the nitrogenous organic matter," J. Chem. Soc. 20(1867): 445-54.

[31] J.A. Wanklyn, "On water and its composition," SR 4(1876): 17-19; idem, "The ammonia process of water analysis," SR 6(1877): 15-16; Wanklyn in disc. of Percy Frankland, "Water purification, its biological and chemical basis," MPICE 85(1885-86): 241.

[32] W.H. Brock, "James Alfred Wanklyn," DSB 14(1976): 168-70. Dugald Campbell showed that Wanklyn's definite proportions did not exist ("A note on Messrs. Wanklyn, Chapman, and Smith's method for determining nitrogenous organic substances in water," CN 16(1867): 139-41). Many felt that a promising technique had been compromised through premature publication ("Chemical society," CN 16(1867): 7). Wanklyn's personality quickly emerged as an issue. The Chemical News review of the first edition of his water analysis manual spoke of a "sort of scientific afflatus, which forms part of the unconscious poetry of these gentlemen's natures, and which impels them to burst forth in paeans at the Chemical Society whenever one of them has a new idea" ("Rev. of Wanklyn and Chapman's Water Analysis," CN 18(1868): 151-53).

wrong.[33] Besides, pure organic substances, as opposed to the mysterious aquatic albumin, gave no definite proportion of their nitrogen as ammonia upon distillation with caustic permanganate, nor was there any reason to suppose organic substances always broke up in the same manner under the influence of such oxidizing agents.[34]

Wanklyn quickly retaliated. Frankland's own data on waters of known composition revealed errors in the combustion process sometimes larger than the amount of organic matter in the water.[35] He listed flaws in Frankland's procedure. First, if combustion were to measure organic nitrogen accurately, all nitrates had to be destroyed during the evaporation that preceded actual combustion. E.T. Chapman, Wanklyn's co-author, found nitrates in residues of samples evaporated according to Frankland's instructions. Second,

---

[33] E. Frankland and H.E. Armstrong, "On the analysis of potable waters," J. Chem. Soc. 21(1868): 99-100. See also Fox, Sanitary Examinations, p. 59, for another example of the rejection of Wanklyn's method on the grounds that its results were discordant with Frankland's. In fact Frankland's argument here is not quite so high-handed as at first seems the case. Frankland was attempting to measure all organic nitrogen; Wanklyn to measure some part of it. Finding that sometimes Wanklyn's process gave numbers larger than his, other times smaller, Frankland concluded, on the basis of about 100 comparisons, that the hypothesis that Wanklyn got a definite proportion was wrong.

[34] Frankland and Armstrong, "Analysis of potable waters," pp. 97-98; Campbell, "Note on Wanklyn's process," pp. 139-41.

[35] This is not so serious as it sounds. A 100% error was still minor in comparison with other processes. Because it measured an artificial and arbitrary substance, albumenoid ammonia, Wanklyn's process was not susceptible to this criticism.

any volatile organic materials would be lost during evaporation and efforts to insure total destruction of nitrates -- as the addition of more sulphurous acid, for example -- might increase volatilization.[36]

The points were well taken. Wanklyn had identified real problems in Frankland's procedures. Other critics of Frankland diligently echoed Wanklyn's objections during the early 1870s; Frankland devoted considerable attention to solving these problems.[37] But substantive criticism was not Wanklyn's style. The _ad hominem_ attacks made on him during the early days of the ammonia process, the ruthlessness and arbitrariness of Frankland's rejection of the process played upon Wanklyn's own paranoia and convinced him that he was a victim of a conspiracy of more established chemists.[38] Chemical objections became secondary, as Wanklyn portrayed Frankland as head of an intellectually empty scientific elite which parasitized the youthful and brilliant, i.e., Wanklyn. Significantly, Wanklyn's

---

[36] J.A. Wanklyn, E.T. Chapman, and Miles H. Smith, "Note on Frankland's analysis," _J. Chem. Soc._ 21(1868): 152-60; "Chemical society," _CN_ 17(1868): 79-80, 127-28; E.T. Chapman to _The Builder_, "Water analysis," _The Builder_ 27(1869): 324.

[37] J.J. Day to H.E. Armstrong, 21 June 1868, 27 March 1869, in Imperial College, Armstrong Papers, C 245-46; E. Frankland, "On some points in the analysis of potable waters," _J. Chem. Soc._ 3rd series 1(1876): 88-89.

[38] Brock, "Wanklyn," pp. 168-70; Colin Russell, et. al., _Chemists by Profession, the Origin and Rise of the Royal Institute of Chemistry_ (Milton Keynes, U.K.: The Open University Press/The R.I.C., 1977), pp. 116-17; Bernard Dyer, _The Society of Public Analysts and other analytical chemists: Some Reminiscences of its first 50 years_ (Cambridge: The Society, 1932), p. 13.

portrayal meshed with the image of Frankland perpetrated by critics grinding other axes -- Crookes, opponent of Frankland's views on sewage treatment, and Henry Letheby and Charles Meymott Tidy, defenders of the London water companies.[39]

Not until 1875 could Wanklyn offer proof that Frankland's system of analysis was not merely unreliable, but dishonest. The final report (1874) of the Rivers Commission included data showing recent improvements in the accuracy of the combustion process. Frankland had analyzed solutions of quinine sulfate of known strength and obtained results within a few thousandths of a gram. On checking this proof Wanklyn discovered that Frankland had used the wrong molecular weight for quinine sulfate and had therefore miscalculated the amount of organic nitrogen in each sample. Calculated and measured levels of organic nitrogen were actually far apart. Triumphantly, Wanklyn wrote to Chemical News of the "happy mischance" that showed that the combustion process gave results "in accordance with the expectation of the analyst rather than with the real composition of the sample."[40] The implication is clear. Frankland had doctored his results to make combustion look good, but done a poor job of it.

Frankland had an explanation. The tests had been done by William Thorp, jr., chief of the Rivers Commission lab. Thorp had reported to Frankland the amounts of organic nitrogen calculated

---

[39] J.A. Wanklyn, "Dr. Frankland's researches," SR 8(1878): 174-75.
[40] J.A. Wanklyn, "The rivers commission," CN 32(1875): 268-69.

and measured for three solutions of differing strengths. From these numbers Frankland had calculated, using the wrong molecular weights, the amounts of quinine sulfate Thorp must have begun with. The error was therefore only clerical.

Wanklyn, without saying why, simply refused to accept Frankland's explanation. The explanation was "curious," the preponderance of evidence was still against Frankland, and "the sinister correspondence between expectation and result still remained."[41] No longer was there pretence that the disagreement was merely scientific. It is unlikely that any explanation would have satisfied Wanklyn; in his view Frankland was a liar. And Frankland's lying was on two levels. First, Frankland had lied about the reliability of the combustion process. As Wanklyn presented it, that deception had hitherto succeeded; now, however, all the accumulated results obtained by combustion would have to be re-examined. Further, if Frankland were guilty of deception in one area of science, he might be guilty in others. Thus, not only were other components of Frankland's water science products of faulty methods, but they might also be wilful deceptions for various purposes.[42]

By the early 1870s Wanklyn had attacked most aspects of Frankland's water science. Wanklyn had allied with the London water

---

[41] E. Frankland, "The rivers commission," CN 32(1875): 279-80; W. Thorp, "The rivers commission," CN 32(1875): 280; J.A. Wanklyn, "The rivers commission," CN 32(1875): 290, 312-13.

[42] J.A. Wanklyn, "On the determination of organic matter in potable water," CN 36(1877): 42-43.

companies. In 1867 he had presented himself to the Water Supply Commission as a radical scientist, whose new found precision would reveal clearly the subtle dangers of Thames water and show self-purification to be an illusion. Frankland turned out to be more radical and by 1868 Wanklyn had declared Thames water satisfactory.[43] In 1872 he ridiculed the concept of previous sewage contamination and accused Frankland of using it for political purposes.[44] In 1872 and again in 1879 Wanklyn began programs of alternative analyses of the London water supply. The government, by employing a chemist who defiantly used "illusory and defective methods," had shirked its duty; Wanklyn would selflessly perform that duty.[45] Not surprisingly, the water Frankland condemned Wanklyn found satisfactory. Wanklyn repeated the analyses of Cumberland and Welsh waters examined by the Water Supply Commission, and found them no better

---

[43] Compare RCWS, Evidence, QQ 5401, 5446-51, 5482-85; with RCWS, Appendix AK, p. 95.

[44] Wanklyn's complaint was that in 1867, when Frankland had been actively campaigning for a northern (Cumberland or Wales) water supply for London, he had listed PSC of the water of northern reservoir-supplied towns. After the 1869 verdict of the RCWS that northern supplies were unnecessary, Frankland had dropped these data from his reports, and had begun championing the Kent Company, whose water he had condemned in 1867 and 1868 (J.A. Wanklyn, "The Registrar General's reports on the London water," CN 25(1872): 169-70). See also "London drinking water," BMJ, 1872, i, p. 671, which from its similar tone is probably also Wanklyn's work.

[45] Wanklyn's reports are in PRO MH 29 3-7. See also J.A. Wanklyn, Water Analysis, 11th ed. (London: Trubner, 1906), p. 210; "Reports on the examination of Thames water," CN 47(1883): 23; J.A. Wanklyn, "Report on the condition of the water supply of London," CN 26(1872): 239, 287; "London drinking water," BMJ, 1872, ii, p. 549.

than the Thames.[46] He said filtration removed all parts of sewage, including typhoid germs, Frankland said it did not.[47] Wanklyn accepted dilution as real purification.[48] Finally, as W.H. Brock has pointed out, Wanklyn never accepted the germ theory, even though he occasionally claimed that his process measured germs. To his death in 1906 Wanklyn viewed germs as a Frankland trick.[49]

It is doubtful that most chemists and sanitarians believed all the nasty things Wanklyn said about Frankland.[50] Nevertheless it was hard to avoid facing the issue as Wanklyn presented it. As a crusader for the abandonment of river water, Frankland was active politically. As we have seen, the political aspects of Frankland's water supply philosophy were inseparable from his methodology, analytical formats, or conceptions of morbid poisons. Frankland and others had dealt shabbily with Wanklyn during the early days of

---

[46] J.A. Wanklyn to BMJ, 1871, i, p. 153.

[47] Wanklyn in disc. of Byrne, "Experiments on the removal of organic and inorganic substances in water," MPICE 27(1867-68): 13; Wanklyn in disc. of Jabez Hogg, "River pollution, with special reference to impure water supply," JRSA 23(1875): 591; Wanklyn, "Water supplies," SR 8(1878): 60. On filtration of typhoid germs see Wanklyn, "On the purification of drinking water by the process of filtration," SR 4(1876): 391-92; which is a reply to J.W. Tripe, "Typhoid germs and their alleged destruction," SR 4(1876): 387-88. See also Tripe's reply (SR 4(1876): 419-20).

[48] J.A. Wanklyn, Water Analysis, 6th ed. (London: Trubner, 1884), pp. 10-11; Wanklyn in disc. of Percy Frankland, "Upper Thames," p. 432.

[49] Brock, "Wanklyn," p. 169; Wanklyn in disc. of Percy Frankland, "Water purification: biological and chemical bases," p. 241.

[50] The vast majority of analysts used Wanklyn's process, however. It was significantly easier, quicker, and cheaper to run than Frankland's. This fact doubtless had an impact on the Wanklyn-Frankland controversy. See "The battle of the waters," SR 8(1878): 52-53.

the ammonia process. What was difficult to resist was the implication Wanklyn tried to draw from these facts, that these personal and political contexts determined scientific validity. Thus, to show political implications of components of Frankland's water science, or to demonstrate Frankland's unfair treatment of Wanklyn, was a sufficient basis for concluding Frankland's water science invalid.[51] Similarly, to suggest that Wanklyn was persecuted implied that his process was valid. It is clear that Wanklyn had successfully discarded the evidentiary bases of scientific reasoning. Instead, verity was to be determined by the scientist's success in portraying himself as a political martyr. Wanklyn's perspective towards the water analysis debate was widely accepted during the early 1870s. Because the business of water analysis involved trying to measure an unknown morbid poison, no evidentiary bases for deciding who was right existed and the alternative basis for decision-making Wanklyn was promulgating was therefore particularly attractive. At the same time, belief that radical water supply politics and self-aggrandizement were the sole bases of Frankland's water science, was immensely attractive to Frankland's other critics.

---

[51] Ibid., p. 52. See also "The water supply of London," SR 4(1876): 55, 216. Here it is suggested that Frankland's opposition to river water is a result of his method.

William Crookes, 1870 onward

As editor of Chemical News, William Crookes was a powerful opponent for Frankland. In 1869 Crookes had championed Frankland's new perspective on water quality evaluation and advocated the combustion process as the best means of water analysis.[52] Growing differences over sewage treatment technology, anti-pollution legislation, and water analysis priorities caused a rift between the two in the early 1870s; by 1872 Crookes was repudiating Frankland's water science wholesale. As with Wanklyn, Crookes' main objections were the unreliability of the combustion process and the issue of Frankland's competence and integrity as a government advisor. Unlike Wanklyn, these complaints were not the basis of Crookes' criticism, however; they were merely the most effective arguments available. Notably Crookes also took the side of the London water companies against Frankland.

The main area of disagreement between Crookes and Frankland was the viability of patent sewage treatment processes in which assorted chemicals were added to sewage to purify it and precipitate sludge fertilizer simultaneously. Frankland believed such schemes were unworkable and thought they were often downright fraudulent. Crookes thought they were necessary and was involved -- as director, publicist, chemical advisor, salesman, and even for a time as chairman -- with the most notorious of these, the ABC process of the Native

---

[52] "Dr. Letheby on the methods of water analysis and on 'previous sewage contamination,'" CN 19(1869): 231-32.

Guano Company. In 1870 Frankland made an example of that very process, hinting that Native Guano tried to fool its manure customers by sending trial samples that were adulterated with artificial fertilizer and that it tried to fool his investigators by diluting effluent to suggest a higher degree of purification than was in fact the case.[53]

During the 1870s it was Chemical News editorial policy to advocate precipitation schemes and at the same time to portray land treatment -- Frankland's alternative -- as a rarely practicable technique advocated by muddle-headed idealists. As filler Crookes used some of the more outrageous irrigationist testimony given to parliamentary select committees in the mid 1860s.[54]

A second basis for opposition to Frankland was Crookes' advocacy of the British chemical industry. Crookes believed the standards for effluents Frankland had proposed were arbitrary, harsh, and unfair to the British chemical industry. He rejected Frankland's contention that industrial wastes could be used profitably if only industry chose to do so.[55] Finally, Crookes objected to Frankland's

---

[53] RPPC/1868/2nd Rept.; E.A. Fournier D'Albe, The Life of Sir William Crookes (London: MacMillan, 1923), pp. 257-70.

[54] CN 28(1873): 121-22, 191, 216; 29(1874): 63, 124, 156, 166.

[55] "A bill to amend the law relating to public health," CN 25 (1872): 145; "Rev. of Higgins, The Law relating to the Pollution and Obstruction of Water-Courses," CN 35(1877): 274; "The pollution of rivers bill," CN 31(1875): 221; "The pollution of rivers bill," CN 28(1873): 219-20, 207-9, 37-39; Crookes in Evidence to S.C. Rivers Pollution, Q 398.

attempt to dominate water analysis. The time, expense, and expertise required for the combustion process would tend toward the centralization of water analysis. The rank and file of analytical chemists who made up the Chemical News constituency had rejected combustion. It was important that water analyses be compatible with one another; therefore in the name of both public health and professional health, Crookes suggested that combustion results be banned from public documents.[56]

As was the case with Wanklyn's objections, these substantive bases of Crookes' criticism were points well taken. By contrast, Crookes used highly unsavory tactics to discredit Frankland's judgment. Crookes was an able polemicist and rarely neglected opportunities to make scornful asides about Frankland's water science. A review of Pettenkofer's book on cholera or of an annual report by the Massachusetts State Board of Health was occasion to snipe at PSC or the Rivers Commission.[57] Moreover, Crookes cleverly employed ambiguities in etiological theory to attack Frankland. If germs best discredited Frankland, germs were causes of diseases. Where specific putrefying nitrogenous chemicals were more effective, these were causes of disease. By 1872 Crookes had decided that the com-

---

[56] "Water analysis," CN 25(1872): 157; "A bill to amend the law relating to public health," p. 145. Compare the former with "Water analysis: statement of the results," CN 3(1861): 285, 315.

[57] "Rev. of Pettenkofer, Cholera: How to prevent and resist it," CN 31(1875): 139-40; "Rev. of the 4th Annual Report of the Massachusetts State Board of Health," CN 27(1873): 311-12.

bustion process was unreliable while the ammonia process was reliable, thereby reversing his 1869 position. For the next five years he championed the ammonia process, sympathized with Wanklyn's outrage at the chemical establishment, and offered Wanklyn space in Chemical News to vent his spleen. The first edition of Wanklyn's Water Analysis had provoked Crookes' most extreme excoriation, the third and fourth (1874 and 1876) elicited his highest praise.[58]

A good example of Crookes' tactics is an 1873 editorial on the effluent standards in the rivers pollution bill then under consideration. Here Crookes maintained that combustion was not only unreliable, but retrograde to the progress of etiological theory. Ignoring Frankland's view that combustion analyses gave valuable information about the kind of pollution a water had received, Crookes contended that "albumenoid ammonia" was far more relevant than "organic nitrogen" because it was only some nitrogenous materials, such as albumen and gelatine, which

> enter readily into decomposition and yield products highly offensive. Others, also nitrogenous, are much more permanent, and, even if decomposed, yield no products injurious to public health; we may instance

---

[58] "Rev. of Wanklyn's Water Analysis," CN 18(1868): 151-53; 29 (1874): 272-73; 34(1876): 204-5. Crookes appears to have been slightly taken aback when Tidy, the defender of the London water companies, decided in 1878 that Frankland could be discredited more effectively by accepting his combustion process and highlighting the irrational (to Tidy) principles with which Frankland interpreted combustion results (see below). Within three months, however, Crookes had fallen into line with Tidy, agreeing that Wanklyn was not the best means for bringing down the Franklandian enterprise. Compare "Rev. of Tidy, The London Water Supply . . . during the last 10 years," CN 37(1878): 100-1; with "Rev. of Water Supply of the State of New Jersey," CN 37(1878): 248.

urea.[59]

The combustion process offered no means of distinction, while the ammonia process did since it measured only the nitrogen of these albuminoid materials.[60] Crookes went on:

> To group under one head two bodies so utterly dissimilar in their sanitary bearings as urea and albumen is to violate the fundamental laws of sanitary classification.[61]

Use of the combustion process in official reports was therefore "a reproach to modern science no less than a misfortune to the nation."

It is clear that Crookes understood very well -- perhaps better than anyone else in Britain -- the philosophy underlying Frankland's analytical procedures. Since sewage-contaminated water might contain resistant, undetectable disease germs, water analysis must be directed toward discovering whether water had been polluted by sewage. Crookes' understanding is evident in the 1869 editorial considered in the preceding chapter. Indeed, Crookes had been one of the founders of the germ theory in Britain. Yet here Crookes employed a much older etiological theory with its complementary justification of traditional water analysis. The purpose of analysis was to measure morbid poisons, and Wanklyn's process purportedly did this; Frankland's did not.

---

[59] "A bill to amend the law relating to public health," p. 145.

[60] Ibid. See also "The pollution of rivers bill," CN 28(1873): 38.

[61] "A bill to amend the law relating to public health," p. 145.

Crookes had not abandoned his germ theory, however. When the rhetorical context demanded, germs replaced the old zymotic theory as the rationale for criticizing Frankland's views. Such was the case with sewage irrigation. Frankland believed the passage of sewage through an oxidizing layer of soil would remove most of the potential for harm from sewage. But if microorganisms caused disease this could not be, Crookes argued: "'intermittent downward filtration' [Frankland's adaptation of sewage irrigation which reduced the area required] must be powerless to deal with sewage containing the germs of disease, and germ-destroyers, such as carbolic and cresylic acids, are imperatively required in the treatment of sewage."[62] Sewage irrigation, Frankland's other recommended mode of treatment, was also unsafe: germs might cling to the crops grown on sewage farms.[63] Chemical treatments, such as the ABC process, were the only reliable means of killing the germs in sewage.

In regard to the London water supply, Crookes' opposition to Frankland developed more slowly. In 1872 he criticized Frankland's public relations techniques, although in connection with the ABC process rather than with the London supply. In 1873 he described self-purification as an important safeguard of water supplies. In 1875 he called the Registrar General's water reports, and specifically

---

[62] "Abstracts from Gazzeta Chemica Italiana," CN 29(1874): 37.

[63] "Rev. of Sewage, suggestions for its utilisation," CN 28(1873): 193; "Rev. of Downes, How to avoid Typhoid Fever and related Diseases," CN 34(1876): 181; "Rev. of 8th Annual Report of the Massachusetts State Board of Health," CN 37(1878): 60.

PSC, "partly rational, but partly foolish, exaggerated, and sensational."[64] After 1877 he published the analyses of the metropolitan supply that Tidy did for the water companies, agreeing with Tidy that the Rivers Commission's condemnation of the supply was unjustified because the commission had included no medical members.[65] In 1880 Crookes himself became (with Tidy and William Odling) one of the companies' water analysts. Thereafter Chemical News frequently published attacks on Frankland's water views.

Finally, Crookes, like Wanklyn, made direct assaults on Frankland's credibility. In 1873 he argued that the effluent standards of the Rivers Commission had been drawn up specifically to aid in the fight for a new metropolitan water supply.[66] Thus, as was the case with Wanklyn, Crookes' message was that the whole of Frankland's water science was fallacious and probably intellectually dishonest.

The London Water Companies, 1868 onward

Significant opposition to Frankland's policies came from advocates of river water supplies, and particularly from the barristers, chemists, and publicists -- Meymott Tidy was all three -- who represented the London water companies.[67] Their criticism too was broadly

---

[64] "Rev. of Pettenkofer, Cholera and how to prevent it," pp. 139-40; Crookes in Evidence, Select Committee on River Pollution, 1873, Q 420.

[65] "Rev. of Tidy, The London water supply during the past ten years," pp. 100-1.

[66] "The pollution of rivers bill," CN 28(1873): 207.

[67] Tidy became a barrister in the mid-1880s. On Tidy see DNB 19,

based, but it focused, first, on the reliability of Frankland's methods, and, increasingly throughout the 1870s, on his integrity. They depicted Frankland's ideas as unorthodox and unfounded, accused him of using devious methods to impose unrealistic standards of pure science in practical issues, suggested his motives were selfish.

From the early 1860s until his death in 1876 the water companies' chief chemical advisor was Henry Letheby, medical officer of health for the City of London, professor of chemistry at the London Hospital. Letheby was succeeded by Charles Meymott Tidy, medical officer of health for Islington, professor of chemistry and forensic medicine at the London Hospital, and, toward the end of his life, a member of the bar. In 1881 Tidy was joined by William Odling and William Crookes to form a triumvirate of consultants. Heretofore monthly analyses of the companies' supplies were made; these three quickly began daily analyses using the combustion process, hoping that in this way they could reveal the meagre grounds on which Frankland's condemnations were based.[68] From 1882 to 1884 they carried on a particularly intensive campaign against Frankland's integrity.

Once again a tacit alliance against Frankland is evident. Letheby accepted neither Wanklyn's nor Frankland's analytical

---

pp. 864-65; J. Chem. Soc. 63(1893): 766-68.

[68] On Letheby see RCWS, Evidence, Q 3861; DNB 11, p. 1010. On Crookes' involvement see Fournier D'Albe, Crookes, p. 315. See also "Reports on the examination of Thames water," CN 47(1883): 23.

processes, yet he treated Wanklyn far more gently than Frankland, and used Wanklyn's arguments to criticize Frankland's process.[69] Letheby and Tidy shared Crookes' abhorrence of sewage farming and advocacy of precipitation sewage treatment.[70]

During the early 1870s the opposition to Frankland by river water advocates echoed the views of the 1867-69 Water Supply Commission. Although its support of the existing supply had been only tepid, the commission's report was welcomed by the companies' representatives because it strongly contradicted Frankland's views on germs, and on self-purification, and chastised him for unfairly dealing with the public by using devious means (PSC, principally) to ignite anti-river water sentiment. The views of William Pole, civil engineering professor at University College, London typify the companies' objections to Frankland's water science.

Pole quickly disposed of the evidence against river water. The quantities of supposedly morbific matter in Thames water were "infinitesimal, . . . so small that we may almost doubt whether they represent anything tangible, and . . . may . . . reasonably hesitate

---

[69] Letheby in disc. of Frankland, "On some points in the analysis of potable waters," CN 33(1876): 105; "Processes of analyses of potable waters," BMJ, 1869, i, p. 379.

[70] "Processes of analysis of potable waters," p. 379; The Sewage Question: comprising a series of reports: being investigations into the condition of the principal sewage farms and sewage works of the Kingdom. From Dr. Letheby's notes and chemical analyses (London: Balliere, Tindall, & Cox, 1872); Letheby in disc. of A. Smee, "Proposed heads of legislation for the regulation of sewage grounds," JRSA 24(1875-76): 156; C.M. Tidy, "The treatment of sewage," JRSA 34(1885-86): 1127-89, 612-25.

to draw from them any practical argument against the wholesomeness of the water."[71] The condemnation of Thames water was the work of rabble-rousers at the Registrar General's Office. The analytical method on which that condemnation was based, the combustion process, was without

> scientific value or . . . practical utility. Disputed both in principle and practice, by every chemist except its inventor, professedly incapable of giving any information on the actual condition of the water, it presents irreconcilable discrepancies and anomalies in the data it pretends to furnish.[72]

Frankland's germs might exist, though no one had seen one. If they did, they would likely require conditions "incompatible with those existing in well-aerated streams."[73] Besides aeration, Pole dragged out the venerable argument that aquatic organisms would dispose of whatever it was that was dangerous:

> streams, independently of their mere chemical oxydizing influence, contain large developments of the higher animal and vegetable forms, and a still greater abundance of healthy infusorial life, all well known and powerful instruments of purification; by their constant and energetic action on organic matter present in water. Hence the persistent endurance, under such unfavourable conditions, of the poisonous vitality of these morbific atoms, if any such exist, is in the highest degree improbable.[74]

By the later 1870s initial outrage of the companies' defenders

---

[71] [Pole], "The water supply of London," Quarterly Review (American edition) 127(1869): 242.

[72] Ibid., p. 245.

[73] Ibid.

[74] Ibid.

had been replaced by a quieter, yet sarcastic toleration of Frankland's ideas. They humored Frankland, refusing to take his ideas seriously enough to refute them. This tone is exemplified in Sir Edmund Beckett's cross-examination of Frankland during the 1878 hearings of the Commons select committee considering the Cheltenham Corporation Water Bill. Beckett, a senior member of the parliamentary bar, regularly represented the London water companies and appeared here on behalf of a company promoting a river water supply for Cheltenham. The testimony suggests that he and Frankland had crossed swords previously.

Beckett treated the cross-examination as a rhetorical exercise. Whenever he and Frankland met the same question had to be asked: "have you got any evidence yet of the germ theory?" The same old theories -- "facts," Frankland insisted -- had to be gone over. Frankland still claimed a spoonful of Oxford sewage might ravage London with cholera, Beckett found the notion patently ridiculous. "You have been preaching this doctrine for ever so long," Beckett remarked, implying that whatever plausibility Frankland's crotchets had possessed had long since vanished, and that Frankland's continual "trying it on" of disproved ideas was becoming tiresome. Frankland said the Rivers Commission reports confirmed his ideas; Beckett countered with the Water Supply Commission report and pressed Frankland with having been the "principal person" on the Rivers Commission. "Oh dear no," Frankland replied.[75]

---

[75] S.C. Cheltenham Corporation Water Bill, Evidence, QQ 2912,

Through continual insinuations about Frankland's judgment, Beckett had lowered the level of discourse from discussion of verifiable scientific concepts to a pro forma exchange of stale speculation. Both he and Frankland spoke their lines, and collected their fees; the proceedings in the committee rooms were unconnected with the pursuit of knowledge. It is unfortunate that the yawns which appear to have punctuated Beckett's questions were not recorded in the transcript.

Tidy's treatment of Frankland complemented Beckett's. In 1879 Tidy made a significant change in the companies' tactics against Frankland. Hitherto, Tidy had employed Letheby's oxygen absorbed (potassium permanganate) process as well as a version of Wanklyn's ammonia process. The contention that Frankland's combustion process was unsound had formed a key part of his and Letheby's criticism.[76] In 1879 Tidy decided that, with some reservations, Frankland's process was acceptable because it gave results concordant with Letheby's process. In part, Tidy's acceptance of combustion probably reflects a move to keep his criticism of Frankland current with the changing views of analytical chemists away from the strongly anti-combustion stance most had taken in the early 1870s.[77] It is also

---

2935-40, 2949-55.

[76] Tidy in disc. of Frankland, "On some points in the analysis of potable waters," p. 105; "Metropolitan water supply," CN 34(1876): 216; C.M. Tidy, "Processes for determining the organic purity of potable waters," J. Chem. Soc. 35(1879): 52-56; idem, "River water," p. 272.

[77] The low point of popularity for Frankland's combustion procedure was sometime in the first half of the 1870s. During the last

likely that Tidy recognized that Frankland's depiction of the London water supply could be combatted more effectively by using the same analytical processes (and ideally analyzing duplicate samples) as Frankland and drawing attention to Frankland's extreme interpretations.[78]

By the late 1870s, therefore, the companies' criticism focused on Frankland's integrity rather than his competence. Frankland was now not simply one working ethically with untrustworthy tools, he was something much more reprehensible: one highly competent, who used his expertise to deceive.[79] This was the theme of the campaign against Frankland that Tidy, Odling, and Crookes carried on in Chemical News and in their monthly reports from April 1882 until the middle of 1884. They objected to Frankland's "dialectical artifice,"

---

half of the decade it was increasingly spoken of with respect, although few still used it. Attempts were made to improve the process on the assumption that "it [combustion] must form the philosophical starting point in an endeavour to solve problems of water analysis" (E.J. Mills, "On potable waters," CN 36(1877): 264). See also W. Dittmar and H. Robinson, "On the determination of the organic matter in potable waters," CN 36(1877): 26-29; C.T. Kingzett and J. Thudichum, "A note on some trials of Frankland and Armstrong's combustion process in vacuo," CN 33(1876): 237; E.W.T. Jones in disc. of A.H. Allen, "On some points in the analysis of waters and the interpretation of results," The Analyst 1(1877): 65; W. Russell in disc. of Frankland, "On some points in the analysis of potable waters," p. 105.

[78] W. Odling, "The chemistry of potable water," CN 50(1884): 205. The paper is also in JRSA 32(1883-84): 930-38.

[79] A. Civil Engineer, The London Water Supply, being an examination of the alleged advantages of the schemes of the Metropolitan Board of Works and of the inevitable increase of rates which would be required thereby (London: E & F.N. Spon, 1878), pp. 25-28; and William F. Rowell in Our Water Supply. Discussion for and against the fitness of Thames and river water for domestic use. Rpted. from the Surrey Comet (London: W. Trounce, 1880).

and to his tables "skilfully made to subserve . . . partisan purposes." They refused even to consider any grounds for questioning the wholesomeness of London's water.[80]

Frankland's Allies, 1870-72

The best indication of the refusal of Frankland's contemporaries to examine the substance of his water science is the response of his allies rather than his opponents. In 1872 the Local Government Board acquired partial sponsorship of Frankland's analyses of the London water supply. George Graham, the Registrar General, who had been paying Frankland's salary, wrote to the Local Government Board on 8 May 1872 regarding Frankland's status.[81]

---

[80] Odling, "The chemistry of potable water," CN 50(1884): 204; Crookes, Odling, and Tidy, "London water supply," CN 45(1882): 180; 47(1883): 241, 181; 48(1883): 103, 159, 197, 297; 49(1884): 111, 185; Crookes, Odling, and Tidy, "London water supply. Annual report for 1882," CN 48(1883): 90.

[81] PRO MH 19 67, Graham to LGB, 8 May 1872 as 26162/72; Graham to LGB, 29 May 1872 as 30440/72. These letters appear to have been part of a more extensive correspondence, other pieces of which have not been located. They appear to be in general reference to the 1872 Metropolis Water Act which transferred responsibility for the water supply from the Railway Dept. of the Board of Trade to the LGB. Graham is asking for instructions on what to do with Frankland's reports so it is likely that the LGB took responsibility for Frankland's work as well at this time. It is clear that in the later 1870s Frankland's fees were paid by the LGB (PRO MH 29 2, 13119 cc/75). It also appears that among LGB officers there was some concern at this time in regard to Frankland's scientific status and activism in water supply politics. A minute by John Simon on the 8 May letter alludes to the necessity of a "re-consideration" of Frankland's reports. Indeed from the tone of Graham's letter it seems likely that the suggestion had been made that Frankland's reports should be edited. The idea of censoring Frankland was certainly considered in 1883 (PRO MH 29 5, Thomas to Dalton, 18 Aug. 1883).

Graham defended Frankland. Frankland had not shrunk "from the duty imposed upon him, but in an unflinching manner conscientiously stated the defects he discovered however unpalatable to the water companies." Frankland's activism had led to improvement of the water supply:

> nothing has had so great an effect upon them [water companies] as Prof. E. Frankland's undisguised description of the impurities discovered, published by the authority of the government, and his comparison of what we are here compelled to drink, with what is supplied to Glasgow, Manchester, etc.[82]

What is important about this is the view Graham, and by implication, the officers of the LGB, were taking of Frankland's science. Graham defended Frankland not because he was convinced that Frankland was right, but because Frankland's sensationalism had been so successful. Graham avoided substantive issues: "I . . . do not effect to have any knowledge of chemical experiments."[83] Further, Graham depicted a scientific establishment governed by the struggle for advancement rather than the pursuit of knowledge. Frankland's practices "incurred the displeasure of rival chemists, who deprecated his conclusions, censured his system of analysis." The water companies had found "philosophic statists and rival chemists glad to dispute the accuracy of the analyses."[84] Among experts such conflicts were to be expected, and it was naive to expect

[82] Graham to LGB, 8 May 1872. See also a similar tone in Arthur Hassall's Food, Water, and Air 1(1871-72): 120-21.

[83] Graham to LGB, 8 May 1872.

[84] Ibid.

agreement: "the higher the individual attacked may be, the greater glory will be their fate who demolishes his reputation."[85]

Thus, Graham was arguing that 1) if Frankland's rhetoric had helped the cause, that, and not its validity, was sufficient justification for sponsoring it; and 2) that all those criticisms of Frankland's methods which had appeared could be ignored because those who uttered them were ambitious malcontents desirous of seizing Frankland's prestige.

The perspective of Frankland's supporters was therefore not significantly different from that of his opponents. Both groups recognized the manifold political implications of his conclusions. Either they accepted these implications without examining their foundations, as with Graham, or they took positions on water issues antithetical to Frankland's as a response to the political power contained in his water science. Such is most clearly the case with Wanklyn. Wanklyn believed his success as a water scientist was inverse to Frankland's. It is apparent that he took positions on issues which opposed Frankland's positions to emphasize what he believed was the incorrectness of the combustion process. That is, Wanklyn tried to show that combustion results were not simply less accurate than ammonia results, but that they led to utterly and dangerously wrong conclusions in contrast to the accurate and safe conclusions of the ammonia process.

---

85 Ibid.

The extent to which the content of Frankland's water science was politicized is profound. As Tidy, doubtless with some justice, pointed out, Frankland's image as an uncompromising champion of pure water was maintained at the government's convenience. The time might come when bold words would no longer be politically desirable, and then, Tidy suggested, Frankland's prestige in water matters would rapidly disappear.[86]

III. The Frankland-Tidy Debates, 1880-84

In 1880 Frankland's self-purification experiments and river studies were finally challenged in detail. Having accepted the analytical methods on which Frankland's self-purification views were ostensibly based, Tidy faced the problem of showing where Frankland had gone wrong.[87] On 18 March 1880, barely a year after endorsing the combustion process, Tidy presented to the Chemical Society his proof of oxidative self-purification. He presented new data showing changes in the concentrations of the organic elements in the Shannon, Severn, and other rivers. He had simulated river conditions with a network of troughs and observed rapid destruction of sewage. These data were embedded in traditional speculations about the

---

[86] Tidy in disc. of Percy Frankland, "Upper Thames," p. 449.

[87] Charles Folkard called upon Tidy to accept the self-purification studies because he accepted the process by which they were done (Charles W. Folkard, "The analysis of potable water, with special reference to previous sewage contamination," MPICE 68(1881-82): 100).

sewage-destroying potency of aquatic biota and traditional descriptions of exceedingly ephemeral morbid poisons measurable by techniques of analytical chemistry.[88]

Frankland took Tidy's challenge seriously. At his request, discussion of Tidy's paper was postponed until 20 May, by which time it had been printed and Frankland had organized his rebuttal. That evening's agenda reflects Frankland's power within the Chemical Society. Two papers were listed: Frankland and Lucy Halcrow on "Peaty Water," and Frankland on "The Spontaneous Oxidation of Organic Matter in Water." "Peaty Water" described experiments showing that peat extracts were not rapidly oxidized.[89] "Spontaneous oxidation," however, was not a report of research, but simply Frankland's commentary on Tidy's paper. Frankland had managed to have his often sarcastic, point-by-point refutation of Tidy's views published as a paper in the society's journal. Further, Frankland had recruited Huxley (in person), and Tyndall (by letter) to refute Tidy's conception of germs. The show of force was apparently effective. The day after the meeting Frankland gloated to Tyndall: "Huxley came. . . . No stick was left for Tidy to stand upon and he really had nothing to say."[90]

---

[88] C.M. Tidy, "River water," J. Chem. Soc. 37(1880); 267-327.

[89] Lucy Halcrow and Edward Frankland, "On the action of air upon peaty water," J. Chem. Soc. 37(1880): 506-17. These studies were not entirely a response to Tidy's March paper, for some of them had been under way for more than a year. The paper did serve to refute Tidy's conclusions regarding the oxidation of peat extracts in the Shannon, however.

[90] Royal Institution, Tyndall Papers, General Correspondence,

Tidy was not cowed by such overwhelming opposition. The following March he delivered a second "River Water." The paper contained nothing new and the Chemical Society did not print it. Still the debate continued. In March 1883, W.N. Hartley, professor of chemistry at the Royal College of Science in Dublin, showed Tidy's studies of the Shannon to have been amateurish and his conclusions premature.[91] Hartley's substantive and well-founded criticisms were not to be the norm, however, and the gutter was finally reached in March 1884, when Percy Faraday Frankland, employed as his father's assistant after receiving his Ph.D. at Würzburg, presented a paper to the Society of Arts on "The Upper Thames as a source of water supply," in which a minimum of new data describing chemical changes in the upper Thames served as a vehicle for an excessively vicious attack on Tidy. For once we may sympathize even with Tidy, who observed that he had "never read a paper so deluged with extravagant partisanship."[92]

With the possible exception of Hartley's paper this debate exemplifies the context of distrust discussed in Chapter Two. Despite Tidy's noble sentiments about uniting with Frankland to fight "the

---

Frankland to Tyndall, 21 May 1880, TS 3964. See also Frankland to Tyndall, 14 May 1880, TS 3964; and Tyndall to Frankland, 16 May 1880, TS 3971.

[91] W.N. Hartley, "The self-purification of peaty rivers," JRSA 31(1882-83): 469-84.

[92] Tidy in disc. of Percy Frankland, "Upper Thames," p. 446. Folkard's "The analysis of potable water, with special reference to previous sewage contamination" might also be considered part of this series of papers.

prevailing heresies on this question which tended so to upset the public mind" neither he nor Frankland sought to resolve the issue. For Tidy, the papers on river water gave the legitimacy of scientific publication to the doctrines he espoused in courts of law and before parliamentary select committees. Nevertheless, the framework of point-by-point contradiction which characterized this debate led to better definition of the issues. Two issues emerge clearly. First, was organic matter in water oxidized, and if so, how? Second, how did morbid poisons respond to stream conditions?

Tidy's studies are best examined in light of Frankland's criticism. The bulk of Tidy's data were taken on the Shannon, in which most of the organic pollution came from peat. This data was therefore of little relevance to the issue of the safety of sewage-polluted water. Tidy found that in the Shannon organic carbon decreased by about 50% in a 35 mile flow. Elsewhere, however, self-purification (presumably due to spontaneous oxidation) was much more rapid. Spontaneous oxidation was to account for a 32% reduction in organic elements in one 30 yard stretch of the Severn.[93] Cleverly altering the language of Frankland's 1870 report on "alleged" self-purification, Tidy wrote:

---

[93] Tidy, "River water," p. 304; E. Frankland, "On the spontaneous oxidation of organic matter in water," J. Chem. Soc. 37(1880); 527; Frankland in disc. of Tidy, "River water II," p. 115. Frankland noted of this example from the Severn that at that rate the river ought to be pure in 100 yards and that "If I believed that some substance in sewage oxidised at this rate, I would endeavour to isolate it and bring it before this [Chemical] society; it would undoubtedly burst into flame when exposed to air" ("River water II," p. 115.

> the oxidation of the organic matter in sewage, when mixed with unpolluted water and allowed a certain flow, proceeds with extreme rapidity, and that it is impossible to say how short a distance such a mixture need flow under favourable conditions before the sewage matter becomes thoroughly oxidised. It is certain, to my mind, that there is not a river in the United Kingdom, but what is many times longer than is required to effect the destruction of sewage by oxidation.[94]

Frankland's main objection was that Tidy's river data showed nothing about the disappearance of organic matter in running water because they consisted entirely of concentrations of pollutants and were uncorrected for changes in river volume. Thus Tidy's conclusion that the sewage of the million inhabitants of the Thames basin above London had been completely oxidized was invalid. Although the Thames at Hampton contained approximately the same concentration of organic elements as the Thames 70 miles further upstream, the volume of the river had increased fivefold in that distance.[95] Tidy had also neglected mixing. The 32% reduction in 30 yards reflected the mixing of raw sewage with the main body of water in the river, rather than oxidation.[96]

Tidy was at least aware of this lapse and offered excuses for it. He admitted that the streams he had studied received tributaries, but claimed that the tributaries were usually more polluted than the rivers receiving them.[97] Further, diminution in the concen-

---

[94] Tidy, "River water," p. 316.

[95] Ibid., p. 304; Frankland, "Spontaneous oxidation," pp. 531-33.

[96] Frankland, "Spontaneous oxidation," p. 528.

[97] Tidy, "River water," pp. 307, 295. On a later occasion Tidy

trations of organic elements between two points on a river might represent only net diminution. If a stream had continuously received organic matter between sampling points, and had continually been oxidizing that organic matter, then a much larger amount of oxidation might have occurred than the analyses indicated. Tidy compared this phenomenon to the impossibility of telling how many passengers had disembarked from a train by subtracting the number still aboard at the end of the line from those on board at the beginning.[98]

As Frankland pointed out, however, apparent reduction might just as well reflect dilution of highly stable organic matter with clean water. Moreover, there was an accepted means of estimating the dilution of sewage. Frankland and R. Angus Smith in England, William Ripley Nichols in America, and others recognized that destruction of sewage could be distinguished from dilution of sewage by comparing the decrease in organic elements with the decrease in chlorine. Sewage was the chief source of chlorine in most inland rivers and there were no natural agencies which removed chlorine from rivers to any substantial degree. Therefore, a reduction in organic matter concentration without a corresponding reduction in

---

claimed that estimation of river discharge was not within the province of a chemist (Tidy in RCMSD, Evidence, QQ 10475-89). On this problem see also the comments of W. Thorp, jr., and H.Y.D. Scott in W. Thorp, jr., "River pollution, with special reference to the late commission," JRSA 23(1874-75): 384, 389.

[98] Tidy, "River water," p. 307.

chlorine concentration indicated destruction of organic matter. On the other hand, dilution with unpolluted water would result in a symmetrical decrease in both organic matter and chlorine.[99] Tidy had tested for chlorine and he had trouble explaining his results. Diminution of chlorine in the Severn was due to the presence of aquatic chlorine-assimilating plants, Tidy suggested, refusing even to consider that dilution offered a much likelier explanation.[100]

Not only was Tidy determined that these data indicated self-purification, he was convinced that spontaneous oxidation was responsible for it. Tidy's faith in spontaneous oxidation ought not to surprise us. As we have seen, the portrayal of oxygen as the "great burner-up of dead organic matter" was fundamental in British physiological chemistry and Liebigian organic chemistry. Tidy alluded to the marvellous destruction by atmospheric oxygen of all organic debris cast into the air, especially those virulent "myriads of dead epithelial scales from every membrane." Air also acted "mechanically, thereby dividing up the noxious particles," thus making them more easily oxidizable.[101] Much the same thing happened

---

[99] RPPC/1868/2nd Rept., pp. 14-15; R. Angus Smith, "On the examination of water for organic matter," Proc. Manchester Literary and Philosophical Society, 3rd series 4(1871): 47, 80-87; William Ripley Nichols, "On the present condition of certain rivers in Massachusetts," in 5th Annual Report of the Massachusetts State Board of Health, January 1874 (Boston: Wright & Potter, 1874), p. 79. See also Fox, Sanitary Examinations, pp. 104-5; Charles A. Cameron, A Manual of Hygiene, Public and Private, and Compendium of Sanitary Laws (Dublin: Hodges and Foster, 1874), pp. 61, 71; and C. Estcourt, "Pollution of rivers: The Irwell," SR 5(1876): 417-19.

[100] Tidy, "River water," p. 306; Frankland, "Spontaneous oxidation," p. 528.

in rivers. Anything that helped to aerate a stream -- rapids, dams, or waterfalls -- aided oxidation by "breaking up . . . the water particles [thus producing] more efficient aeration, and necessarily more complete actual contact of the organic matter with the oxygen necessary for its combustion."[102] The motion of water also helped break solids into more easily oxidizable portions, while the respiration of aquatic plants yielded an unusually active form of oxygen.[103] There was nothing new in this explanation and Frankland met it with an equally venerable response: Brodie's 1866 assertion that pure organic substances were singularly resistant to the depredations of oxygen.[104]

Much the same repetition of stale speculation characterized discussion of the fate of morbid poisons in rivers. As in earlier discussions of self-purification, models of morbid poisons were adopted which either resisted or succumbed to stream conditions and therefore confirmed the conclusions of their adopters: either that polluted water remained permanently polluted, or that self-purifica-

---

[101] Tidy, "River water," pp. 301-2. Frankland suggested that if Tidy thought atmospheric oxygen expeditiously destroyed all cast off organic matter he ought to go home and look on the top of his bookshelf. If Tidy were right a great deal of money could be saved on servants (Frankland, "Spontaneous oxidation," p. 543).

[102] Tidy, "River water," p. 301.

[103] Ibid.; and C.M. Tidy, Handbook of Modern Chemistry, Inorganic and Organic, for the use of students (London: J. & A. Churchill, 1878), p. 493.

[104] Frankland, "Spontaneous oxidation," pp. 517-18; Frankland in disc. of Tidy, "River water II," pp. 113-15.

tion occurred rapidly and reliably. The discovery of germs of disease had a considerable impact on the range of plausible morbid poisons, however. No longer were morbid poisons simple chemical entities, no longer could it be facilely assumed that virulence would disappear in proportion to the destruction of nitrogenous organic matter. As Percy Frankland observed in 1884, "the question of [oxidative] self-purification was, from a sanitary point of view, a wholly idle one," since "if there was no guarantee . . . that this materies morbi was destroyed, what mattered it whether the other organic matter was removed or not."[105] Tidy opposed this trend. Morbid poisons obeyed the same laws as dead organic matter, he maintained, and his oxidation studies were therefore relevant to the issue of drinking water quality. Perhaps in desperation, he asserted that one had to make such an assumption if chemical analysis -- in his view the only guide available -- were to have any meaning.[106]

Since Tidy was the most ardent defender of the use of polluted river water during the 1880s, his understanding of morbid poisons is worth examining in some detail. In 1878 Tidy acknowledged that

---

[105] Percy Frankland, "Upper Thames," p. 565.

[106] Tidy declared that if Frankland's views were correct "a certain terrible doubt is cast over any opinion we may base on analytical results, and, which doubt, as a chemist, I cannot allow. In fact, if such be true, I fail to understand the value of water analysis at all. Admit on the other hand that the materies morbi disappears together with the disappearance of the organic pollution (as I contend the facts prove), it is manifest that the chemist is strengthened when he pronounces a water wholesome and fit to drink" (Tidy, "River water," p. 326; see also "River water II," pp. 113-14).

germs existed but were destroyed in running water. Water was therefore a disinfectant. Its "endosmic action" was "inimical to the corpuscular structure of many specific contagia, for, by bursting their cell-like envelopes, it destroys their vitality."[107] In "River Water I" the same idea appeared not as a statement but as a speculation. If there were such things as germs, it was likely, suggested Tidy, that their envelopes endosmically burst. Complementing this notion were equally scholastic constructions. If there were germs, probably they were only marginally alive and therefore behaved more like dead than living matter. And if germs were really alive, then they would multiply, and the incidence of water-borne disease would be directly proportional to the distance downstream from the source of infection -- clearly not the case. Thus, by manipulating assumptions, Tidy arrived at the conclusion that "be this <u>materies morbi</u> . . . organised matter or chemical poison, it is subject to the same laws of destruction as the ordinary organic matter of sewage."[108]

With the help of Huxley and Tyndall, Frankland ridiculed this conception of germs. The endosmic bursting of bacterial envelopes was ridiculous. A living morbid poison would not multiply indefinitely since its numbers would be limited by the food available in the environment.[109] In 1881 and 1884 Frankland made further claims

---

[107] Tidy, <u>Handbook of Modern Chemistry</u>, p. 493.

[108] Tidy, "River water," p. 326.

[109] Frankland, "Spontaneous oxidation," p. 542.

for the resistivity of germs. Frank Hatton, a Frankland disciple, found that bacteria (no species given) were unaffected by cyanogen atmospheres and similarly harsh chemical environments. Frankland alluded to Hatton's work repeatedly, complementing it with the observation that spores resisted prolonged boiling. The implication was that if germs could withstand such treatment they could surely withstand the gentler conditions of a running stream.[110]

In 1880 Tidy had argued that if germs existed they obeyed the same laws as ordinary organic matter. Since his opponents rejected his description of germs, Tidy decided that germs did not exist. He abandoned the notion of bursting envelopes after 1880, and took up an intransigent positivism: until somebody could show him a cholera germ or a typhoid germ, he was unwilling to consider the characteristics of such entities. He admitted that splenic fever and pig typhoid germs had been discovered, but that made no difference. Any discussion of the characteristics of germs (in which Tidy had been quite willing to indulge so long as it had led to the conclusion that they could not survive in river water) was unwarranted. It was simply "diving deeper and deeper into mere speculation, when we discuss the laws governing the life of bodies, the very existence of which at present is unproved."[111] The adoption of positivism gave

---

[110] Frank Hatton, "On the action of bacteria on gases," <u>J. Chem. Soc.</u> 39(1881): 247-58; Frankland in disc. of Tidy, "River water II," pp. 113, 115; in disc. of Percy Frankland, "Upper Thames," p. 445.

[111] Tidy, "River water," p. 321. See also Tidy in disc. of Folkard, "The analysis of potable water," pp. 79-82, and Folkard, p. 100

Tidy the benefit of not having to replace his exploding germs. He continued to claim, however, that whatever morbid poisons were, they obeyed the laws of dead organic matter.[112]

It is clear that Tidy was here practicing much the same kind of science as Frankland had in promoting the germ theory in 1868. Both men had developed hypotheses about the nature of morbid poisons and proceeded to advise the public as if those hypotheses were proved. The main difference was that Frankland had been ahead of his time, while Tidy was behind his. Both men responded to the public's demand for certainty from the scientists who investigated issues of water quality.

IV. Artificial Rivers and Sewage-Eating Germs

Both Frankland and Tidy had done experiments on self-purification. In Frankland's experiments organic matter didn't oxidize, in Tidy's it did. Experimental design was therefore a key issue in the 1880-84 debates. Frankland's 1868-69 experiments had been done on bottles of sewaged water. In some cases these had been allowed to stand for long periods, in some cases they had been agitated and/or aerated periodically. Frankland had looked for changes in the amount of organic matter as well as changes in dissolved gases and changes in the overlying atmosphere. The experiments on the oxidation of

[112] Tidy in disc. of Percy Frankland, "Upper Thames," p. 446, and Percy Frankland, p. 565.

peat extracts, done with Halcrow in 1879-80, perpetuated this design. Even continuous agitation failed to produce substantial oxidation. From these experiments Frankland concluded that: 1) oxygen was removed from the overlying air only slowly in the case of sewaged waters, and virtually not at all in the case of peaty waters; 2) small amounts of CO_2 were discharged into the atmosphere during these experiments, but CO_2 was also discharged by a control sample of water presumably free from oxidizable organic matter; 3) organic content dropped slowly in the case of the sewaged waters, virtually not at all in the case of the peaty waters.[113]

Tidy repeated Frankland's 1868-69 experiments and accepted Frankland's results. But while these "were doubtless most interesting and satisfactory experiments as regards shaking fluids up in bottles, . . . they did not represent the flow of a river."[114] Tidy's own experiments were quite different. He had constructed in series, twenty V-shaped troughs, each ten feet long, each with a drop of one foot. Dilute sewage flowed from a cistern at the top, through the series of troughs, which were joined so as to create a small cascade between each trough, to a cistern at the bottom, whence it was pumped back to the top. Thus Tidy could simulate the effect of a certain number of miles of flow on a certain quantity of

---

[113] Halcrow and Frankland, "Peaty water," pp. 506-16; Frankland, "Spontaneous oxidation," p. 526.

[114] Tidy in disc. of Frankland, "Spontaneous oxidation," CN 41 (1880): 248. See also Tidy, "River water," p. 316; idem, "River water II," pp. 113-14.

water.

Tidy had trouble making his apparatus produce the numbers he wanted. For a long time his running water would not purify. The main problem, he believed, was that sooty London air was contaminating his artificial river. He went to great pains to cover the apparatus, and lined the troughs with glass as well. While admitting to the Chemical Society that he had included only successful data -- those which showed purification, or in Tidy's view those which had not been contaminated by the atmosphere -- Tidy still maintained that he had experimentally demonstrated the self-purification of sewaged rivers. He noted that the smell of sewage disappeared after a few runs, and that organic matter decreased rapidly at first, slowly thereafter.[115]

Frankland was skeptical. He distrusted outdoor experiments -- all manner of complications might intervene. He distrusted experiments in which the experimenter kept struggling with the apparatus until it confirmed the experimenter's preconceptions. Frankland asked for repeatability, which Tidy's data did not reveal, nor his apparatus apparently offer. He asked the identity of Tidy's "occult influence," or "oxidising demon" -- whatever it was in water flowing in a river that didn't operate when water was shaken in a bottle or stirred by an agitator.

In short, this mysterious influence which so favours

[115] Tidy, "River water," p. 308.

> the oxidation of polluted water running in rivers with numerous unpolluted affluents appears always absent when the water is put under conditions admitting of the application of accurate experimental tests.[116]

Frankland had three explanations for Tidy's success. First, thin sheets of water flowing in V-shaped troughs would have an unrealistically large surface to volume ratio which would greatly aid aeration. Second, some sewage matter might be settling out in the apparatus. Third, and most importantly, the apparatus might be home for bacteria which would aid in the destruction of organic matter. Clearly familiar with the multitude of fermentation studies done in the previous twenty years, Frankland suggested that *micrococcus ureae* lived in the apparatus and were responsible for changing organic urea into inorganic ammonium carbonate. This bacterial hypothesis explained "the failure of the experiments *at first*, and their success *at last*; for if the troughs were clean, . . . the sewage would pass through unchanged, but a marked diminution of organic carbon and nitrogen would ensue when the accumulation of bacteria and their germs provided the conditions necessary for rapidly transforming urea into carbonate of ammonia."[117]

Thus Frankland had admitted that at least some self-purification was the work of bacteria. We might expect this casual observation to lead to an explicit acknowledgement of biological self-

---

[116] Frankland, "Spontaneous oxidation," pp. 538-39.

[117] Ibid., p. 539.

purification. We might see it as a classic example of the dialectical workings of science. Opposites had clashed; from between grinding gears of contradiction had fallen an insight which could lead to the asking of much more fruitful questions about the chemical dynamics of polluted rivers.

Frankland did not recognize that the excuse he used to explain away Tidy's experiments was the key to resolving the issue about which they were arguing. It is worth asking why he did not. Two possible reasons may be quickly rejected. First, it might be suggested that Frankland was unwilling to countenance mechanisms tending to the conclusion that rivers were purified. This is not the case. Frankland acknowledged the apparent self-purification of rivers and ascribed it to dilution and subsidence. He had done considerable research showing how the joining of streams with different sorts of chemical loads might produce a general settling-out of impurities and result in substantial purification.[118] Here Frankland directed attention to _M. ureae_ as another mechanism of self-purification. In fact, the issue had never been self-purification _per se_, but spontaneous oxidation. As Frankland explained in the introduction of his paper on that subject, polluters and water companies had "always contended that the most abominable of organic rubbish is destroyed by oxidation, in fact utterly 'burnt up' during a flow

---

[118] RCWS, Appendix BE, p. 111 (Frankland and Odling, "Supplementary Report on the chemical quality of samples of water collected in the basin of the Thames").

of a few miles."[119] It was this doctrine that he had long combatted by arguing that spontaneous oxidation was virtually non-existant and further that the "most abominable" matters were exceptionally resistant to oxidation. Frankland meant bacteria to be an alternative to oxidation. The change from urea to ammonium carbonate was "effected by the mere assimilation of two molecules of water by one molecule of urea, and . . . has nothing to do with oxidation."[120]

Second, it might be argued that the idea of bacteria purifying sewage was totally new to Frankland, that he was unfamiliar with the ideas of Pasteur, and therefore missed the significance of the biological mechanism he had broached.[121] While it is true that on oc-

---

119 Frankland, "Spontaneous oxidation," p. 517.

120 Ibid., p. 540. In general, Tidy shared Frankland's perception that the issue was one of spontaneous oxidation. In testimony to the RCMSD in March 1883 he admitted that bacteria might aid in the oxidative process (RCMSD, Evidence, Q 11556), but in 1884 he was still maintaining a vision of self-purification in which oxidation was effectively spontaneous (In disc. of Percy Frankland, "Upper Thames," p. 448).

121 It may be that British chemists continued to take a peculiar perspective toward Pasteur's early work. Hartley, for example, obtained from Pasteur the concept that there was virtually no such thing as spontaneous oxidation, but not the corallary of that concept, that in fact an immense amount of oxidation was accomplished by the metabolic processes of aerobic bacteria. Hartley emphasized changes in bedrock as having a profound influence on oxidative self-purification. Except for such changes, Hartley implied that organic matter was essentially stable in much the same manner as Brodie had asserted it was. Hartley wrote: "Those chemists who are well acquainted with the classic researches of M. Pasteur, must have experienced a feeling of surprise when informed in the pages of the Journal of the Chemical Society that organic matter of a particularly stable character was oxidized and destroyed by the oxygen dissolved in water, under the influence of comparatively low temperature" (Hartley, "Self-purification of peaty water," pp. 469-70).

casion Frankland exhibited discomfort with the application of biological principles to matters of water quality, it seems impossible that one in regular contact with Tyndall and Huxley would be unaware of Pasteur's notions of the significance of bacteria as the mechanisms running the cycles of the elements.[122]

The chief reason for Frankland's failure to follow up the notion of bacteria as purifiers was that he regarded bacteria as the quintessential impurities. Frankland would not tolerate a world whose inhabitants were quite so morally ambivalent towards the human enterprise. In refuting Tidy's claims that bacteria must either burst their envelopes or multiply indefinitely, Frankland had pointed out that bacteria required "congenial molecules capable of yielding active energy by transformation into new molecules."[123] The congenial molecules were, of course, from sewage, the process

---

[122] Examples of Frankland's discomfort with biological principles are: 1) In 1881 Frankland suggested that consumption of solid parts of sewage by fish was not truly purification because the fish also removed dissolved oxygen from the water. This represents a failure to realize that oxygen inhaled by a fish oxidizes organic matter in exactly the same proportion as would spontaneous oxidation (In disc. of Tidy, "River water II," p. 114); 2) In 1872 Frankland gave evidence to the Select Committee on the Birmingham Sewerage Bill suggesting that plants assimilated organic matter directly, obviating the need for an intermediary process of decomposition (House of Lords Record Office, Minutes of Evidence, House of Commons, 1872, Vol. 5 [Birmingham Sewerage], 29 April 1872, pp. 184-85, 138); 3) Frankland's 1885 paper on "Chemical changes in their relation to microorganisms" was justly criticized for employing outdated concepts of the distinction between plants and animals and for simplistic biological methodology ("Chemical Society," CN 51(1885): 79-80; 53(1886): 82; E. Klein, "Bacteriological research from a biologist's point of view," J. Chem. Soc. 49(1886): 197).

[123] Frankland, "Spontaneous oxidation," p. 542.

yielding active energy was oxidation, and the new molecules produced were harmless products of complete oxidation. Yet the microbes Frankland was discussing in this passage were not to be encouraged as the purifiers of sewage-polluted streams; instead, Frankland was discussing disease germs, the ultimate enemies of humanity.[124]

The perspective that Frankland represents was sketched at the close of Chapter Four. Frankland and his associates, Tyndall and Huxley, preached a doctrine of anti-bacterial militancy. Bacteria were a menace to humanity; science could combat that menace successfully, if unhampered by anti-vivisection laws. Tyndall concluded an 1876 lecture with a call to arms:

> We have been scourged by invisible thongs [sic], attacked from impenetrable ambuscades, and it is only today that the light of science is being let in upon the murderous domain of our foes.[125]

The zymotic pathology, still very much in force in the mid-1880s, downplayed the distinction between those bacteria which caused disease and those which recycled matter. Putrefaction remained the archtypal pathological process. Speaking at Frankland's request on May 20, 1880, Huxley told the Chemical Society that the human body

---

[124] On a later occasion Frankland admitted these bacteria were purifiers, but thought they accelerated putrefaction and were therefore undesirable (RCMSD Evidence, Q 10914). In 1892 Frankland spoke of "wicked" bacteria being eaten by harmless ones. The choice of words is interesting (Royal Commission on Metropolitan Water Supply [RCMWS], Evidence, P.P., 40 pt. 1, 1893-94, [C.-7172.], QQ 4484-85).

[125] John Tyndall, "Fermentation and its bearings upon surgery and medicine," in Essays on the Floating Matter of the Air and its Relation to Putrefaction and Infection (New York: D. Appleton, 1888), p. 274.

was a putrescible substance, and that bacteria, invisible to chemistry, were "as deadly to the human body as prussic acid."[126] Owing to the immaturity of bacterial taxonomy there was some doubt -- and again Huxley was one of the doubters -- whether bacteria could be separated into pathogenic and non-pathogenic classes. Bacteria might be the resolvers of dead organic matter, but if that process of resolution were the essence of disease, and if there were no way to restrict bacteria to dead organic matter, then the might of science had to be mustered against bacteria.[127]

Perhaps a dialectical scientific context forced the expression of the hypothesis that M. ureae, and by implication, other bacteria, were responsible for river self-purification. Yet that context also obscured the significance of that hypothesis. There was no room in the arguments of Frankland and Tidy for an entity possessing both benign and malignant aspects. Frankland was a crusader, and believed that the tools of science had to be continually employed if mankind were to win in the struggle against nature, e.g., the germs of disease. By contrast, Tidy had faith in nature's benevolence. He made his differences with Frankland explicit: "He [Tidy] did not believe in the bewildering consciousness of ever-present peril, and that all things around and within them were working together for

---

[126] Huxley in disc. of Tidy, "River water," CN 41(1880): 248.

[127] William Bulloch, The History of Bacteriology (London: Oxford University Press, 1938), pp. 188-89, 195-96.

evil."[128]

At least one British chemist was appalled by his colleagues' utter obliviousness to the Pasteurian paradigm. This was Robert Warington, jr., son of the inventor of the aquarium. On 11 June 1880, three weeks after the first of the great self-purification debates, Chemical News published a letter from Warington, an agricultural chemist and associate of Gilbert and Lawes, under the title, "Observations upon Dr. Tidy's paper on 'River Water.'"

Warington was surprised that spontaneous oxidation was still a subject of debate:

> Modern investigations plainly teach us that in nature organic matter is resolved into inorganic matter not by mere contact with oxygen, but by the agency of organic life.[129]

He could not believe Frankland and Tidy didn't know this, yet the "primary importance of organic life for the destruction of sewage has been strangely absent in their discussion of the question." The destruction of sewage, Warington went on, was the work of a series of organisms, each with a specific chemical function, among them the "innumerable army of bacteria, embracing many families of similar physical structure, but endowed with very different chemical powers." Point-by-point Warington explained how the microbial hypothesis resolved issues of the Frankland/Tidy debate.

---

[128] Tidy in disc. of Percy Frankland, "Upper Thames," p. 448. For a similar statement see Baldwin Latham in Disc. of Folkard, "The analysis of potable water with special reference to previous sewage contamination," p. 77.

[129] Robert Warrington, "Observations upon Dr. Tidy's paper on

Tidy was correct that self-purification occurred, but only in ideal circumstances where biological action was not "excluded by addition to the water of chemical refuse." British streams had become so "artificial" that effective self-purification was rare. The significant differences in the experiments of Frankland and Tidy were clarified:

> the conditions of laboratory experiments cannot easily be made comparable with those of a natural river, if organic life rather than mere exposure to air is the true agent in the destruction of sewage. In any experiments it would seem advisable to employ natural river water and not an artificial mixture, as the needed organisms would probably occur in the natural fluid.

Warington explained seasonal variations in river data. Hardness, for example, varied inversely with plant growth since plants abstracted the CO_2 that held minerals in solution. Pollution and dissolved oxygen were high during the winter, because organisms were least active then. In summer the river was busiest; organic contamination was low, but so was dissolved oxygen, and CO_2 was high. Sharp increases in organic contamination in autumn were the result of plants dying.

In a single short letter Warington had obviated pages of contradiction exchanged by Frankland and Tidy. Again, however, the biological gambit was declined. There were no comments on Warington's letter. Two years later, in the discussion of Hartley's "Self-Purification" paper Warington repeated his criticism, this time

---

river water," CN 41(1880): 265.

in person, and again got no response.[130]

The silence that met Warington's observations is an indication of how dreadfully inflexible the self-purification issue had become. To those preoccupied with the effects of scientific doctrine on water supply policy, this business of good bacteria which under some circumstances cleaned rivers seemed so much noise and confusion, a bunch of ideas not wholly worked out, with implications still unclear. Even Warington would not say if his organisms made water safe to drink, which was, after all, the chief issue.

This episode, Warington's observations and the response they received, represents a collision between two contexts of science. Warington's chemical practice didn't involve attack and defense of sewage or water supply policies. Instead, he was exclusively an agricultural and manufacturing chemist, a protégé of J.B. Lawes and a member of the Rothamstead group. His chief interest was in soil chemistry, and particularly in nitrification. Warington had been one of the first to recognize the importance of the 1877 proof by Schloesing and Muntz that nitrification was a microbial process. When he commented on Tidy's paper, Warington was in the midst of a study of the conditions under which nitrifying bacteria worked and an attempt to isolate them in pure culture.[131]

---

[130] R. Warington in disc. of Hartley, "Self-purification of peaty rivers," p. 482.

[131] E. John Russell, A History of Agricultural Science in Great Britain, 1620-1954 (London: George Allen and Unwin, 1966), p. 160; Richard Aulie, "Boussingault and the Nitrogen Cycle," Diss. Yale University, 1968, pp. 310 ff.

It is clear that as an enterprise for understanding nature, the water science of the 1870s was malfunctioning. Despite the enormous number of scientist-hours devoted to the study of problems of river pollution, water quality, and sewage disposal, the understanding of what purification was and how it happened remained as obscure in 1880 as it had been in 1869. Admittedly, there had been empirical advances in water filtration, sewage treatment, and preventive medicine. Yet the money spent -- and Frankland's Rivers Commission was the most expensive royal commission, and equally the most expensive piece of government-sponsored science to date -- had not produced an understanding of purification.[132] We have examined the reasons for this failure in this chapter. Science had made no progress because scientists refused to look upon one another's data as faithful representations of nature or to regard one another's conclusions as honest, impartial generalizations. Very often this assessment was correct. Yet the assumption that everyone was lying eliminated the necessity of substantive criticism. There was neither motive nor means for attempting to extract bits of truth from the rhetorical productions of Frankland and Tidy, and when, as with the *M. ureae* episode, insights emerged, they were ignored. It took Warington, an outsider, to show that this continual banging of heads was leading nowhere; not surprisingly, the head-bangers ignored

---

[132] *Return of all Royal Commissions issued from the year 1866 to the year 1874*, *P.P.*, 81, 1888, (426.). The Rivers Commissions of 1865 and 1868 spent almost £40,000 total.

Warington. This context of distrust perpetuated itself. Water experts were paid to deceive, the assumption spread among them that because they were deceiving, others -- whether they agreed or opposed -- must also be deceiving.

Warington's perspective would not long be ignored by sanitary scientists. By the early 1880s his work on the conditions required for nitrification in soil was being applied to the design of sewage irrigation works.[133] Warington's insight was not the chief route by which biology was restored to water quality and sewage treatment discussions, however. Instead, when biology was finally re-introduced into sanitary science, it was incorporated into the existing rhetorical and adversarial structures of discourse. The reappearance of biology and the eventual acceptance of biological purification is the subject of the next two chapters.

---

[133] Henry Robinson, "River pollution," TSIGB 8(1886-87): 175-79; idem, "Sewage disposal," SR n.s. 6(1884-85): 150.

IX. and X. BIOLOGY ACQUIRES A CONSTITUENCY

Between 1880 and 1895 the issue of self-purification became two issues. In regard to water supply self-purification was a matter of the conditions under which those newly-discovered causative organisms of enteric diseases survived or perished in rivers. Those concerned primarily with sewage treatment were interested in the capacities of rivers to accept sewage without development of a nuisance. A pathogen-free effluent was only a secondary goal.

Biology was incorporated in both branches of self-purification debate during this period. Re-introduction of biological considerations into water issues was due chiefly to the triumph of the germ theory of disease on the one hand, and to the 1877 discovery of bacterial nitrification by Schloesing and Muntz on the other. But while these events made biology the idiom of water quality matters in Britain, they failed to sponsor thorough study of the ecological aspects of water quality. To be sure, ecological questions were raised -- about conditions required by pathogens, or by aerobic saprophytes -- but no paradigm of systematic ecological study of polluted rivers emerged until the first decade of the twentieth century.[1]

---

[1] F.T.K. Pentelow suggests the "first scientific study" of polluted rivers was begun by the West Riding of Yorkshire Rivers Board, founded in 1894. A survey of the Board's reports reveals little concern with biological questions until the end of the first decade of the twentieth century and only then after Board staff had learned of German ecological studies of polluted streams. See Pentelow in H.B.N. Hynes, The Biology of Polluted Waters (Liverpool: Liverpool Univer-

Instead, biological principles became predominant in water quality debate because they justified particular water-use policies and not because they supplied answers to long-standing questions about water purity. The next two chapters, in roughly parallel fashion, examine this acquisition of constituencies by biological principles between 1880 and 1895 in regard to drinking water quality first, and sewage treatment second.

Although the chapters are ordered differently, both examine the struggle by sanitarians to understand the significance of the new biological discoveries for water quality matters, the intellectual feedback mechanisms between politics and science that characterized the research careers of two water scientists, Percy Frankland and William Dibdin, and finally the incorporation of biology in adversarial water politics as manifested in the proceedings of the Royal Commission on Metropolitan Water Supply of 1892-93 and the Royal Commission on Metropolitan Sewage Discharge of 1882-84.

---

sity Press, 1960), p. xiii; and West Riding of Yorkshire Rivers Board Reports (in library, Yorkshire Water Authority, Leeds), #98, H. Maclean Wilson and H.T. Calvert, "Report on a visit to Germany," Sept. 1906; #101, H. Maclean Wilson, "Report on a visit to Germany in connection with the 14th International Congress of Hygiene, Sept. 1907," Nov. 1907; #113, H. Maclean Wilson, "Report on the biological work in the laboratory to 31 March 1910," October 1910 (A biological laboratory was set up in June 1909); and #114, J.W.H. Johnson, "Report on the biological examination of the River Wharfe," Dec. 1910 (first of a series of river studies). Also see West Riding of Yorkshire Rivers Board, 17th Annual Report, 1 April 1909-31 March 1910.

IX. BIOLOGY ACQUIRES A CONSTITUENCY I: ORGANISMS AND THE POLITICS OF WATER SUPPLY

> We are not dealing -- as mostly in analytical chemistry -- with matter unalterable in quantity. Our matter is capable of self-multiplication; and further, amongst microphytes, the same fight is going on for supremacy or existence which we find amongst the highest of animals.
>
> -- Gustav Bischof[2]

Between 1880 and 1895 water quality became the province of biology. The techniques and interpretive principles of water analysis were transformed; new understandings of the natural and artificial mechanisms of purification emerged. The return of biology to water quality matters occurred in three phases. First was the return of traditional natural history to water examination. This was a sociological phenomenon. Disenchanted with the London analytical chemists, local sanitarians, chiefly medical officers and public analysts, adopted an eclectic water analysis which included an attempt to relate types of microscopic life to water quality. Biological examinations were ideal for local use -- they were quick and easy, required minimal equipment, conformed to and amplified traditions of amateur natural history, could be performed at some level even by those without formal training. Unfortunately the problem that had vexed able biologists in the 1850s still haunted

---

[2] Gustav Bischof, "Notes on Dr. Koch's water test," J. Society of Chemical Industry 5(1886): 116.

local sanitarians in the 1880s. Microorganisms in water purified and indicated impurity. Were they a good sign or a bad one? Was there really basis for associating species with dangerous pollution?

This eclectic water analysis was short-lived, rapidly overwhelmed by bacteriology. A second phase of the resurgence of biology began in 1884 when Koch's plate culture technique was introduced in Britain. This procedure quickly became popular, and by the late 1880s the number of bacteria in a water had become the chief measure of its purity, despite nearly universal acknowledgment that in regard to bacteria, type, and not number, was what mattered.

A third phase is apparent in the early 1890s, when it was becoming clear that purity was not simply measured in biological units, but determined by the interplay of biological processes. Whether a pathogen lived or died in a river was a question of its physiological requirements and competence in the "struggle for existence" against other stream organisms. Here was fertile ground for speculation. By the early 1890s the experts' explanations for the inevitability or impossibility of bacterial self-purification were almost entirely drawn from biology.

Recognition that water quality was a biological issue did nothing to change the character of the scientific institutions within which scientists dealt with water quality problems. Scientific discussion of water quality remained closely linked to water politics and political conditions had not changed. In London, for example, the struggle between public and private sectors for control of the water supply intensified in the decade preceding public takeover in 1901. Only the language of debate had changed. The question of whether disease germs inevitably exploded, debated by Tidy and Frankland in 1880-81, was typical of the topics of expert contradiction during the hearings of the 1892-93 Royal Commission on Metropolitan Water Supply. Experts now happily hypothesized about bacterial ways of life. One envisioned bacteria marching 1000 abreast between the pores of a sand filter, another described an osmotic force that would draw them to the surfaces of sand particles.[3]

This chapter traces the three phases of the biologicization of water quality issues in Britain and examines the incorporation of biology in water politics.

I. Populism in Water Analysis, 1875-1885

To Frankland and Wanklyn improvement in water analysis had meant more precise measurement and establishment of definite criteria for separating safe from harmful waters. In Wanklyn's best-selling manual of water analysis chemists were told that the ammonia process

---

[3] Charles Folkard, "The analysis of potable water with special reference to previous sewage contamination," MPICE 68(1881-82): 68; and James Mansergh in Royal Commission on Metropolitan Water Supply, Minutes of Evidence, P.P., 40 part i, 1893-94, [C.-7172.-I.], Q 10336, Q 10641 (E. Ray Lankester). [Cited as RCMWS Evidence.]

gave an accurate determination of the harmfulness of a water, and that thousandth gram differences in free and albuminoid ammonia truly distinguished good from bad water. Frankland appeared to be doing much the same thing -- claiming immense precision for the combustion process, promulgating standards for classifying waters.

By the late 1870s many sanitarians had rejected these goals in favor of a broadly based approach to water quality evaluation. They recognized that bad water had no characteristic chemical signature, believed opinions regarding safety should be based on environmental circumstances, qualitative chemical and physical observations, microscopical and biological examination. This new program for water analysis had nothing to do with the germ theory. Indeed, its practitioners shared no common model of morbid poisons. Instead, rather than marking an application of scientific discoveries in pathology and bacteriology, the new water analysis represented a retreat from science toward empiricism.[4] It reflected a change in the make-up of the community of water analysts, and concommitant changes in the situations in which water analysis was done and the expectations of what analysis ought to show.

No longer was sanitary expertise the monopoly of London consultants; by the late 1870s local sanitarians were becoming increasingly confident in their abilities and increasingly involved in water

---

[4] Cornelius J. Fox, Sanitary Examinations of Water, Air, and Food. A Handbook for the Medical Officer of Health (Philadelphia: Lea and Blakiston, 1878), p. 152.

analysis. Often these sanitarians were medical officers of health or public analysts (despite the fact that water analysis was not among the legal responsibilities of either position), employed by any of the manifold sanitary authorities set up under successive public health acts.[5] Increasingly local sanitarians were trained specifically for their jobs and were members of professional organizations. Diploma programs in public health medicine were available at the University of Dublin beginning in 1871, at Oxford and Cambridge beginning in 1875. The peripatetic Sanitary Institute of Great Britain, a sort of British Association for sanitary matters, and the Society of Public Analysts were both established in 1877.

The model for this de-centralized sanitary science came from John Simon's Local Government Board Medical Department. Simon had rejected Chadwick's universal sanitary engineering in favor of preventive medicine practiced by local sanitary officers experienced in epidemiology and sanitary inspection. Simon recognized quite different uses for water analysis than had the London chemists. Usually local sanitarians were not concerned about the general quality of a water supply but with fluctuations in water quality on a single street, perhaps even in a single house, which might mark the

---

[5] Ibid., p. 2; and "Water analysis," SR 10(1879): 27-28. On Medical Officers of Health see Jeanne Brand, Doctors and the State: the British Medical Profession and Government Action in Public Health, 1870-1912 (Baltimore: Johns Hopkins University Press, 1965), pp. 109-15. On Public Analysts see Ernst W. Stieb (with the collaboration of Glenn Sonnedecker), Drug Adulteration in 19th Century Britain (Madison: University of Wisconsin Press, 1966), pp. 181-84.

infection of a normally safe water. Local sanitarians had to know the normal quality of the water and the pollutions to which it might occasionally be subject. They needed tests which could quickly reveal changes.

Cornelius J. Fox's Sanitary Examinations of Water, Air, and Food (1878) illustrates the new perspective in water analysis. The book was written by a medical officer of health and for medical officers of health. Fox explained that his job demanded immediate evaluation of water complaints. His evaluations usually required less than 40 minutes, in sharp contrast to the "two or three months" a London expert [Frankland] would require.[6] Fox's methods were eclectic. The odor a sample gave when heated and the appearance and odor of its residue when ignited were important. Frankland and Wanklyn scorned such techniques.[7] Fox used Wanklyn's ammonia process but only to supplement other examinations. Regarding ammonia results alone as a measure of water quality was analogous to making a medical diagnosis from a single observation, Fox suggested.[8] Frankland's combustion process was certainly "beautiful," but com-

---

[6] Fox, Sanitary Examinations, pp. 139, 143. Reference to "foolscap papers of formidable appearance, . . . incomprehensible to all but experts" is likely to Frankland.

[7] Fox, Sanitary Examinations, pp. 14-15, 98-102; SR 1(1874): 199-200.

[8] Fox, Sanitary Examinations, pp. 46, 78. Fox and Wanklyn exchanged letters on this point (SR 6(1877): 15-16; SR 2(1875): 330, 348; M.F. Anderson, "Water analysis," SR 6(1877): 77-78; and J.A. Wanklyn, "Water analysis," SR 6(1877): 96).

pletely impractical for use by local sanitarians. Not only did a combustion analysis take two days, but the equipment cost a phenomenal 13 guineas and, Fox estimated, six months would be required for a novice in chemistry to learn to run it properly.[9] Fox agreed with Frankland that inorganic nitrogen compounds, as well as phosphates and excessive chlorine, usually represented previous sewage contamination, although he doubted PSC could be reliably quantified. PSC was not an indicator of resistant sewage germs, however, but important because one could never be sure that putrefaction would always have been completed before the water reached the consumer.[10] Fox believed that mineral constituents, particularly sulfates, were often responsible for water-borne illness and was appalled that Frankland and Wanklyn had paid so little attention to these.[11] Microscopical, though not necessarily microbiological, observations were also important. As Hassall had shown, an experienced microscopist could recognize in mud the stuff of sewage -- such things as meat fibers that had passed unscathed through the digestive tract.[12] Fox con-

---

[9] Fox, Sanitary Examinations, pp. 48-49.

[10] Ibid., pp. 70-78.

[11] Ibid., pp. 111-12, 143.

[12] Ibid., pp. 124-25. See also J.H. Timms, "On water analysis for sanitary purposes," SR n.s. 3(1881-82): 216-17; and Edmund A. Parkes, A Manual of Practical Hygiene, 6th ed., ed. by F.S.B. Francois de Chaumont, 2 vols. (New York: Wm. Wood, 1883), Vol. I, p. 70. Interest in such methods was probably sparked by Henry Clifton Sorby's quantitative microscopical method for distinguishing natural mud from sewage mud. See Sorby's testimony in Thames Navigation Act, 1870. In Arbitration. The Thames Conservators and the Metropolitan Board of Works, Minutes of Proceedings, Report, and Determination of Arbitrators, 1878, QQ 10924 ff.

cluded with two cardinal rules for interpreting water analyses: judgments must never be made from a single analytical character; and judgments must never be made without knowledge of the circumstances of the water -- whether it came from well, lake, river, etc.[13]

Between 1875 and 1885 such views of water analysis were widely espoused. They could be found in public health manuals by George Wilson, medical officer of health for mid-Warwickshire, and Edmund Parkes, professor of hygiene at the Army Medical School at Netley. In 1878 the chemist George Wigner, a prominent member of the Society of Public Analysts, developed a point system, in which many of the characteristics Fox had included were assigned numerical values, and water judged by the number of points it accumulated. Wigner gave values for 30 different observations, including total solids (1 point for every 5 grains), color (2 points for pale yellow, 4 points for yellow-green, 6 points for urine yellow), taste (2 points if like decayed leaves, 1 point if slightly saline, etc.), smell (2 points if flat, 6 points if urinous), as well as values for nitrates, nitrites, free and albuminoid ammonia, chlorine, oxygen absorbed, loss of weight on ignition, lead, and copper. Microscopical observations got the highest values, the presence of muscular fibers warranting 18 points. Fewer than 35 points meant a water was first class, more than 75 meant it was sewage.[14] While many quibbled with the number

---

[13] Fox, Sanitary Examinations, p. 155.

[14] George Wigner, "On the mode of statement of the results of

of points assigned to certain observations, there was general agreement that Wigner had the right idea. In modified form the scale was used for several years by members of the Society of Public Analysts.[15]

In 1881 the perspective of Fox and Wigner received scientific reinforcement as a result of two studies of the validity of chemical water analysis. Under sponsorship of the Local Government Board, R. D. Cory studied the ability of the ammonia, combustion, and permanganate processes to distinguish water contaminated by the excreta of typhoid sufferers. All the processes failed this test. In the same year even more exhaustive tests of these processes were done in the United States by J.W. Mallet under commission from the U.S. National Board of Health. Mallet too found none of the processes distinguished infected water. Mallet went beyond Cory and measured

---

water analysis and the formation of a numerical scale for the valuation of the impurities in drinking waters," The Analyst 2(1878): 215; also in SR 8(1878): 235-38, 337-39. On Wigner see R.C. Chirnside and J.H. Hamnence, The Practising Chemists. A History of the Society for Analytical Chemistry, 1874-1974 (London: The Society, 1974), pp. 15, 65-66. A photograph of Wigner follows p. 20. Also see Parkes, Manual of Practical Hygiene, Vol. I, pp. 65-100; and George Wilson, A Handbook of Hygiene and Sanitary Science, 2nd ed. (London: Churchill, 1873), pp. 153-62; and 4th ed. (London: Churchill, 1879), pp. 165-88.

[15] Wigner, "On the mode of statement," pp. 216-20; Chirnside and Hamnence, Practising Chemists, pp. 65-66; Fox, Sanitary Examinations, pp. 158-59; Wigner, "On the valuation of the relative impurities of potable waters," The Analyst 6(1880): 111-25; "Analysis of public water supplies in England. Instructions for analysis prepared by a committee appointed by the Society of Public Analysts," The Analyst 6(1880): 127-39; Wigner in Sanitary Record, SR 7(1877): 256; Wigner, "On the outbreak of typhoid fever at Baxenden and Accrington," SR 7 (1877): 262-64; Wigner, "Water analysis," SR 8(1878): 27, 78-79, 126-27; G.B. Longstaff, "Water analysis," SR 7(1877): 402; R. Barnes

the accuracy of the processes by various methods. He studied the amount of variation when the same sample was analyzed by different chemists, the effects of dilution on error, and the proportion of organic matter really being detected. Both investigators came to similar conclusions: processes for analyzing organic matter in water might provide useful information but should never be the sole basis for judging water.[16]

The studies of Mallet and Cory were enthusiastically received by local sanitarians. They were taken as proof that the domination of water analysis by Frankland and Wanklyn had ended. Recondite methods and impressive quantification had been shown to be inconsequential. Henceforth sanitarians might trust their own judgment to make decisions on water quality.[17]

Austin, "Water analysis," SR 8(1878): 63, 110-11.

[16] J.W. Mallet, "The determination of the organic matter in potable water," CN 46(1882): 63-66, 72-75, 90-92, 101-2; R.D. Cory, "On the results of the examination of certain samples of water purposely polluted with excrements from enteric fever patients, and with other matters," in 11th Annual Report of the Local Government Board. Supplement containing the report of the Medical Officer for 1881, P.P., 30 part ii, 1882, [C.-3337.-I.], App. B-1, pp. 127-65. See also Buchanan's introductory remarks, pp. xvii-xxi. Both studies were widely reported: "The sanitary requirements of water analysis," SR n.s. 4(1882-83): 346-47; "Water Analysis," SR n.s. 4(1882-83): 364-65; "Special scientific investigations on sanitary subjects in America," SR n.s. 4(1882-83): 403-4; "Water examination," JRSA 31(1882-83): 215.

[17] Alfred Ashby, "The fallacies of empirical standards in water analysis as told by the story of a polluted well," SR n.s. 5(1883-84): 533; Charles E. Cassal and B.H. Whitelegge, "Remarks on the examination of water for sanitary purposes," SR n.s. 5(1883-84): 427-29, 479-82; Timms, "On water analysis for sanitary purposes," pp. 216-17; G. Gore, "Analysis of drinking water for ammonia," CN 50 (1884): 182-86; Jabez Hogg in disc. of William Anderson, "The Antwerp water works," MPICE 72(1882-83): 61; "The fallacies of water

Ironically, the perspective of local sanitarians in the late 1870s was the same perspective Frankland had been advocating since 1866: that since harmful matters in water could not be measured directly, the analyst must watch for indications of danger. This new attitude among water analysts was manifested in rejection of Wanklyn's contention that his process directly measured morbid poisons in water, but it was not manifested in an approbation of Frankland. The reasons for this were social. Despite compatibility of interpretive principles, local sanitarians regarded Frankland as representing a concept of water analysis antithetical to that which they hoped to establish. In opposing the founding of the Institute of Chemistry the Sanitary Record noted that the problem with water analysis had not been a lack of competence to provide useful information, but a concentration of water analysis practice in the hands of a very few expert chemists who made their livings contradicting one another in courtrooms. Frankland was clearly included in that group.[18]

Thus, by the late 1870s local sanitarians had declared independence from the metropolitan experts. Necessarily new methods would be adopted. A battery of less sophisticated qualitative techniques available to those with only minimal chemical training would replace the delicate and expensive techniques of the experts. Frequent

---

analysis," SR n.s. 6(1884-85): 406; "Rev. of Fox, Sanitary Examinations," Lancet, 1878, ii, p. 663.

[18] "The Institute of Chemistry," SR 4(1876): 232-33.

examinations of changing conditions would replace the expert's one-time verdict. Because samples could be examined when fresh, biological examinations were practical.[19]

II. Biology Returns to Water Analysis, 1875-1885

Fox, Wilson, Parkes, and Wigner all stressed biological examination as one of the factors to be included in an inventory of water quality characteristics. In the 1850s biological examinations of water had held equal status with chemical examinations. There had been good reasons for rejecting biological examinations. Although some biological studies had been done in impressive detail, they had usually been partisan -- linked to campaigns to get water supplies changed. Their apparent purpose had been to nauseate the water-drinking public with visions of swallowed insects. As sanitarians became convinced that their quarry was some obscure chemical substance or process, biological argumentation appeared only so much sound and fury, an obfuscation of the real issues at hand. By 1872 even Arthur Hassall, champion of biological water assay during

[19] In one of the rare cases in which biological examinations of water were utilized to a significant extent, the research undertaken by the Massachusetts State Board of Health, it is noteworthy that arrangements were made for samples to be analyzed within 24 hours. See Thomas M. Drown, "The Chemical Examination of waters and the interpretation of analyses," in Massachusetts State Board of Health, _Report on Water Supply and Sewerage, part 1. Examinations by the State Board of Health of the Water Supplies and Inland Waters of Massachusetts, 1887-1890_ (Boston: Wright and Potter, 1890), pp. 522-23.

the 1850s, had expressed confidence in chemical water analysis, though he still believed microscopical studies should be done. Hassall wrote: "this is certain, that when the chemist and the physicist [actually an engineer, Francis Bolton, who was charged with inspecting the London water works] assure us the water is pure, we know it is salubrious."[20] Hassall regarded as irrelevant Burdon Sanderson's studies of water-borne germs which appeared in the early 1870s.[21]

Biology was slow to shake off sensationalism. During the mid 1870s the opthamologist and microscopist Jabez Hogg campaigned against the London water supply, using arguments and demonstrations almost identical to those used by Hassall, Redfern, and Lankester 25 years earlier. Hogg even allied himself with the sponsor of the 1852 studies of Lankester and Redfern, engineer Samuel Homersham, who still advocated an alternative supply for London. Like his predecessors, Hogg used strong language to describe aquatic life. Almost every living thing in the water was "noxious": "filariae and larvae of the most noxious kinds"; "small fish, eels, and numerous noxious animals."[22] Aquatic plants did not aid in the oxidation of

---

[20] "The Registrar General on the water supply of London," Food, Water, and Air 1(1872-73): 119.

[21] "Dr. Sanderson's experiments on the growth of microzymes in water," Food, Water, and Air 1(1871-72): 8; and "Dr. Tyndall's discourse on dust and disease," Food, Water, and Air 1(1871-72): 47.

[22] Jabez Hogg, "River pollution with special reference to impure water supply," JRSA 23(1874-75): 581; and A Microscopical Examination of Certain Waters submitted to Jabez Hogg and a Chemical Analysis by Dugald Campbell, with introductory notes by S.C. Homersham (London: Trounce, 1874), pp. 21-24.

sewage, but simply increased "mud accumulations, afford[ed] protection to fish, and [became] the breeding ground of animal life of all kinds." Infusorians were pathogenic, Hogg declared, citing Swedish claims of pathogenic paramecia. These infusoria, however, did not lessen the threat from still "more subtle and dangerous organisms . . . held in solution or suspension," or from unorganized decaying matter itself. Where a hypothetical morbid poison lent itself to the condemnation of river water, Hogg uncritically employed it.[23] There was no incentive to narrow the list of poisons that could be in polluted water, for Hogg was not seeking an answer to a scientific problem but crusading for a political change. Hogg maintained that all water examinations should include microscopical examination, but it is clear that the microscopy Hogg advocated would not be objective.

In his monthly reports to the Registrar General on the London water supply, Edward Frankland made similar, if less brazen use of biological examinations. Among the techniques with which Frankland undermined confidence in London water was the listing of "living and moving organisms" in each company's water. Frankland was not concerned with these organisms as indicators of aquatic conditions, nor, with a few exceptions, did he characterize or identify them.[24] The ostensible reason for listing the presence of organisms was to demonstrate which of the companies' filters were working well enough

---

[23] Hogg, "River pollution," pp. 580-81, 584-87.

[24] But see BMJ, 29 January 1870, p. 110.

to catch hypothetical pathogenic organisms. The effect of these reports was quite different; "living and moving organisms" was more unsettling even than "previous sewage contamination" in spurring Londoners to complain to newspaper editors and members of parliament.[25]

Fox, Wilson, Parkes, and Wigner saw biological evidence in quite a different manner. Unlike Hogg, who apparently knew what biological examination would indicate before any organisms were collected, they were unsure what to make of biological evidence. They advocated biological examination because it was one of the ways of characterizing water. The central premise underlying their biological water assay was that life indicated impurity. As Fox put it:

> the existence of animal life in a water affords good evidence in itself of the presence of a very sensible amount of organic matter, alias filth. . . . These little creatures flourish on what we call organic matter, and in perfectly pure water they cannot live.[26]

---

[25] RPPC/1868/6th Report (The Domestic Water Supply of Great Britain), P.P., 33, 1874, [C.-1112.], p. 4; Frankland, "Annual Report for 1869," in Reports on the Analysis of Waters supplied by the Metropolitan Water companies during 1869, 1870, and 1871, by Professor Frankland; Copy of his letter to the Registrar General, dated 10th July 1869, and analyses of Metropolitan Water Supply for October 1871 and January 1872, P.P., 49, 1872, (99.), p. 2. For problems with the concept see Copy of Reports made to the Board of Trade by the Water Examiner appointed under the Metropolis Water Act, 1871, P.P., 49, 1872, (88.), p. 4; "Impure water supply," SR 2(1875): 151; BMJ, 10 April 1875, p. 486; "Report by Mr. J. Netten Radcliffe on the turbidity of the water supplied by certain London companies," in 12th Annual Report of the Medical Officer of the Privy Council for 1869, P.P., 38, 1870, [C.-208.], App. 5, pp. 141-67; and in BMJ, 22 Jan. 1870, pp. 88-89; "Miniature aquariums," The Builder 35(1877): 770; SR 7(1877): 141.

Life needed food; at the bottom of the food chain that food was conceived to be organic impurity. This assumption was made clear in the furor that followed the discovery of eels in the mains of the East London water company. As the Morning Post noted, "the fact that . . . animalculae are not discovered by the chemists the company employs does not reckon for much . . . indeed, one visible eel is a more trustworthy proof of the existence of living organisms than a dozen reports of analysts who declare that no such organisms are in the water."[27] Wigner suggested that the mass of life was proportional to the insalubrity of the water. In his point system, bacteria warranted three points, "other similar growths in greater quantity" got four, and "few living organisms" six.[28]

Beyond this, sanitarians hoped to correlate species with habitat. Thus *Daphnia pulex* and *Cyclops quadricornis* were thought to be indicators of spring water.[29] It was assumed that these habitat distinctions might have hygienic significance. The filth sustaining *Daphnia* and *Cyclops* was usually safe, while the diet of the amoeba and the worm anguillula was dangerous.[30] Reaching a consensus on

---

[26] Fox, Sanitary Examinations, p. 125. For similar views see A. Wynter Blyth, "Water," A Dictionary of Hygiene and Public Health (London: Charles Griffin, 1876), p. 634; and J.D. MacDonald, A Guide to the Microscopical Examination of Drinking Water (London: Churchill, 1883), p. ix.

[27] Morning Post, 9 October 1886, in PRO MH 29 9.

[28] Wigner, "On the mode of statement," p. 215.

[29] MacDonald, Guide, p. 1; Fox, Sanitary Examinations, p. 129; Parkes, Manual of Hygiene, p. 73; W. Ivison Macadam, "Animal life in fresh water reservoirs," SR 7(1877): 222.

[30] Blyth, Dict. Hygiene, p. 637; Parkes, Manual of Hygiene, p. 72;

such classifications was elusive, however: A. Wynter Blyth regarded paramecia as reliable indicators of sewage, Fox did not. William Crookes observed that many regarded cress as an indicator of pure water, yet in France it was grown in sewage.[31]

The scavenger paradox that had so troubled sanitarians in the 1850s persisted. Organisms indicated impurity because impurity was their food. Their metabolization of impurity constituted purification, but their own life processes might produce a new type of impurity. Ought one to rejoice at the presence of purifiers or to despair that purification was necessary? Could scavengers be relied upon to have finished the job before the water was drunk? Attempting to resolve these questions, J.D. MacDonald, the Navy's professor of hygiene and author of A Guide to the Microscopical Examination of Water, suggested that a fat ciliate be regarded as a danger sign. In purer waters, with insufficient filth for ciliates to thrive, "the transparency and leanness of their bodies, and the restlessness of their search for aliment, will show that they are in a half-starved condition."[32] Fox too was unable to resolve the issue. He noted that on the one hand organisms indicated filth, and suggested they might carry morbid poisons; on the other that they were "scavengers that assist, like plants in . . . purification."[33]

Fox, Sanitary Examinations, p. 126.

[31] "Rev. of Fox, Sanitary Examinations," CN 38(1878): 219.

[32] MacDonald, Guide, p. 50.

[33] Fox, Sanitary Examinations, pp. 125-28. For the continued existence of the notion of organisms as purifiers see C. Meymott Tidy,

It is clear that acceptance of the germ theory was not responsible for the interest of local sanitarians in biological water analysis. Parkes and Fox recognized the existence of specific disease poisons which sometimes contaminated water but did not believe these were independent organisms. As late as 1883, Parkes' Manual still regarded putrefying matter as the dangerous substance in water; bacteria were significant as an indicator of putrefying matter.[34] Fox too thought that bacteria were not harmful, but that "poisons of several zymotic diseases find a congenial soil amongst such organisms, which act as carriers, to which they attach themselves."[35]

Instead, the biology that MacDonald, Fox, and Parkes advocated was a poorer version of what Hassall, Edwin Lankester, and Redfern had been doing in the 1850s: an attempt to associate habitat with purity. As Crookes, who advocated bringing biology into water analysis, observed, biological knowledge was not sufficient for this to be done with confidence; at best correlations were tentative, requiring broad categories of habitat and equally loose conceptions of purity and impurity.[36] In the 1850s biological water analysis had

---

"River water," J. Chem. Soc. 37(1880): 301; and Arthur Angell, "The discolouration of Southampton water," SR 7(1877): 334-35.

[34] Parkes, Manual of Hygiene, p. 71.

[35] Fox, Sanitary Examinations, p. 126.

[36] Most successful work in this direction was done by Alphonse Gerardin in France ("Altération, Corruption et assainessment des rivières," Annales d'Hygiène Publique et de Médecine Légale, 2nd series 43(1875): 5-41). In the view of Gerardin and J.A. Dumas, the famous chemist, the ability of a water to support higher life forms was the chief measure of its salubrity (Gerardin, "Altération," p. 6; Dumas to Frankland, 26 April 1872, in RPPC/1868/4th Rept. (The

not simply been advocated, but frequently used. Such appears not to have been the case in the late 1870s and early 1880s. Although writers of public health manuals were excited about biological water analysis, there is scant evidence of sanitarians practicing it.[37] This application of ecology to water quality evaluation was short lived. By the late 1880s a very different sort of biology had replaced it. This was bacteriology, specifically the counting of bacterial colonies on a culture plate.

III. Percy Frankland and the Politics of Bacteriological Water Analysis

During the late 1880s the deluge of bacteriological research, which had risen in France and Germany, swept across Britain. Bacteriological water analysis was a profoundly different scientific enterprise from water analysis based on natural history. Bacteriology encouraged professionalism. A good bacteriologist had extensive methodological training, laboratory facilities, was able to read foreign languages, had access to foreign colleagues. Despite early attempts to fit bacteriology into the populist water analysis of Fox

---

Rivers of Scotland), P.P., 34, 1872, [C.-603.], pp. 300-1). In Britain, Crookes admired and publicized Gerardin's work ("Rev. of Water Supply of the State of New Jersey," CN 37(1878): 248; "Rev. of Fox, Sanitary Examinations," CN 38(1878): 218-19).

[37] Fox mentions Wigner condemning a well on microscopical (and presumably in part microbiological) grounds which would have been acceptable on the basis of chemical tests (Fox, Sanitary Examinations, p. 128).

and Parkes, water bacteriology quickly became the province of experts. In the late 1880s many London water chemists underwent metamorphosis and emerged as water bacteriologists. Crookes, Odling, Tidy, and Edward Frankland dabbled in water bacteriology; Percy Frankland became a first-class bacteriologist.[38]

The techniques for culturing bacteria developed by Robert Koch were introduced to the British scientific public in 1884. The International Health Exhibition, held that summer at South Kensington, included a biological laboratory depicting the techniques of Koch and Pasteur, arranged by the surgeon Watson Cheyne. The Lancet was particularly intrigued and treated its readers to a "deluxe" catalog of the exhibit, including illustrations of many bacterial species as they appeared in culture tubes and under the microscope. The Lancet also gave directions for culturing water bacteria. A drop of sample water was added to molten sterile gelatin in a test tube. After shaking the mixture, the analyst poured the contents onto a level glass plate (hence the term "plate culture"), about five by six inches, which he immediately covered. Colonies appeared in one or two days. Instructions were also given for isolating species by taking further samples from individual colonies.[39]

---

[38] H.E. Armstrong thought it a good thing that chemists were becoming bacteriologists. Biologists would look at matters "from too one-sided a point of view," and, he implied, did not have good enough laboratory technique (in disc. of Percy Frankland, "New aspects of filtration and other methods of water treatment; the gelatine process of water examination," J. Soc. Chemical Industry 4 (1885): 709).

[39] "The biological laboratory at the Health Exhibition," Lancet,

Several British sanitarians, notably Robert Angus Smith, had advocated culturing techniques of various sorts prior to 1884, but such practices had not become widespread.[40] In 1884 times were right for such techniques. Germs were no longer a theory, Koch was a celebrated authority, chemistry had proven impotent, and the plate culture technique was seemingly easy. Many sanitarians, Smith among them, thought they had witnessed the millenium in water analysis. Bacteriology would provide simple and objective procedures to end years of "wretched wrangling."[41]

Early optimism proved unfounded, despite widespread appreciation of the potential of bacteriology and employment of Koch's techniques. Bacteriology in the 1880s was not the monolithic institution automatically grinding out answers to sanitary problems that some had envisioned. Instead, it was an immature science, in flux methodologically, conceptually, socially. It was not clear who would be using bacteriological methods. Some, like Francis Bolton, the London water

---

1884, ii, pp. 251-52, 332-33, 380-81, 557-58, 609-10, 705-6, 751-52, esp. 705-6; "Scientific aspects of the Health Exhibition," Lancet, 1884, ii, p. 24.

[40] John M. Eyler, "The conversion of Angus Smith: the changing role of chemistry and biology in sanitary science, 1850-1880," BHM 54(1980): 230; Charles Heisch, "On organic matter in water," J. Chem. Soc. 23(1870): 371-75. Some discussion of Koch's techniques had appeared in the preceding year and a half (Cassall and Whitelegge, "Remarks on the examination of waters," pp. 479-82; R. Angus Smith, "Development of living germs in water," SR n.s. 4(1882-83): 344-47; "The new method of testing water," SR n.s. 4(1882-83): 360, 521; J. Hogg in disc. of William Anderson, "The Antwerp water works," pp. 62, 69). An improved version was published by C.J.H. Warden, who was working in Koch's laboratory ("The biological examination of water," CN 52(1885): 52-54, 66-68, 73-76, 89, 101-4).

[41] Ashby, "The fallacies of empirical standards," p. 533.

examiner, saw plate culturing as the central technique in the management of water works and believed it could be used by engineers. Others, recognizing the technical dexterity required for accurate results, conceived (correctly) that use of bacteriological methods would be the preserve of highly trained experts. At a variety of levels there was disagreement about what bacteriological examinations signified. Even in the early 1890s some bacteriologists still doubted the integrity and even the reality of bacterial species. If supposedly harmless forms could at any time become pathogens there was no point in determinative bacteriology.[42] In Britain, bacteriological water analysis was largely confined to counting the colonies that grew on a gelatine-coated plate, but that technique itself was fraught with ambiguity. Some colonies might represent more than one bacterium; some bacteria did not grow in gelatine, others grew only slowly. The number of bacteria found in a sample was affected by the time that had elapsed between sampling and culturing.

Thus, far from liberating water analysis from partisan wrangling, bacteriology became the dominant mode of discourse of such wrangling. By the late 1880s the question of the validity and utility of bacteriological water analysis was as much a political issue as the validity of Frankland's combustion process had been in the 1870s. The alignment of positions on the issue was similar to that which had characterized debate on water analysis as far back as 1830. On

---

[42] E.E. Klein, Microorganisms and Disease, 3rd ed. (London: MacMillan, 1886), p. 222.

one side were those who found bacteriological water analyses a meaningful measure of salubrity, on the other those who felt that vastly reducing the number of bacteria in water was still insufficient purification and that bacterial counts did not provide trustworthy basis for proclaiming sewage-polluted water safe. Plate cultures showed that sand filters removed a large proportion of water bacteria, usually more than 95%, and the technique was quickly pressed into service in defense of the London water companies.

Much of the responsibility for politicizing bacteriological water analysis lies with Percy Frankland. Percy (1858-1946) was Edward Frankland's second son. He was educated at the School of Mines and obtained a Ph.D. from Würzburg in organic chemistry in 1880. For most of the 1880s Percy Frankland worked as a demonstrator at the School of Mines and as a partner in his father's water laboratory. In 1888 he was appointed chemistry professor at Dundee and in 1894 moved on to a similar post at Birmingham. With further training at Koch's laboratory in Berlin and aided by his biologist wife Grace Toynbee Frankland, Percy Frankland became one of the leading late nineteenth century British bacteriologists.[43]

Percy Frankland apparently discovered plate culture techniques at Watson Cheyne's 1884 exhibition.[44] In the beginning of 1885 he

---

[43] W.H. Garner, "Percy Faraday Frankland," J. Chem. Soc., 1948, pt. iii, pp. 1996-98. The only reference to work in Koch's laboratory I have found is Percy Frankland's statement in disc. of Bischof, "Notes on Dr. Koch's water test," p. 120.

[44] Colin Russell, "Percy Frankland: the iron gate of examination," Chemistry in Britain 13(1977): 425.

began using plate cultures to test the efficacy of filtration by comparing the number of bacteria in water before and after filtration. In May he read a paper to the Royal Society on the subject. He noted that sand filters were initially highly effective but lost effectiveness with time; that animal charcoal, thought excellent by chemists, was not, while spongy iron filters were, as his father had maintained, excellent removers of bacteria.[45] Still he was skeptical of plate culture results. Koch's methods were "exceedingly beautiful and ingenious tests for ascertaining the number of individual organisms present in a given water," but were of little value since pathogens were presumably only exceptional inhabitants of water, and "their absence on a given occasion [could] afford no permanent security whatever."[46] Despite his reservations, Percy Frankland continued to use plate culture techniques. In November 1885, at the request of Francis Bolton, the water examiner, Percy Frankland began submitting to the Local Government Board reports on the bacterial efficiency of the water companies' sand filters. For

---

[45] Percy Frankland, "The removal of micro-organisms from water," CN 52(1885): 27-29. James Dewar believed E. Frankland had money in filtration schemes, and he was probably referring to Bischof's spongy iron filter scheme (Imperial College, Lyon Playfair Papers, General Correspondence, C231, J. Dewar to L. Playfair, 15 Jan. 1887). For Frankland's surprisingly strong advocacy of spongy iron see his remarks in disc. of William Anderson, "The Antwerp water works," pp. 45-47. Compare Percy Frankland's 1885 remarks with his 1887 views (Percy Frankland, "Recent bacteriological research in connection with water supply," J. Soc. Chem. Ind. 6(1887): 319).

[46] Percy Frankland, "The selection of domestic water supplies," SR n.s. 6(1884-85): 549.

the next three years these were published in the monthly water reports, along with his father's chemical analyses and those done for the water companies by Crookes, Odling, and Tidy. Organisms in filtered water were "strikingly less" than those in river water, he noted.[47]

Here politics entered. The 1880s were a time of anti-germ hysteria. Announcement that the "unknown and covert enemy" lurked in the water supply horrified the public. Following publication of the first filtration report, Percy Frankland and Bolton wrote to the Times to reassure the public that microorganisms were ubiquitous, and that those which got through the filters were neither necessarily from sewage nor necessarily dangerous.[48]

But if the bacteria in the water were not cause for concern, what did Percy Frankland's results signify? Did 95% reduction in bacteria mean water was 95% safer, as the water companies implied? Did filters remove all types of bacteria equally well? Did the gelatin medium accurately reflect the bacterial population of a water? These questions were quickly raised and were to dominate water analysis for the next decade.

Formidable skepticism toward plate culture results came from within the Local Government Board itself. Percy Frankland's reports

---

[47] Percy Frankland to Local Government Board, 7 Nov. 1885 in PRO MH 29 8. See also draft October report, PRO MH 29 8, #110229/85.

[48] William Anderson, "The interdependence of abstract science and engineering," MPICE 114(1892-93): 278-79; Percy Frankland and Francis Bolton to Times, Times, 4 December 1885; PRO MH 29 8.

were submitted to Bolton, the water examiner, whose job included insuring that the companies "effectively" filtered river water. Bolton saw bacteria counts as an objective means for checking filtration efficiency. On the other hand, George Buchanan, the Board's medical officer, was concerned that substantial reductions in bacteria would be improperly regarded as having hygienic significance. Noting that gelatin had not been proved an effective medium for all microbe species, he objected to the phrase "relative freedom from organic life." Percy Frankland accepted this censorship, but he was not contrite. Whatever its faults, he argued, bacteriological analysis certainly gave a better indication of relative salubrity than chemical analysis.[49] Defying Buchanan in spirit if not in letter, Bolton and Percy Frankland put their case before the public. The December report contained Percy Frankland's more cautious discussion of filtration results. However, it also contained his defiant acquiesence to Buchanan's criticisms and an introduction to the filtration data by Bolton even bolder than Percy Frankland's original language. The effect was to counteract completely the minor revisions Buchanan had requested and to make the Board's officers appear as censors of science as well. This airing of the Board's dirty linen suggests that Bolton was committed to making bacteria counts central in water evaluation.[50]

---

[49] Bolton to Owen, 15 December 1885, PRO MH 29 8.

[50] "Report for December 1885," PRO MH 29 9, #8074/87; Percy Frankland, "Bacterial report for December 1885," PRO MH 29 9.

Buchanan's response was to consult the Board's consulting bacteriologist, E.E. Klein, on the significance of plate culture results. Klein (1844-1925), "father of bacteriology in Britain," was a Yugoslav histologist brought to England to run the Brown Institution. Although his bacteriology was self-taught, Klein, as professor of advanced bacteriology at St. Bartholomew's Hospital, was one of the main British teachers of bacteriology.[51] Klein shared Buchanan's skepticism. Some microbe species did not grow in gelatine, and it was not proved that that medium was preferential for pathogens. Plate cultures were often poorly done and therefore gave no reliable indication of the number of microbes actually in the sample. Finally, the procedure, when done by water company engineers, "might tend to erroneous and mischevious conclusions." Klein concluded:

> The method of gelatine plate culture is excellent, if it is required to determine which of several samples of water contains more organisms capable of growing in gelatine, provided the gelatine cultures are made at the same time and place. It gives no notion of the absolute numbers of organisms, or of their character, i.e. septic or pathogenic.[52]

On the basis of Klein's letter Bolton was mildly rebuked, and he agreed to print bacteriological results "without comment."[53]

---

[51] "Edward Emanuel Klein," Proc. Royal Society of London, Biological Series 98(1925): xxv-xxix.

[52] "Internal correspondence, memos, etc., c. February 1886," PRO MH 29 10, # after $12665k_2$/87. Klein to Buchanan, 29 Jan. 1886.

[53] Ibid.; "Bolton to L.G.B., 6-8 February 1886," PRO MH 29 10.

The fear of Klein and Buchanan that plate culture results might lead to "erroneous and mischevious" conclusions shows their recognition of the great public relations impact bacteria counts could have in the struggle between the water companies and those attempting to make the supply public. By the end of 1885 bacteriology had already been politicized, used by the water companies with considerable success to enhance the reputation of Thames and Lea water. Crookes, Odling, and Tidy pointed to the great reduction in bacteria due to filtration and quickly began studies of their own which showed that anthrax bacilli could not live long in London water. Even though the existence of reliable techniques for isolating pathogens was a matter of debate, they regarded absence of pathogens as significant. In December 1885 these three correctly predicted that "the attempt from the biological standpoint to condemn London water will. . . be as conspicuous a failure, . . . as the attempt to condemn it from the chemical standpoint of organic carbon and nitrogen."[54] As Percy Frankland pointed out a decade later, bacteriology quickly became a friend to the water companies.[55]

Thus, from the outset, Koch's techniques were subjected to the stresses and strains of British water politics. Plate cultures became another of the public relations weapons used in the battles for

---

[54] "Metropolitan water supply," CN 52(1885): 296; CN 53(1886): 91; CN 54(1886): 44-45, 183, 211-12, 269, 318; William Odling, "Micro-organisms in drinking water," J. Soc. Chem. Ind. 5(1886): 544.

[55] Percy Frankland, "London's water supply," SR n.s. 19(1897): 504.

control of water supplies. Yet the politicization of bacteriology was not simply an application of scientific tools for political purposes. As with the self-purification controversy, certain bacteriological techniques and concepts had inherent political implications. Again, Percy Frankland's career exemplifies this. Between mid-1885 and mid-1887 Percy Frankland's ardent opposition to domestic use of the sewage-polluted Thames and Lea turned to equally strong support. This switch in loyalty resulted from his research on the bacteria-removing capabilities of sand filters.

During these two years Percy Frankland was trying to understand what his filtration results signified. In March 1886 he still saw them primarily as Bolton had. By showing when filters needed cleaning they aided in the management of water works. It was tempting, however, to consider the "wider significance," to think in terms of true purification. Already, in February 1886, he had vigorously defended his studies against H.E. Armstrong's contention that bacteriological tests were so immature that their results should not be publicized. The studies showed that "sand filtration, hitherto . . . regarded by most authorities as of little value, was really an exceedingly important process in rendering river water more fit for domestic use."[56]

The 1886 results on the London filters showed even higher rates

---

[56] Percy Frankland and Sir Francis Bolton, <u>Lectures on the Collection, Storage, Purification, and Examination of Water</u> (London: Harrison, 1886), pp. 27-28; Percy Frankland in disc. of Bischof, "Notes on Dr. Koch's water test," p. 121.

of baoterial removal than in 1885, about 98%. In April 1886 he spoke to the Institution of Civil Engineers on the bacteria-removing properties of various methods of purification. Unstated, but implied, was the notion that the degree of bacterial removal was the degree of purification. The implication was challenged. Jabez Hogg, doubtless astonished at the facility with which biology could defend pollution, maintained that since bacteria could multiply, "the one or two colonies, or even one or two individuals any process left behind, was a danger which could not be contemplated without a shudder."[57] Gustaf Bischof, a German-born consulting chemist who had learnt culture techniques in Koch's laboratory, asserted that in bacteriology there was no relation between number and insalubrity. Further, Bischof claimed the organisms Percy Frankland was culturing were chiefly aerobic bacteria, indicators of well-aerated, good water.[58] Percy Frankland replied that he had never claimed hygienic significance for his results and that

> investigation could be carried out without any reference to the influence of those micro-organisms upon health, the problem being simply to ascertain whether and to what extent the various processes of purification had the power of removing micro-organisms in general.

There was a major proviso, however. Since pathogens were presumably physically indistinguishable from other bacteria, it could be assumed that they were removed in the same proportion as "micro-organisms in

---

[57] J. Hogg in disc. of Percy Frankland, "Water purification: its biological and chemical basis," MPICE 85(1885-86): 229.

[58] G. Bischof in ibid., pp. 224-26.

general."[59]

A year later Percy Frankland again tried to explain what his results meant:

> I wish it to be clearly understood that no conclusions whatever, as to the relative excellence of the various waters, are to be drawn from their greater or less freedom from micro-organisms, any more than it is possible, on the strength of chemical composition, to say that one water is more wholesome than another.[60]

This was no help since there was enormous disagreement among sanitarians regarding the utility of chemical water analysis. More explanation led to more ambiguity.

> On the other hand, these determinations undoubtedly do indicate what would be the probable fate of any harmful organisms gaining access to the sources of supply, and what is the relative chance of their reaching the consumers; _for that method of treatment which abolishes the largest proportion of organisms of all kinds is also the most likely to abolish any pathogenic forms should they be present_.[61]

Predictably, Bischof reiterated his objection that Percy Frankland was attaching great significance to meaningless numbers.[62]

Slowly, Percy Frankland was gravitating toward a philosophy of water quality evaluation in which his filtration studies assumed greatest importance. Regarded from the assumption of purificatory nihilism advocated by Edward Frankland, filtration studies had little

---

[59] Percy Frankland in ibid., p. 244.

[60] Percy Frankland, "Recent bacteriological research in connection with water supply," p. 319.

[61] Ibid. Compare with G. Bischof, "Dr. Koch's gelatine-peptone water test," _CN_ 53(1886): 206, to which Percy Frankland's statement was probably a direct response. Italics orginal.

[62] Bischof in Percy Frankland, "Recent research," pp. 323-24.

significance. Edward Frankland had always admitted that purification processes improved water, but maintained that sewage-polluted water supplies must be rejected because it was impossible to tell when purification was complete. The view was incapable of disproof since it was an axiom about water quality which could be undermined neither by Percy's filtration studies nor by any empirical demonstration. Viewed from this perspective the filtration studies showed London water safer, yet not safe. As late as March 1886 Percy Frankland appears to have shared this perspective.

Assume on the other hand that sewage-polluted water could be purified and the filtration results acquired vast importance. It was tempting to assume that filtered water was virtually pure, to focus on the organisms removed rather than the few not removed. In such a perspective Percy Frankland was the discoverer of the scientific basis of water purification and of the explanation of the safety of London's water supply.

By mid-1887 Percy Frankland's conversion was complete and the efficiency of sand filters in removing bacteria had become his claim to fame. In the next decade he addressed gatherings of sanitarians frequently, regaling audiences with the story of how sand filters had been regarded as ineffective (by chemists) prior to his momentous discovery.[63] He upheld the interests of the water companies

---

[63] Percy Frankland, "The bacterial examination of drinking water," in Seventh International Congress of Hygiene and Demography, Reports of the Meetings and Discussions held in London, August 10-17, 1891 (London: The Society of Medical Officers of Health/E.W. Allen, [1891]), p. 168; idem, "The filtration of water for town supply,"

and was employed by them as a consultant. In October 1886 he carried out a study for the East London company which purported to prove that dead eels discovered en masse in the company's mains caused no harm to water drinkers. In this study he was associated with Tidy, whom he had so viciously scorned two years earlier as a dupe of the companies. Percy Frankland's report on the eels utilized neither the new bacteriology nor chemical analysis. He attributed typhoid in part to emanations from garbage. Buchanan correctly regarded the report as "red herrings drawn across the scent" and assigned his own investigators to study the problem.[64]

After 1887 Percy Frankland increasingly took positions on water issues which either supported the water companies or enhanced the importance of his own research, and usually did both. He praised the water companies' engineers for having developed successful filters despite ridicule from chemists. Removal of 95% of bacteria was impressive in itself, but he hinted that this number even did not reflect the full extent of purification. The few microbes in filtered water might not have come through the filter, but represent new, post-filter contamination. Water mains, after all, were not

---

TSIGB 8(1886-87): 276-84; idem, "On the application of bacteriology to questions relating to water supply," TSIGB 9(1888): 369-77, 394; idem, in SR n.s. 15(1893-94): 141; idem, "London's water supply," SR n.s. 19(1897): 504-5; idem, in Royal Commission on Metropolitan Water Supply, Appendices, P.P., 40 pt. ii, 1893-94, [C.-7172.-II.], Appendix C 63, pp. 462-66.

[64] Percy Frankland, "The upper Thames as a source of water supply," JRSA 32(1883-84): 565-66; PRO MH 29 9, #92059/86; and A. deC. Scott and W.H. Power to the L.G.B., PRO MH 29 10, #67728/87.

sterile. Those bacteria that reached the consumer, therefore, might not be sewage inhabitants, but only harmless air-borne species.[65] He suggested that pathogenic bacteria were typically extremely susceptible to the elements and likely to die in greater proportions than normal river bacteria under almost any set of conditions. He argued that a critical mass of pathogens was required to cause illness; filtration made it unlikely that enough pathogens would be swallowed.[66] In what seems a gratuitous attack on his father's convictions, he cast doubt on the wisdom of obtaining deep well water for London. As Edward Frankland had long claimed, percolation through hundreds of feet of limestone did result in virtually sterile water. After reaching the surface, however, such water quickly became recontaminated and its bacterial population underwent particularly rapid expansion. That a similar, though less striking increase in bacterial numbers occurred after filtration of river water Percy Frankland failed to point out.[67] Still, unlike Tidy, Percy Frankland did not see nature as unambiguously aligned on the side of the water companies. He refused to accept the doctrine that predatory aquatic organisms could be relied upon to destroy pathogens or to trust any proposed mechanism of self-purification.[68]

---

[65] Percy Frankland and Mrs. Percy Frankland, Micro-organisms in Water. Their Significance, Identification, and Removal (London: Longmans, Green, & Co., 1894), p. 142.

[66] Percy Frankland in RCMWS, Appendices, p. 466.

[67] Ibid.

[68] Percy Frankland, "Recent research," p. 322; idem, "The present state of our knowledge concerning the self-purification of rivers,"

Percy Frankland's handling of his filtration results reflects the operation of a feedback system involving analytical methodology, concepts of purification, water supply politics, and scientific prestige. Having tentatively adopted plate cultures as a method for investigating purification techniques, Percy Frankland increasingly placed confidence in that method because it gave such striking results. Percy Frankland's status as a scientist was linked to the importance of these results and he adopted positions on water supply and water purification which confirmed and amplified those results. Those who emphasized quantitative bacteriological water analysis often discovered impressively large proportions of bacteria being removed by natural and artificial operations. It is noteworthy that in testimony to the 1892-93 Royal Commission on Metropolitan Water Supply bacteriologists representing the water companies relied chiefly on numerical reduction of bacteria as proof of purification, despite their acknowledgment that numbers meant nothing. Those opposing the companies tried to show that all methods of bacteriological water analysis were untrustworthy, and that switching from chemistry to bacteriology had not lessened the hazard of sewage-polluted water. Not surprisingly, Percy Frankland was one of the companies' chief witnesses.

---

in Seventh International Congress of Hygiene and Demography, Reports, p. 147.

IV. The Royal Commission on Metropolitan Water Supply and the Politics of Biological Purification

In 1892-93 a Royal Commission on Metropolitan Water Supply collected vast amounts of contradictory testimony on the safety of the river water supplied to London. Dominating the commission's deliberations was an ecological issue: the response of pathogenic bacteria to natural and artificial environments encountered between sewer and household tap. Purification, it seemed, was also a biological question.

The Commission was to consider the adequacy of London's water supply, the same issue considered in the inquiries of 1850-52 and 1867-69. Despite the presence of cholera on the continent, water quality was a shadow issue in 1892. Pressure for inquiry came from interests (chiefly the London County Council) desirous of taking over the privately owned supply; a negative report, it was felt, would spur parliament to force sale of the water works at a reasonable price. Unlike the largely non-professional 1867-69 commission, professions concerned with water supply were represented on the 1892 commission. Its members were Archibald Geikie, a geologist, James Dewar, a chemist, William Ogle, a physician, and three civil engineers: George Barclay Bruce, George Henry Hill, and James Mansergh.[69] Lord Balfour of Burleigh was chairman. In light of much

[69] Of the commissioners only Mansergh had expertise in water matters. Dewar, who appears sympathetic to the companies during the hearings, later joined Crookes as one of the companies' analysts.

of the testimony, it is noteworthy that none of the commissioners was a bacteriologist. To aid it in the evaluation of scientific evidence, the commission required statements from its scientific witnesses describing testimony to be given and relevant studies done.

Like earlier inquiries the 1892-93 hearings were a partisan affair. Witnesses upheld one of two well-defined positions: either the water was safe or it was not. Supporters of the companies included William Crookes and William Odling as chemists (Tidy died in March 1892 but his evidence was included in the report), the engineer Thomas Hawksley, defender of the companies since 1850, and Percy Frankland and E. Ray Lankester, son of Edwin Lankester and editor of the Quarterly Journal of Microscopical Science, as bacteriologists. Appearing for the London County Council (L.C.C.) were Edward Frankland as a chemist, and E.E. Klein and German Sims Woodhead, editor of the Journal of Pathology and Bacteriology and director of the joint laboratories of the Royal Colleges of Physicians and Surgeons, as bacteriologists.[70]

As in earlier inquiries testimony was contradictory. In 1892, however, it is clear that experts on both sides were blatantly distorting bacteriological knowledge. Such had not been the case previously. In the late 1860s chemists had drawn conclusions about purification from a science permeated by uncertainty. Contradiction

---

[70] "German Sims Woodhead," Proc. Royal Society of Edinburgh 42 (1921-22): 394-95.

was a true reflection of science; the identity of morbid poisons and the means by which they were destroyed were questions for which a wide range of legitimate answers could be considered. Bacteriology had supplied an empirical basis for investigating purification. Causative organisms of typhoid and cholera had been isolated. Conditions required for their survival and multiplication as well as natural and artificial processes which destroyed them had been studied in detail. Unfortunately the quantity of research was inverse to the amount of agreement. In some experiments the comma bacillus of cholera survived only a few days, in others, as long as 392 days.[71] Bacteria, as Woodhead noted, were "fickle" and refused to conform to the scientist's desideratum of repeatability.[72] While bacteriology offered much relevant data, it provided no clear answer to the question of whether London's water was safe, and when expert bacteriologists tried to draw unambiguous conclusions for or against London water they were distorting contemporary bacteriological knowledge. In 1869 contradiction had resulted from ignorance; in 1892 it resulted from conscious distortion.

The case of Percy Frankland is a prime example. Percy Frankland was an able witness and gave the commissioners a highly plausible account of pathogens succumbing to a variety of destructive agencies in the river and the water works. He was also a bold witness.

---

[71] Frankland and Frankland, Micro-organisms in water, pp. 297-300, 218.

[72] Woodhead in RCMWS, Appendices, p. 492.

His statement to the commission contained the same data on which he had based a bitter attack on "The Upper Thames as a source of water supply" in 1884. In eight years Percy Frankland had become an ally of the water companies and these data had become unimportant -- "a few analyses of unfiltered river water taken at long intervals of time."[73]

Percy Frankland's statement to the commission of June 1892 contrasted sharply with a mid-May report on water bacteriology he and H. Marshall Ward, professor of botany at the Royal Indian Engineering College, had prepared for the Water Research Committee of the Royal Society. That report included bibliographies on many controverted issues the commission would consider -- the survival of pathogens in rivers, for example -- and it reflected the real uncertainty in contemporary bacteriology. Percy Frankland admitted to the Royal Society that some studies showed that bacteria which deposited on stream bottoms were not necessarily destroyed and might in future re-infect the stream, yet no such qualification was made to the commission.[74] In the Royal Society report the series of purifying agencies appeared simply as factors affecting pathogen

---

[73] Percy Frankland in RCMWS, Appendices, pp. 462-63. Compare with Percy Frankland, "Upper Thames," pp. 435-37; and Percy Frankland, "The cholera and our water supply," 19th Century 14(1883): 346-55.

[74] Percy Frankland and H. Marshall Ward, "First report to the Water Research Committee of the Royal Society, on the present state of our knowledge concerning the bacteriology of water, with especial reference to the vitality of pathogenic schizomycetes in water," Proc. Royal Society of London 51(1892): 200.

survival rather than reasons for complacency toward a sewage-polluted water supply.

Those witnesses who appeared before the commission to defend the water companies envisioned a series of barriers to pathogens as sewage made its way down river to the companies' intakes, through reservoirs and filters, and finally into the cisterns and stomachs of consumers. These barriers made the likelihood of water-borne disease infinitesimal. Dewar, who sympathized with this view, listed the odds as infinity to one.[75]

The first threat to the hapless pathogen was sewage treatment itself. The vast majority of pathogens were destroyed when sewage was filtered through soil or by use of Clark's lime precipitation process (in which water was softened by the addition of lime which removed the carbonic acid which held excess calcium in solution), Percy Frankland pointed out.[76] By 1892 nearly all towns on the upper Thames and Lea had sewage treatment plants.[77]

Germs that survived sewage treatment faced several hazards in the river. There was the well-known phenomenon of attenuation. In certain physical and chemical environments pathogens lost virulence. In the hands of E. Ray Lankester and Percy Frankland attenuation became a principle to be employed in the planning of water supplies.[78]

---

[75] RCMWS, Evidence, Q 4283.

[76] RCMWS, Appendices, p. 467.

[77] Ibid., pp. 61-126.

[78] Ibid., pp. 452, 468.

Pathogens might succumb to oxygen, traditional nemesis of organic pollution in rivers. Odling and C.E. Groves, the Thames Conservancy's chemist, suggested that pathogens were typically anaerobic and therefore unable to tolerate well-oxygenated rivers.[79] Pathogens might fall to the bottom of the stream. In apparent contravention of the laws of physics, E. Ray Lankester argued that subsidence removed bacteria more effectively from flowing water than from still water.[80] It was assumed -- incorrectly -- that those germs which fell to the bottom ceased to exist. Finally, the "straining and adhesive action of weeds" might remove bacteria.[81]

The most popular mechanism for getting rid of pathogens was other stream organisms. Debates on the metropolitan water supply in 1830 and in the early 1850s had focused on the removal of morbid matters by scavenging organisms created for that purpose. By 1893 metaphor but not results had changed. Now pathogenic bacteria were vanquished in the "struggle for existence" with harmless organisms. Company defenders described two distinct modes of struggling for existence. First, pathogens presumably competed with native stream species for food and, being ill-adapted for life in streams, were "starved out" because they could not compete successfully. Lankester searched diligently for a species which competed with the typhoid bacillus. He mixed pure cultures of several fluviatile species with

---

79 RCMWS, Evidence, QQ 3768-69, 10123, 10208-16.

80 Ibid., QQ 10641-45, 10840-42.

81 RCMWS, Appendices, p. 455.

cultures of B. typhosus but found no species with which typhosus would not coexist harmoniously.[82] In fact, food requirements of pathogens were not clear and the argument probably would have been more popular had there not been substantial evidence of pathogens multiplying successfully in distilled water.[83]

Pathogens also struggled for existence against aquatic predators. This concept was less a product of the new bacteriology than a reaffirmation of an old faith in benevolent scavengers. William Anderson's 1885 discussion of pathogen-eating microbes, for example, retains much of the analogical flavor of the natural theology-based scavenging arguments of the 1850s:

> just as small birds preyed upon insects, and so did more good than they did mischief, it was quite conceivable that some harmless species of microbes preyed upon the injurious species, and it was by that means that water which had been contaminated by the germs of zymotic diseases gradually cleared itself and became inoffensive.[84]

There was, however, evidence to support removal by predation. Bacteria multiplied rapidly in filtered waters and when pure cultures were added to sterile natural waters. Lankester argued that sterilization or filtration destroyed or removed predatory organisms that normally limited bacterial numbers.[85]

---

[82] Ibid., p. 458; Evidence, Q 10676.

[83] Frankland and Frankland, Micro-organisms in Water, p. 232; RCMWS, Appendices, pp. 453-55; Evidence, QQ 10122, 10128, 10146.

[84] Anderson in disc. of Percy Frankland, "Water purification: its biological and chemical basis," p. 233.

[85] RCMWS, Appendices, p. 455.

Additional ordeals awaited, "should these pathogenic forms . . . succeed in reaching the intakes of the water companies."[86] Storage of water in reservoirs substantially reduced the bacterial population, presumably through subsidence and continued competition with native forms.[87] Next were the sand filters whose "astonishing" prowess in removing bacteria Percy Frankland had demonstrated. By 1892 the importance for bacterial removal of a gelatinous film on the sand was recognized, although its action was conceived primarily as physical rather than biological. Various company defenders alluded to "adhesion" to this layer, or to its "osmotic" or "colloidal" attraction for bacteria. Thomas Hawksley claimed "filter beds act . . . by what I may call the principle of universal attraction, which is the same thing as gravity."[88]

Bacteria were present after filtration, but company representatives suggested that these were peculiarly innocuous. Percy Frankland argued that the general incompetence of pathogens outside mammalian bodies was grounds for supposing that filtration would be preferentially effective in removing them. Odling, who believed pathogens were typically smaller than harmless forms, thought "the more minute these particles are the more liable they are to be laid hold of, so to speak, by the attracting jelly."[89] Many of the post-

---

[86] Ibid., p. 467.

[87] Ibid., pp. 456, 467; <u>Evidence</u>, Q 10778.

[88] RCMWS, <u>Evidence</u>, QQ 7204, 10130, 10641, 11111.

[89] RCMWS, <u>Appendices</u>, p. 466; <u>Evidence</u>, Q 10133.

filter bacteria might not have crossed filters, but result from new contamination. Ignoring the species of these post-filter bacteria, Percy Frankland argued that it was unreasonable to suppose filtration efficiency varied much, so variations in the number of post-filter bacteria must be due to varying degrees of new contamination. Thus post-filter bacteria, no matter how much they might subsequently multiply, posed little hazard to health because they were normal occupants of the atmosphere and the water works, not morbid poisons brought down with sewage.[90]

Finally, despite manifold obstacles, some pathogens might reach the consumer. Percy Frankland and Lankester reassured the commissioners that a few swallowed pathogens were not sufficient to produce disease. Besides, within the human body were further lines of defense. On their way to the stomach typhoid bacilli would be "brought into competition with a host of other micro-organisms."[91]

These multiple safeguards explained the unexpectedly good health of Londoners, and the companies' representatives noted that properly filtered water had never been proved to have caused an epidemic. As in earlier inquiries, the companies' defenders had put forward a highly plausible and appealing account of how nature worked. The biological and bacteriological principles and processes they pointed to were real enough but were used to support the unwarranted conclu-

---

[90] RCMWS, Evidence, QQ 11080-81.

[91] RCMWS, Appendices, p. 466; Evidence, QQ 10661-65.

sion that London's water was always safe. The assumption that nature benevolently looked out for humanity still underlay the companies' case. In words that recall the natural theology of the 1850s, Odling spoke of the "continuous alternate transformation by natural processes of sweet into foul and of foul into sweet."[92]

The main argument against river water was the same as it had been since the late 1860s, that there could be no assurance that sewage-polluted water would always be safe. The old-style nihilism of Brodie characterized the testimony of Alfred Ashby, George H. Fosbroke, and George Turner, medical officers of health the L.C.C. had commissioned to catalog all the pollutions the Thames and Lea received. They found many minor pollutions, demonstrating that two decades of policing by conservancy boards had not ended pollution. Ignoring recent concepts of disease transmission by specifically contaminated sewage, these three perpetuated etiological concepts of the 1850s. They granted to tannery refuse and animal wastes the same hygienic status as human sewage. They denied the relevance of chemical and bacteriological analyses, claimed there was no proof water did not cause disease in London, and discounted any form of self-purification. "We cannot rely on the . . . flow of a river for doing everything," Ashby observed.[93]

The nihilism of Ashby, Fosbroke, and Turner was no longer the

---

[92] RCMWS, Appendices, p. 444.

[93] RCMWS, Evidence, Q 3886; Appendices, pp. 179-200.

nihilism of Edward Frankland. Frankland told the commission that although he appeared at L.C.C. request he did not represent the council; his opinions were his own. His views on the suitability of Thames water had moderated over the years. In 1891 he had begun a series of plate culture counts of bacteria in filtered and unfiltered London water, similar to that done by Percy Frankland in the mid-1880s. These analyses were financed directly by the water companies, and the companies also hired Frankland to suggest ways of improving filtration.[94] By 1892, therefore, Edward Frankland was no longer the sharp-tongued critic of the companies' practices, a fact some of his colleagues on the Local Government Board probably resented. As London water quality became even more a political issue, Frankland became less active in water supply politics.

Denying twenty years' medical progress was not a satisfactory basis for opposing river water, especially when bacteriology itself contained such vast potential for a nihilistic perspective. The thrust of bacteriological testimony against river water was a bacteriological analytical nihilism, the contention that no mode of bacteriological analysis was reliable. Unlike the companies' bacteriologists, whose denial of the meaningfulness of bacteria counts was at variance with their reliance on the technique, L.C.C. representatives assiduously avoided bacteria counts. Klein and Woodhead were also skeptical of methods of determinative bacteriology. Both

---

[94] PRO MH 29 19, East London Water Company to L.G.B., 16 Feb. 1894, 24 November 1893.

raised doubts that the causative organisms of typhoid and cholera had really been discovered. They suggested that B. typhosus, rare in water, was simply a mutation of more common and morphologically similar B. coli communis. Therefore, at any time, B. coli-containing water might cause typhoid.[95]

Even if the comma bacillus and B. typhosus abdominalis were true species and the causes of cholera and typhoid, techniques for determining their presence were untrustworthy. First was the problem of dilution. Because pathogens did not appear in the culture of a given cubic centimeter of water was no proof that they might not exist in another cubic centimeter. Next, there was no guarantee that even if pathogens were in the cubic centimeter chosen for analysis they would be discovered. Woodhead noted that it was difficult to culture the bacilli of cholera and typhoid.[96] At present bacteriological conclusions had to be "very guarded," since the science was too new to deliver certainty. Klein noted that as determinative techniques were improved, it would become easier to find pathogenic bacteria. His implication was that eventually searches for pathogens in Thames water -- in 1892 bacteriologists on both sides had looked for them in vain -- would be successful. Therefore, just because pathogens had never been found in London water was no grounds to assume that they weren't there. In Klein's hands negative

[95] RCMWS, Evidence, QQ 11009-11, 13325-27.

[96] RCMWS, Appendices, pp. 202, 491; Evidence, QQ 10918-22, 11004-7, 12619.

results -- failure to find pathogens -- assumed the same significance positive results would have.[97]

In their discussions of analytical and determinative bacteriology Klein and Woodhead had drawn plausible but extreme interpretations from the bacteriological literature. It was true that determinative bacteriology for water-borne pathogens was difficult; Percy Frankland admitted as much in his report to the Royal Society. Still, both cholera and typhoid bacilli had been found in river water on several occasions. Klein and Woodhead had transformed occasional failure into total unreliability.

The purificatory nihilism that complemented the analytical nihilism of Klein and Woodhead was less far-fetched. Both admitted that several obstacles to pathogen survival separated the excrement of upstream inhabitants from the London water supply. River conditions, however, were not invariably beneficial as the companies' bacteriologists had maintained. Oxygenated water was exactly what some pathogens required, Klein noted.[98] The vanquishment of pathogens by fluviatile species could not be relied upon; sometimes cholera bacilli adapted to life in rivers and acquired considerable resistance.[99] Storage reservoirs might be places for pathogens to multiply.[100] Certainly filtration improved water but there was "no tittle

---

[97] RCMWS, Evidence, QQ 13064, 13077, 13093, 13248-49.

[98] Ibid., Q 10990.

[99] Ibid., Q 10927; Appendices, pp. 491-92.

[100] RCMWS, Evidence, Q 10993.

of evidence to show that a pathogenic organism can not pass through a filter."[101] Much larger objects than pathogens came through filters and Frankel, a German bacteriologist, had demonstrated that comma bacilli successfully crossed sand filters.[102] Moreover, while filtration might substantially reduce the number of bacteria, there was great opportunity for pathogen multiplication after filtration since the filter could be presumed to have blocked passage of larger microscopic species which preyed on pathogens. This post-filter multiplication was an observed phenomenon. As Woodhead pointed out, there was little point in counting bacteria immediately after filtration, since their numbers might vastly increase between the filter and the consumer.[103] Finally, while it might be true that a single pathogen was insufficient to initiate disease, both the virulence of pathogens and the resistance of victims varied enormously, so discussion of the number of germs required to kill a person was not fruitful.[104]

The vision of multiple safeguards was therefore misleading. Some of these safeguards were not safeguards, and even if pathogens were rare inhabitants of drinking water, multiplying some tiny

---

[101] Ibid., Q 4588.

[102] Ibid., QQ 10947, 11012-29, 12876-79; Appendices, p. 494.

[103] RCMWS, Appendices, pp. 493-94.

[104] RCMWS, Evidence, QQ 10951, 10972, 10974, 13149. Edward Frankland and George Fosbroke still regarded a single germ as sufficient (QQ 4144, 4561-65).

probability of their presence, say 1 in 10,000, by the population of London (5,000,000) made it likely that some Londoners were victims of water-borne disease.[105]

As in 1867-69 the royal commissioners found London's water satisfactory. It is likely that the most convincing evidence had nothing to do with the contradictions of rival bacteriologists. Those assailing the water had sputtered doubts that it might not be safe and had made excuses for their failure to prove it unsafe. Much of their argument had been based on sentiment, that it was wrong "to take the sewage of a large population and then pass it through a sand filter and depend on that."[106] The commission happily noted that no water-borne epidemics had occurred in London since 1868 and observed that even Edward Frankland, "no sparing critic of the London water," had come to accept river water as safe.[107]

It is remarkable that the replacement of chemistry by biology brought no greater degree of confidence to the identification of pure water and the understanding of purification. Local sanitarians who turned to biology in desperation as one data category to be included in an inventory of characters which might reflect danger, found that although they could identify organisms there was little basis for associating species with hygienic properties. Notably MacDonald's

---

[105] Ibid., QQ 10981-89.

[106] Ibid., Q 12885.

[107] RCMWS, Report, pp. 62, 67-68, 69.

Guide, the authority for this strategy of water quality evaluation, failed to tell analysts what various species signified.

More significant is the failure of bacteriology to resolve questions of drinking water purity and purification. The prowess of late nineteenth century bacteriology has been a hallmark of medical history. As Shryock observes, bacteriology made doctoring respectable again.[108] Practical success, however, was despite profound intellectual uncertainty. As the Percy Frankland/Marshall Ward reports to the Royal Society and the reams of contradictory evidence bacteriologists gave to the Royal Commission on Metropolitan Water Supply illustrate, nearly every fundamental issue of water bacteriology was still unsettled in the early 1890s.

The failure of the commission to resolve the long-standing dispute regarding the quality of the London water supply was not mainly due to the immaturity of bacteriology, however. Instead, the dispute was not resolved because it was, and always had been, couched in unrealistic terms, terms defined and legitimized by politics, not science. As in 1867-69, in 1850-52, and even in 1830, the public demanded certainty from scientists. In the 1890s the bulk of British water science was still done within a political context and scientific pipers played the tune they were paid for. What emerged from the commission's hearing were two contradictory pretenses of certainty, each distorting the unsettled condition of bacteriology.

---

[108] Richard H. Shryock, The Development of Modern Medicine (rpt., Madison: University of Wisconsin Press, 1979), pp. 336-37.

The employment of bacteriological arguments in political disputes regarding water supplies is less a reflection of the success of the bacteriological revolution in public medicine than it is a reflection of the utility of biological and bacteriological methods and principles as a format for disputing water quality issues. As the use by the water companies' bacteriologists of the Darwinian metaphor of a "struggle for existence" and Percy Frankland's perspective toward his plate culture results both exemplify, principles and methods acquired constituencies in water supply politics.

X. BIOLOGY ACQUIRES A CONSTITUENCY II: ORGANISMS AND THE POLITICS OF SEWAGE TREATMENT

> . . . the whole mystery of Nature's method of purification is made plain. In this we see but one more instance of the power of science to clear up mystery after mystery. A few years ago it was stoutly denied that rivers had the power of purifying themselves! Then we knew practically nothing of Nature's method. Now that this has been so far revealed to us, it is declared with equal force that not only are effete matters rendered innocuous, but even disease-producing microbes are themselves voraciously devoured by others of like kind, and the formerly much-dreaded bacteria are -- and properly so -- considered the best friends of man.
>
> -- William Dibdin[1]

Between 1880 and 1900 biology also replaced chemistry as provider of the scientific basis for sewage treatment and for understanding what processes caused or prevented nuisances in rivers. Bifurcation of water purification discussions led to the setting of new, more realistic priorities for sewage treatment. Prevention of nuisance, rather than production of potable water, became the primary goal of sewage treatment. Although sewage works managers no longer lost sleep worrying about pathogenic effluents, public health justifications of sewage treatment remained. Sanitarians worried that the gases generated in putrefaction, while not themselves pathogenic, might convey pathogenic bacteria from sewage to the atmosphere.[2]

---

[1] William Dibdin, The Purification of Sewage and Water, 3rd ed. (London: Sanitary Publishing Co., 1903), pp. 287-88.

[2] Charles Meymott Tidy, "The treatment of sewage," JRSA 34(1886): 1149; James Edmunds in disc. of Tidy, "The treatment of sewage,"

More importantly, despite the germ theory, nuisances remained a public health concern.

The new priorities for sewage treatment clarified the scientific questions involved in self-purification and fostered new methods for investigating that phenomenon. In the 1882-84 hearings of the Royal Commission on Metropolitan Sewage Discharge, for example, the kinetics of oxygen diffusion and of sewage oxidation emerged for the first time as central issues. Since no one drank the water of the Thames estuary (where the Metropolitan sewage was discharged), these questions could be considered without the complicating factor of the oxidation-resistance of disease germs, the issue with which Frankland had so successfully muddied the waters of self-purification debate. A similar simplification was achieved when the commission's witnesses considered the issue of the purifying capabilities of aquatic organisms. The question was simply whether organisms destroyed organic matter, rather than the relationship between their activity and the safety of the water for drinking. As with water questions, re-definition did not alter the traditional partisan context of sewage science. Contradiction still characterized the testimony given to the 1882-84 commission. Nevertheless, the re-definition of self-purification was a progressive step, blocking some

---

JRSA 35(1886-87): 41; John Burdon Sanderson in disc. of Percy Frankland, "Some of the conditions affecting the distribution of organisms in the atmosphere," JRSA 35(1886-87): 496. In the 1890s there were still traces of old-style miasmatism in British sanitary literature (Thomas Wardle, On Sewage Treatment and Disposal [Manchester: John Heywood, (1893)], p. 10).

avenues of obfuscation, forcing scientific adversaries to consider more fruitful questions.

During the 1880s the role played by aerobic bacteria in the destruction of cast-off organic matter came to be widely appreciated by British sanitarians. Despite an enormous quantity of continental research describing step-by-step the various hydrolytic and oxidative reactions that took place during decomposition, British sanitarians continued to regard anaerobic action as illegitimate until the late 1890s, when the concept of septic sewage treatment was developed and popularized.[3] Interest in aerobic bacteria was less a result of the dissemination of Pasteur's ideas than a response to the 1877 discovery by the French chemists Theophile Schloesing and Achille Muntz that soil nitrification was caused by microorganisms. British sanitarians fit this discovery into the moralized view of natural economy which had traditionally underlain their ideas of purification. Yet because distinctions between good and bad bacteria remained unclear and because technologies for encouraging the former while eliminating the latter were untrustworthy, there was widespread anxiety about whether sanitary engineers should attempt to harness aerobic bacteria. This problem too was resolved in the mid-1890s, as it became clear that the same processes that encouraged good

---

[3] There were exceptions. See Samuel Rideal, Sewage and the Bacterial Purification of Sewage (London: Sanitary Publishing Co., 1901), 2nd ed., pp. 83ff; H.E. Armstrong, "The alkaloids, the present state of knowledge concerning them and the methods employed in their investigation," J. Soc. Chemical Industry 6(1887): 482-91.

bacteria usually destroyed bad bacteria.

As had hitherto been the case, ideas about biological purification developed in a political context and because they had dialectical value. William Dibdin, chemist for the Metropolitan Board of Works, the organization responsible for disposing of London's sewage, is noteworthy as one able to transcend that context. Between 1887 and 1895 Dibdin transformed a rhetorical tool into a scientific principle, using the concept of biological purification as basis for investigation of the Thames estuary ecosystem and for experimentation on biological sewage filtration.

In retrospect, Dibdin saw himself as founder of a scientific revolution in sewage treatment that had occurred during these years. In his view the principle of biological purification had liberated sewage issues from commercial and political partisanship. A revolution certainly occurred, but it was not primarily the scientific phenomenon Dibdin imagined. Instead, on a practical level, it was a manifestation of trial-and-error technological development by experienced sewage works managers, such as Donald Cameron, the Exeter surveyor and inventor of the septic tank, while on a verbal level it was an endorsement by sanitarians and popularizers of a technology that was cheap and that reaffirmed an outlook pervasive in British sanitarianism -- that nature, working properly, was beneficent.

This chapter examines the articulation of new questions about oxidation and the activities of aquatic organisms during the

1882-84 hearings of the Royal Commission on Metropolitan Sewage Discharge, the discussions about the propriety of encouraging bacteria by British sanitarians in the late 1880s and early 1890s, and the nature of the revolution in sewage treatment and self-purification that William Dibdin believed he had led.

I. Oxygen and Organisms, 1882-84

Among the forums in which self-purification was discussed was a series of hearings held between 1882 and 1884 by the Royal Commission on Metropolitan Sewage Discharge, established in 1882 to consider the wisdom of the Metropolitan Board of Works' policy of dumping London's sewage, raw, into the Thames estuary. In 1855 the Metropolitan Board had been established with the primary fuction of solving London's sewage problem. By the mid-1860s it had completed a system of intercepting sewers which led to twin outfalls at Barking and Crossness, on opposite sides of the river about ten miles below central London. There the sewage was to be stored in reservoirs until it could be carried off to sea by the outgoing tide.[4]

The system merely relocated pollution. Much of the sewage did not slip quietly off to sea but oscillated with the tides. Large black patches floated on the water; low tide uncovered stinking black mudbanks. Complaints by fishermen, shippers, and local residents

---

[4] Royal Commission on Metropolitan Sewage Discharge, First Report, P.P., 1884, 41, [C.-3842.], pp. xv-xxx. [To be cited as RCMSD, 1st Report.]

occasioned major government inquiries in 1870 and 1878, but these failed to resolve the issue. When a cruise ship, the Princess Alice, sank near the outfalls in 1878 many attributed deaths from the accident not to drowning, but to immersion in only slightly diluted sewage. Some even felt the Board should be charged with murder. The 1882 commission was a response to that accident, to a petition by 13,000 lower Thames residents, and to agitation by the Port of London Sanitary Authority (responsible for health all along the estuary) and the residents of the town of Erith, located across the river from the northern outfall.[5]

George William Wilshire, Lord Bramwell, chaired the otherwise highly professional commission: civil engineers C.B. Ewart, Sir John Coode, and James Abernethy; chemist A.W. Williamson, and physicians F.S.B. Francois de Chaumont and Thomas Stevenson. (One other commissioner, Sir Peter Benson Maxwell, a politician, was absent for much of the proceedings.) Their mission was aspecific: "to inquire into and report upon the system under which sewage is discharged into the Thames, . . . whether any evil effects result therefrom, and in that case what measures can be applied for remedying or preventing the same."[6] From July 1882 to July 1883 the commission considered whether evil effects resulted. Having decided they did, and that sewage treatment was necessary, it held further

---

[5] F.R. Conder, "The question of the Thames," Fraser's Magazine n.s. 18(1878): 726-27; RCMSD, 1st Report, Evidence, p. 153.

[6] RCMSD, 1st Report, p. iii.

hearings on methods of treatment from May to October 1884 before recommending chemical deodorization as a short term, and sewage irrigation as a long term solution.

The commission's hearings mark the reappearance of the biological defense of pollution in Britain. The interplay of social and ideological factors was important in bringing forth that concept. Because the issues under consideration were inherently of an adversary nature, the commission to a great extent allowed counsel from each side to control the direction of inquiry. Because the issues at stake were so large -- £40,000 were spent to mount the show -- both sides were staffed with the best legal and scientific talent available.[7] The relationship between organisms and pollution as well as other fundamental aspects of pollution and purification were brought forth within a context which exemplifies characteristics of adversary science: glibness, flexibility, persuasiveness, and arrogance on the part of witnesses; searching questions about philosophical and methodological underpinnings of scientists' beliefs, and attempts to exploit the many ambiguities in sanitary science on the part of cross-examining barristers.[8] So thorough was this working-over of the content of ideas of pollution and purification that issues raised in the commission's hearings -- in particular

[7] Greater London Council Record Office, London County Council Main Drainage Committee Presented Papers, Vol. 4, for Jan. 23, 1890, #5, R. Ward to Ctte., p. 1. [Cited as GLCRO/LCC/MD Cttee. Pres. Papers.]

[8] See "The whole duty of a chemist," <u>Nature</u> 33(1885-86): 74.

the question of the purifying abilities of organisms -- were transferred from this litigious context to the less blatantly partial context of scientific societies after the hearings.

One issue occupying much of the commission's attention was the actual damage to local health caused by existing sewage-dumping practices. Chadwick and the sanitarians of the 1850s had assumed that polluted rivers were harmful to health even when their waters were not drunk because the air above them became tainted by putrefying water. By the early 1880s the germ theory, a more definite understanding of the routes of disease transmission, and greater trust in the specificity of diseases had made direct causation of zymotic diseases by a pestilential lower Thames atmosphere doubtful, but sanitarians still worried about the effect of such unpleasant surroundings as predisposing causes of disease. During the 1882-84 hearings the health effects of sewage-dumping were debated in terms of statistics, rather than with germ theories or through epidemiological demonstrations. Statisticians on opposite sides contradicted each other successfully, and the commission concluded that sewage-dumping had no effect on health.[9]

More fundamental were the ambiguous natures of basic concepts such as purification and pollution. In the perspective of Victorian sanitarians there was a sense in which both terms referred to the same phenomenon. By polluted water, sanitarians generally meant

---

[9] RCMSD, 1st Report, pp. xlix-l, lxvii.

either water which was unpleasant, presumably because it contained some putrefying substance, or water which chemical analysis showed to contain excessive organic matter. In essence they were objecting both to existing and to potential putrefaction. The concept of purification was similarly ambivalent. In an ideal hygienic world neither form of pollution existed because oxygen speedily destroyed refuse organic matter in benign fashion. In practice, purification meant either an attempt to foster those ideal oxidative conditions, such as was done when sewage was filtered through soil, or an attempt, by antiseptic treatment, to prevent any change whatever, the famous application of carbolic acid to the sewage of Carlisle that inspired Lister being an example.

In the view of the Port of London Sanitary Authority's witnesses and lawyers both modes of purification were occurring naturally in the estuary and both were failing, resulting in a river polluted according to both definitions given above. Antiseptic action in the river was objectionable because it left water polluted with organic matter from sewage, while decompositive processes in the river were also objectionable because they happened so rapidly that the river was left deaerated, stinking, and fishless.

The exploration of this paradox appears most clearly in the testimony of Charles Meymott Tidy, ironically appearing here as an opponent of self-purification and a comrade of Edward Frankland. Tidy argued that the Thames estuary was not able to purify itself like an inland stream due to its brackishness and lack of plants.

Although Tidy was here recognizing the antiseptic action of the river, he was not regarding this action as true purification because it did not destroy impurity.[10] Elsewhere, however, Tidy described a river purifying far too fast. Citing a multitude of dissolved oxygen determinations done on the lower Thames, Tidy noted "the water has been trying to keep itself pure, it had a struggle in the reach between Vauxhall and Deptford [central London], and after it had passed that reach it could not do any more, its power was exhausted, it could not do any more."[11]

That same paradox was reflected in the Port of London Sanitary Authority's case generally. Acting for the Authority, engineer Baldwin Latham had performed experiments in which floats were put into the river near the sewage outfalls and followed as they oscillated with the tides. Assuming that sewage would behave in the same manner as the floats, Latham and other engineers concluded that the residence time of sewage in the estuary was about one month.[12] In cross-examination the assumption underlying this calculation was brought out: float experiments and impressive calculations of how thick the estuary must be with London sewage had meaning only if self-purification did not occur. Noting that during that month in the lower Thames the sewage underwent a flow of nearly 1500 miles, the Metropolitan Board's counsel suggested that surely this was

---

[10] RCMSD, 1st Report, Evidence, Q 10325.

[11] Ibid., Q 9958. See also Q 9918.

[12] Ibid., Q 9215.

distance enough even for the most confirmed opponents of self-purification.[13]

While the Authority's engineers built their case on the assumption of the permanence of sewage, its chemists built theirs on the assumption of permanent putrefaction. They argued that oxidative purification was happening, but too quickly for oxygen diffusion to keep up with sewage input and yet too slowly to result in appreciable diminution of the organic matter in the river. J.T. Way worried that the oxygen demand of London's sewage was so vast that if oxidation occurred at the rate of sewage input "the air of the neighbourhood would be irrespirable; all the oxygen would be gone." He admitted the data did show a great "eating up of oxygen" taking place in the river, "but where the next lot of oxygen is to come from is not so easy to see."[14] In Way's view, however rapidly organic matter was being oxidized, the rate was not sufficient to destroy a significant quantity of sewage. Moreover, the very low amounts of oxygen typically found in the lower river could be misinterpreted as representing amounts of oxygen available for combustion, rather than, in effect, amounts left in an oxygen reservoir which was being filled and emptied at some rapid rate, which they really were. Thus, Charles W. Heaton, lecturer in chemistry at Charing Cross Hospital, contrasted the enormous quantity of oxygen required by the sewage

---

[13] Ibid., QQ 9320-23.

[14] Ibid., QQ 12289-90.

with the insignificant quantity actually found in the river.[15]

While the Authority's chemists bemoaned the emptiness of the oxygen cup, the Board's chemists celebrated the thirst that had been quenched by its contents and observed that the cup was continually being refilled. The themes of their case were the permanence and rapidity of oxidation and re-aeration. Low oxygen levels, in their view, were proof of the immense amount of oxidative purification actually occurring in the lower Thames. As chief counsel G.P. Bidder, jr., put it, the river was "doing its work."[16] More importantly, underlying the Board's view of the oxidation issue was the recognition that the problem must be seen in kinetic terms. Dissolved oxygen levels naturally equilibrated at some fraction of saturation which reflected the rate at which oxygen from the air was diffusing into the water and the rate at which dissolved oxygen was combining with organic matter. The lower the oxygen level, the more rapid was the combustion. One of the Board's engineers, Henry Law, pointed out that it was meaningless to compare the oxygen demand of sewage with the oxygen present in the water without knowledge of rates of re-aeration. Law noted that deoxygenated water "sucks [oxygen] from the air with avidity, just as a sponge sucks water." He also recognized that the re-aeration rate increased the lower the dissolved oxygen level got.[17] Low oxygen levels therefore

---

[15] Ibid., QQ 11440, 11473-75.
[16] Ibid., QQ 10262, 10256.
[17] Ibid., Q 13877.

reflected very rapid oxidation.[18] There was plenty of oxygen available -- 46 times the amount necessary in a one-foot layer above the river, according to Law.[19] Yet while they recognized the problem at hand was a kinetic one, and understood in principle how it must be set up, the Board's scientists were unable to solve it. They simply argued that a much larger quantity of oxidation went on than the Authority's chemists admitted. In fact the same data (showing oxygen saturation levels of approximately 20%) proved both sides right by demonstrating that substantial oxidation was occurring and confirming that putrefactive conditions existed.

This debate about the meaning of dissolved oxygen data marks an important change in the idiom of self-purification controversies. Although dissolved oxygen measurements had been made in connection with earlier inquiries, they had been made only sporadically and D.O. data had been little used. One reason such data were so important in the 1882-84 hearings was the availability of the Schutzenberger process for measuring dissolved oxygen which was easy and could be done on river. Tidy, who had used this process for about two years prior to the hearings, claimed he had made 4000-5000 determinations by it.[20] The importance of oxygen data was also a

---

[18] Ibid., Q 16066.

[19] Ibid., Q 13876.

[20] Ibid., Q 10187. The Schutzenberger process was based on the spontaneous oxidation of sodium hyposulfite to sodium bisulfite. A water sample was dyed with a solution of Coupier's analine blue, compatible with the bisulfite but not with the hyposulfite. Hyposulfite was added until the solution was decolorised. The amount of hyposulfite was an indication of the amount of dissolved oxygen.

phenomenon of the bifurcation of self-purification debate. So long as the poisoning of water by organic matter and the vitiation of the atmosphere by putrefaction had held equal and associated hygienic significance, the measurement of organic matter (or organic nitrogen or albuminoid ammonia) had held primary status in discussions of self-purification. Because the fraction of saturation of dissolved oxygen in water is a far better index of the general condition of the river as regards potential nuisance than the amount of organic matter, dissolved oxygen determinations quickly became a central method for answering the self-purification questions of interest to sewage works managers, just as bacteria counts were to become the central method for answering drinking-water quality questions.

The 1882-84 hearings reveal sanitary scientists trying to make sense of this new category of data, a struggle made more difficult by the fact that each side's counsel offered a version of the significance of that data that contradicted his opponent's version. The utility of dissolved oxygen studies in biological investigation, and the utility of those biological investigations for understanding self-purification were not here appreciated. As Law cavalierly observed, low oxygen levels in the river did make the

---

The process was known chiefly for its ease rather than its accuracy ("New process of estimating free oxygen," (abstract), CN 26 (1872): 204; Christopher Clarke Hutchinson, "On Schutzenberger's process for the volumetric estimation of oxygen in water," CN 38 (1878): 184-87; Dibdin, Purification of Sewage and Water, pp. 355-57).

river fishless, but the fish were simply sacrificing themselves for the good of civilization.[21]

Biology entered the commission's hearings, not on the coat-tails of the Schutzenberger process, but through another door. The assimilation of sewage by estuarine life was the second of the Board's principal grounds for denying that the lower Thames was a nuisance. Bidder, in introducing his case, noted:

> It is not a mere question of the oxygen in the water, which is the purifying agent, but there is a cause which has been overlooked on the other's side, and which our researches show is a very important one. An enormous quantity of this sewage is consumed by minute organisms of [sic] animals which feed upon it, so that it is their meat, and it ceases to be sewage.[22]

Bidder was right that the biological excuse had surprised the opposition. Although the excuse was not new as a defense of pollution, in recent years it had been out of fashion. The Authority had retained a biologist among its experts, Lionel Smith Beale of University College, London, but Beale's chief task was to confirm the Authority's assertion that the Thames estuary was essentially a stagnant pool rather than a flowing stream by showing that marine diatoms existed far upstream. He was also to do microscopical studies in search of sewage components in Thames mud.[23]

Bidder introduced the biological excuse during his cross-examination of Edward Frankland, one of the Authority's main chemical

---

[21] RCMSD, 1st Report, Evidence, QQ 13878, 16300, p. 453.

[22] Ibid., p. 458.

[23] Ibid., QQ 11649-65, 11576-77, 11560-62.

witnesses. Earlier, Frankland had been asked by Dr. Thomas Stevenson, one of the commissioners, if it were not true that "the power of self-purification in a river depends on what are called micro-organisms in it?" Frankland had agreed but had not elaborated. Asked if these organisms could tolerate impure waters, he had replied that if they were causers of putrefaction they probably could. There the issue had been dropped.[24]

Bidder began his cross-examination by inquiring about Frankland's objections to Tidy's trough experiments on self-purification. Frankland initially pleaded loss of memory, then recalled his suggestion that bacteria flourished in the troughs and these were "very active in destroying sewage matters." He added that bacteria "would induce putrefaction of the sewage, and . . . such matters purify much more rapidly by putrefaction than by oxidation."[25] This was a strategic qualification. Frankland's implication was that bacterial action was illegitimate and ought to be prevented because it was linked exclusively to undesirable putrefactive purification. Frankland was here exploiting the same sanitary paradox as Tidy had -- admitting that purification occurred, but suggesting that it was inevitably of the unacceptable, putrefactive variety; nothing more, indeed, than a different form of pollution. The conclusion to which Bidder was trying to lead him, that microbes improved the Thames, was therefore incorrect.

---

[24] Ibid., QQ 10872-73.

[25] Ibid., Q 10913.

Bidder persisted in his original line of questioning:

> Is not it the case in a river, where sewage is mixed with large quantities of river water, there is a great development of living organisms which play a very important part in dealing with the sewage elements?

Frankland agreed this was "probable," and said he was "not personally familiar with the subject."[26] Here Bidder let the issue drop and set his hypothetical purifying organisms on other rhetorical game. Frankland's analyses showed relatively large amounts of organic matter in the lower Thames. Since, by Frankland's own admission, chemical analysis could not distinguish living organic matter from dead organic matter, might it not be, Bidder suggested, that "what you may be recording in your analyses as organic pollution may be . . . your organic scavengers?"[27]

If he had failed to extract from Frankland a confession that organisms purified streams, Bidder had nevertheless raised that possibility in the commissioners' minds. To a question from Dr. de Chaumont about the oxidation of sewage by intermittent filtration through soil, Frankland admitted that in that case, oxidation was "at all events, helped by bacteria."[28] Sir John Coode and Dr. Stevenson also brought up the issue: might not uneven spots in the river bed, just like the corners of Tidy's trough apparatus, harbor

---

[26] Ibid., Q 10921.

[27] Ibid., QQ 10924-25. MBW counsel used the same suggestion in cross-examining C.W. Heaton (QQ 11493-95, 11506-9).

[28] Ibid., QQ 10965-66.

sewage-destroying bacteria? Frankland agreed that was possible, stressing again the association of bacteria with putrefaction. The mud of "rivers in a putrescent condition . . . would be found to swarm with bacteria."[29] Tidy, doubtless tired of having repeatedly to explain why his self-purification experiments were not applicable to the lower Thames, also accepted the suggestion. He had not observed any bacteria in his troughs, but agreed that if they had been there, they were likely in the river as well, "and if they assisted oxidation, they would assist it in the river just as they would in these troughs."[30]

To lay this troublesome ghost once and for all, the Authority's counsel asked its biologist, Lionel Smith Beale, to consider the question. Beale portrayed the contention that bacteria purified the Thames as a notion plausible to the lay mind but ridiculous to the expert:

> I am afraid they will not grow fast enough, particularly at this time of the year [March]. If time were allowed and if the temperature were sufficient for them to grow very rapidly indeed, they would decompose the sewage and reduce it to something perfectly innocuous, but that would involve an enormous number and enormous time, and a very thin layer of the organic material spread over an enormous surface.[31]

In essence, the bacterial explanation was simply not a practical view to take of London's sewage problem.

---

[29] Ibid., QQ 11231-32. See also Q 11274.

[30] Ibid., Q 11556.

[31] Ibid., QQ 11676-78.

In cross-examination of Beale, W.H. Michael, another of the Board's lawyers, claimed his side had never suggested bacteria destroyed sewage; but that instead "there are much larger organisms developed in matter of an organic character which tend to destroy it." Again Beale was scornful: "I was not aware of any organisms which would go up and down the Thames and destroy the sewage."[32]

Larger organisms were indeed the main line of the Board's biological defense, a defense presented chiefly by Henry Clifton Sorby, well-known as an innovative microscopist. Sorby's appearance as a witness for the Metropolitan Board was not surprising since he had testified for the Board on the use of microscopy for distinguishing sewage mud from ordinary Thames mud in the 1878 inquiry.[33] What was unexpected was Sorby's biological testimony. Sorby had spent two months during the summer of 1882 and another month in spring 1883 on the lower Thames, studying mud.[34] In the mud he found large numbers of fecal particles from entomostraceae, a microscopic crustacean. Sorby noted that these excreta were present in inverse proportion to traces of human excreta and explained this fact by

---

[32] Ibid., QQ 11681-82.

[33] In the 1878 inquiry the source of the mud was a central issue. It was an arbitration between the Thames Conservancy and the Metropolitan Board of Works on the question of whether three mud banks were caused by metropolitan sewage discharge (Thames Navigation Act, 1870. In Arbitration. The Thames Conservancy and the Metropolitan Board of Works, Minutes of Proceedings, Report, and Determination of the Arbitrators. Evidence, QQ 10924 ff.

[34] RCMSD, 1st Report, Evidence, QQ 16593-95.

arguing that these crustaceans lived on human excrement. He experimented, keeping entomostraceans in a tank and feeding them excrement: "I have kept a number of them for six weeks, and you could see that they are healthy and happy and in good spirits; you could see them eating human excrement all the time."[35] The river contained an enormous number of these animals, Sorby observed, and their activities were supplemented by other organisms, particularly worms. He suggested that what Darwin had discovered of the action of earthworms in gardens probably also applied to worms in Thames mud.[36]

As with the dissolved oxygen issue, Sorby's evidence represented a new theme in self-purification debate, a theme with ambiguities that perplexed the commissioners and that were exploited by the lawyers. The first problem was where to fit organismic action in the panorama of self-purification mechanisms. Bidder, probably intentionally, implied that organisms were a supplement to oxidation rather than the mechanism by which oxidation occurred. His opening statement (above, p. 500, n. 22) was extremely crafty in that regard. Bidder did not list oxidation and organisms as two separate self-purifying mechanisms, but spoke instead of it not being "a mere question of the oxygen in the water," and of a "cause" (not "another cause"), "overlooked on the other's side." The deception was probably successful. No one suggested that Sorby's organisms

---

[35] Ibid., Q 16625.
[36] Ibid., QQ 16631-32.

were simply the mechanism of oxidative purification and the Board's chemists strategically discussed oxidation and organismic action separately. When, during cross-examination, the suggestion was made to Sir Frederick Abel, one of the Board's chemists, that organisms simply were sites for the oxidation of sewage, Abel agreed, but refrained from expanding on the significance of the statement.[37] The impression remained that organismic action was independent of the amount of oxygen in the Thames, a sort of magical purification mechanism which could plausibly account for any degree of hypothetical purification.

Ignoring the oxidation issue, the commissioners raised doubts about whether assimilation of sewage by crustaceans represented any gain. This was a pervasive concern among sanitarians and had been a central issue in the London water controversy of the early 1850s. For example, Charles Kingsley had suggested that organisms fed on sewage were a worse impurity than sewage itself. At issue was the fate of the excreta and dead bodies of the assimilating organisms. Liebigian pathology was based on the harmfulness of putrefaction and it seemed plausible that entomostracean excrement might putrefy in as dangerous a manner as human excrement. In reassuring the commissioners Sorby pointed out that while effete matter of crustaceans might indeed be objectionable, there was far less of it than there had been of the original food which had been converted into

---

[37] Ibid., QQ 16256-58, 16326-28.

it. This was, after all, the essence of the metabolic process -- a conversion of food partly into excrement, but chiefly into inorganic combustion products.[38]

Such fundamental confusion about physiological principles helps explain why introduction of biological methods and investigations into self-purification debates in the 1880s and 1890s did not immediately lead to an appreciation of biological processes of self-purification. With regard to drinking water quality, the confusion had centered on the question of the meaning of purification when the impurity, an organism, could multiply. At issue here was the fundamental nature of the process of purification.

More than other self-purification arguments, Sorby's application of biology brought home to sanitarians the extraordinary potential for applying science to deceive or to make excuses for pollution. In 1885, Henry Robinson, professor of engineering at Kings College, London, agreed with Sorby that "organisms and plants may be regarded as serving to remove the normal pollution to which any river may be subjected," but "ought not to be claimed as factors removing the abnormal pollution to which our rivers are exposed." Robinson saw "a danger that the knowledge which has of late years been acquired with reference to the chemical and biological changes which go on in a river will be used to justify, or at all events minimise, the mischief arising from sewage pollution."[39] Beale, with considerably

---

[38] Ibid., QQ 16681, 16799, 16735, 16323; Commission Report, p. lxii.

more warmth, took the same view in a January 1884 paper to the Royal Microscopical Society on the "Constituents of the sewage mud of the Thames." Beale regarded conditions on the lower river as disgusting, if not actively unhealthy. He accepted the facts the Board had produced, rejected the implications drawn from them.

> It may be said that all the animal and vegetable tissues and other constituents of feces with actual fecal matter present, do no harm because they are constantly being disintegrated, while many low animal and vegetable forms live at their expense and grow and multiply exceedingly and consume them. It may be said that by these means and by oxidizing and other disintegrating processes, all the organic matters present in sewage are gradually resolved into substances which are not in the least degree deleterious either to fishes or other organisms living in the water of the Thames or to the inhabitants of the houses near the river. Such statements may be made and supported with facts. Arguments telling in the same direction may be freely admitted without the strong objection to the presence of these things in a tidal river being in the slightest degree diminished, much less removed.[40]

Beale went on to observe that bacteria were responsible for the final resolution of organic matter, and that they performed this in a series of stages.[41]

But while Beale objected to Sorby's politicization of the biological processes in the Thames, he too found it impossible to avoid confronting the inherent political implications of this biological principle. Concluding his paper with a section on "Practical Con-

---

[39] Henry Robinson, "River pollution," SR n.s. 6(1884-85): 394.

[40] Lionel Smith Beale, "The constituents of the sewage mud of the Thames," J. Royal Microscopical Society, 2nd ser., 4(1884): 11.

[41] Ibid., pp. 14-16.

siderations," Beale noted that the problem with organismic action in the Thames was that it was anaerobic. Sewage was too concentrated there; "compounds of a most offensive and deleterious nature" were produced.[42] Were the sewage properly diluted on the other hand, by being poured directly into the sea,

> it is probable that for miles round, at a considerable distance from the outlet, organisms of many kinds would grow and multiply in vast numbers. Many of these would become the food of fishes, which in turn would be taken and help support the population which had supplied their sustenance.[43]

It is impossible to say whether Beale was speaking generally or of a particular solution to the London sewage problem here. It is noteworthy, however, that during the next few years a scheme for extending the sewers to the sea was the most seriously considered alternative to existing practices and that this scheme aroused the intense debate that typically characterized London sewage issues.

Sorby too recognized scientific significance in his studies of organismic decomposition in the Thames estuary. In August 1884 he read a paper on his research to the Society of Arts. Unlike Beale's paper, Sorby's was apolitical. He described results, discussed their implications, but forbore from moralizing about the beneficence

---

[42] Ibid., p. 18.

[43] Ibid., pp. 18-19. The notion that sewage was recycled in fish was quite prevalent at this period. Huxley attributed the flavor of Thames whitebait to their being fed on sewage (in H. Robinson, "Some recent phases of the sewage question with remarks on ensilage," [London: Spon, (1885)], p. 210). J.B. Lawes took a similar view (RCMSD, 2nd Report, Evidence, P.P., 1884-85, 31 [C.-4253-I], QQ 18259-74).

of nature in destroying pollution. The theme of Sorby's wide-ranging and prophetic paper was that "the removal of impurities from rivers is more a biological than a chemical question; and that in all discussions of the subject, it is most important to consider the action of minute animals and plants, which may be looked upon as being indirectly most powerful chemical reagents."[44] Sorby correlated the numbers of different types of organisms with the condition of the water. From a study of entomostraceae, rotifers, and insect larvae, he found

> that the number per gallon and percentage relationships of these mark in a most clear manner, changed conditions in the water, the discharge of a certain amount of sewage being indicated by an increase in the total number per gallon, or by an alteration in the relative numbers of different kinds.[45]

As he had before the commission, Sorby noted that the quantity of entomostracean excrements was an indication of the enormous amount of sewage they consumed. He discussed the importance of plants in oxygenating streams and even suggested that scavengers might destroy germs of disease, and thus be the mechanism of drinking-water self-purification.[46]

Sorby's perspective is remarkable. A relative newcomer to sewage science, he had transcended the conceptual limits of sewage

---

[44] Henry Clifton Sorby, "Detection of sewage contamination by the use of the microscope, and on the purifying action of minute animals and plants," J. Royal Microscopical Society, 2nd ser., 4(1884): 991. The paper is reprinted from JRSA 32(1884): 929-30.

[45] Ibid., p. 989.

[46] Ibid., p. 990.

politics, and drawn from a partisan context a basic principle which had been neglected in the chemists' acrimony. It is noteworthy, however, that neither Beale nor Sorby fully recognized the role bacteria played in the process of decomposition. Sorby was concerned chiefly with larger organisms; Beale, like many of his contemporaries, on the one hand spoke of decomposition as a spontaneous process, on the other, associated bacteria with putrefaction. During these years appreciation that bacteria were also responsible for oxidation was also spreading, due most directly to the 1877 discovery of bacterial nitrification by Theophile Schloesing and Achille Muntz and its popularization by Robert Warington.

II. Learning to Trust Bacteria, 1877-1893

In Britain, appreciation of the beneficial sanitary functions of bacteria came as a result of the 1877 discovery by the French agricultural chemists Theophile Schloesing and Achille Muntz that soil nitrification, the conversion of ammonia to nitrates, was caused by bacteria. Schloesing and Muntz had filtered sewage (containing ammonia compounds) through an aquarium filled with sterile sand and made slightly alkaline with lime. No nitrates were present in the filtrate until after twenty days of filtration, but thereafter nitrification increased rapidly. When the filter was sterilized with chloroform nitrification stopped and resumed only after it had been washed and reinoculated with garden soil. The incubation time for the filter, the gradual increase in nitrification, and the effect of sterilization convinced Schloesing and Muntz that they were looking at a microbial process. In Britain their results were quickly confirmed by the agricultural chemist Robert Warington, who had long been interested in soil nitrification. In 1890 Serge Winogradsky finally isolated the organisms responsible, ending a hunt as intense and competitive as the search for pathogens. Percy and Grace Frankland and Robert Warington were among the unsuccessful competitors.[47]

Besides its agricultural significance, the discovery also had profound implications for sanitary science. Because nitrates could not exist under putrefactive conditions, nitrification had from the late 1840s onward been regarded as the ultimate process of purification. When sanitarians extolled the purifying capacities of agricultural soils they were referring to the same phenomenon Schloesing and Muntz had been investigating. Indeed, essentially the Schloesing-Muntz experiment had been performed at least three times in Britain prior to 1877: by Robert Angus Smith in the late 1840s, by J.T. Way in the mid-1850s, and by Edward Frankland in 1869. Each had been puzzled by the increase in nitrifying capacity; none had sug-

[47] Theophile Schloesing and Achille Muntz, "Sur la nitrification par les ferments organises," Compts Rendus de l'Académie de Sciences 84 (1887): 301-3. The paper is reprinted in English translation in Raymond Doetsch, ed., Microbiology -- Historical Contributions from 1776 to 1908 (New Brunswick, N.J.: Rutgers University Press, 1960), pp. 103-7. On the importance of the Schloesing-Muntz work see Richard Aulie, "Boussingault and the Nitrogen Cycle," Diss. Yale University, 1968, pp. 319-24.

gested microorganisms were responsible for it.[48]

Prior to his collaboration with Muntz, Schloesing had been investigating Paris sewage treatment and in that context had stated the problem of nitrification: what was the nature of the mysterious quality of terre vegetale that caused nitrification?[49] In Britain too the importance of the discovery for scientific sewage treatment was immediately clear. In the discussion of Warington's presentation of the French results to the Chemical Society in December 1877, Edward Frankland suggested that "some experiments should be made to try and assist the action of these industrious, inoffensive mycoderms." In particular, he thought knowledge of their habits could be used to increase the rate of sewage filtration through soil.[50]

The immediate importance of the Schloesing-Muntz discovery was symbolic, however, rather than practical. The route to health,

---

[48] R. Angus Smith, "On the air and water of towns," 18th Report of the B.A.A.S. (Swansea) for 1848 (London: John Murray, 1849), pp. 24-25; J.T. Way, "On the composition of waters of land drainage and rain," J. Royal Agricultural Society 17(1856): 148-49; E. Frankland in RPPC/1868/1st Report (Mersey and Ribble Basins), P.P. , 1870, 40, [C.-37.], pp. 61-70. In testimony before the 1872 Commons select committee on the Birmingham Sewage Bill Frankland tried to explain his filtration results (House of Lords Record Office, Minutes of Evidence, HC, 1872, Vol. 5 [Birmingham Sewerage], 29 April 1872, pp. 259-64). Compare this with [Edward Frankland], Sketches of the Life of Edward Frankland (n.p., n.n., 1901), pp. 145-46.

[49] T. Schloesing, "Assainissement de la Seine. Épuration et utilisation des eaux d'egout. Rapport fait au nom d'un commission," Annales d'Hygiène Publique et de Médecine légale, 2nd ser., 47(1877): 212-15.

[50] E. Frankland in disc. of R. Warington, "On nitrification," CN 36(1877): 263.

British sanitarians believed, lay through a certain set of natural processes; that to disease through another set. This dualistic perspective fostered assignment of moral values to natural phenomena: putrefaction, for example, was the quintessential pathological process, while oxidation constituted both the literal breath of life as well as the safe resolution of organic matter. The antithesis was perpetuated in bacteriology. Septic and pathogenic bacteria were regarded as closely associated if not identical. The nitrifying bacteria, although they remained unidentified for more than a decade, symbolized the completion of that dichotomy: they were aerobic organisms responsible for purification.[51] As sanitary engineer Douglas Galton pointed out in 1885, bacteriology had simply placed "laws of health and disease" on a new scientific footing.

> the study of these creatures, of their habits, of the methods by which they can be increased or destroyed, is, therefore, as it were, a study of the laws of health and disease. . . . if these laws . . . were fully ascertained, the sanitarian would learn how to get rid of refuse matter easily and effectually in a manner beneficial to the community; whilst on the other hand the art of the physician would be raised from dependence on empirical observation to the position of an exact science.[52]

Despite Pasteur's early researches on industrial ferments, the

---

[51] Nevertheless there remained an underlying concern that nitrifiers might be pathogenic. See J.W. Mallet, "The determination of organic matter in potable water," CN 46(1882): 101. Lister was bothered that no attempt was made to check pathogenicity when denitrifying organisms were discovered by Percy and Grace Frankland (Richard Fisher, Joseph Lister 1827-1912 [New York: Stein and Day, 1977], p. 286).

[52] Douglas Galton, "Prevention is better than cure," TSIGB 7(1885-86): 351.

Schloesing-Muntz discovery was seen as the first demonstration that microbes were responsible for enormously important and unambiguously benevolent natural processes. Nitrifiers took on a significance for British sanitarians similar to that which Newton's discoveries held for the philosophes of the eighteenth century. They were the counterweight to the germ theory, the proof that God was a sanitarian.

One manifestation of the importance of the discovery is the tendency among the sanitarians of the 1880s to use the concept of nitrification to refer to the beneficent activities of bacteria in general. Thus Henry Robinson, University College, London, professor of engineering, in speaking of organisms that "consume impurities and convert them into nitrates," concluded with the remark: "the action of living agents thus brings about the oxidation of organic matter in the soil."[53] Similarly, Dr. Thomas Stevenson noted in 1885 that the organisms of Schloesing, Muntz, and Warington were the oxidizers of organic matter. For Stevenson, the goal of sewage treatment was to bring clarified sewage "as speedily as possible under oxidizing influences, and nothing more effectively does this than the nitrifying organisms."[54]

From the early 1880s onward, Victorian sanitarians frequently extolled the wonders of bacterial purification. But could this axiom of sanitary natural theology serve as a principle for the design of sewage-treatment plants? Beneath celebratory rhetoric lay a

---

[53] Henry Robinson, "Sewage disposal," SR n.s. 6(1884-85): 150.
[54] Thomas Stevenson, "Sewage disposal," SR n.s. 6(1884-85): 516.

profound anxiety about the wisdom of encouraging microbes. The potency of saprophytic action and the potential utility of harnessing it were appreciated, but there were doubts that it could be controlled, doubts augmented by uncertainties about the natural history of bacteria. For example, some refused to recognize any genetic distinction between malign and benign bacteria, believing instead that environment, not heredity, controlled bacterial function. This was persistent bacteriological heresy. In 1884 W.B. Carpenter argued that pathogens were saprophytes grown in a soil "whose energies were not turned to good account."[55] He accused bacteriologists of irresponsibly inventing new species in defiance of the Darwinian edict of variability.[56] Frederick J. Faraday, a protégé of Angus Smith, took a similar view. The phenomenon of attenuation and the association between pathogenic and septic action suggested to Faraday that the same microbe might not only "tear oxygen as a saprophyte from dead organic matter, but as a parasite from living tissues."[57] Faraday envisioned a continuum of microbial environments. On one end -- the bad end -- the microbe was an anaerobic parasite; in the middle it was an anaerobic saprophyte; at the good end it was an

---

[55] W.B. Carpenter, "The germ theory of disease considered from a natural history point of view," 19th Century 15(1884): 321.

[56] Ibid., pp. 333-34. See also "A.A.," "Disease germs," SR n.s. 19(1897): 103.

[57] Frederick J. Faraday, "Pasteur and the germ theory," CN 50 (1884): 74. See also idem, "On some recent observations in microbiology and their bearing on the evolution of disease and the sewage question," CN 53(1886): 31. Both papers were originally given to the Manchester Literary and Philosophical Society.

aerobic saprophyte.[58] Edward E. Klein, Britain's most famous bacteriologist, carefully examined alleged instances of saprophytes becoming pathogens before rejecting the concept, but Klein pointed out how serious were the consequences for diagnostic bacteriology and for human well-being, if that could happen.[59]

Even for those who accepted the reality of bacterial species there remained the problem of how to encourage good species while simultaneously discouraging bad ones. The obvious answer was to exploit the oxidation/putrefaction dichotomy, to treat sewage under well-aerated, non-biocidal conditions. From the mid-1880s until the mid-1890s, Walter E. Adeney, demonstrator in chemistry at the Royal College of Science in Dublin, and W. Kaye Parry, a civil engineer, experimented in exactly that direction, treating sewage with oxygen-supplying chemicals.[60] Similar experiments, made in London by August Dupre and William Dibdin, will be considered in the next section. Prior to the mid-1890s, however, results of this

---

[58] Faraday, "Pasteur," p. 90.

[59] Edward E. Klein, Micro-organisms and Disease. An introduction to the study of specific micro-organisms, 3rd ed. (London: MacMillan, 1886), p. 222. See also C.T. Kingzett in disc. of Percy Frankland, "Recent bacteriological research in connection with water supply," J. Soc. of Chemical Industry 6(1887): 324-25; and C.W. Kimmins, The Chemistry of Life and Health (London: Methuen, 1892): p. 112.

[60] W.E. Adeney, "The chemical bacteriology of sewage: its hygienic aspect," J. State Medicine 1(1892-93): 78-89; W. Kaye Perry, "A new method of sewage purification," J. State Medicine 1(1892-93): 90-97; Adeney, "The course and nature of fermentative changes in natural and polluted waters, and in artificial solutions, as indicated by the composition of dissolved gases," Scientific Transactions, Royal Dublin Society 5(Sept. 1895): 539-620.

research were too indefinite for widespread technical application. As the Lancet observed in 1893, while it was likely that certain bacteria would power the sewage treatment plants of the future, for the present it was legitimate, even necessary to try to obliterate the lot of them.[61] Thomas Wardle, whose treatise on sewage treatment occasioned the Lancet's remarks, was indeed perplexed by "the good microbe." Wardle was greatly impressed with the nitrification discoveries and all they symbolized. He repeated Duclaux' vision of the clutter of dead and cast off animal matter which would cover the earth but for the beneficial activity of microbes.[62] He wrote of nitrifiers,

> it is not a little singular and wonderful that this smaller and unseen world of plants act as chemical agents in altering the composition of organic and inorganic substances into such states as to enable the greater vegetable world to assimilate them as food from the soil.[63]

Admirable in principle, nitrification was suspect in application. Wardle described the Adeney-Kaye Parry experiments on treating sewage with chemicals that encouraged aerobic organisms, and reiterated Henry Roscoe's observation that "it was idle to discuss the relative value of different chemicals as agencies for the purification of sewage until the precise part . . . played by micro-

---

[61] "Review of Wardle, Sewage Treatment," Lancet, ii, 1893, pp. 1065-66. See also J.W. Slater, Sewage Treatment, Purification, Utilisation (London: Whittaker, 1888), p. vii.

[62] Wardle, On Sewage Treatment and Disposal, p. 19.

[63] Ibid., p. 296.

organisms had been thoroughly investigated." All the same, encouraging microbes was playing with fire: "To rapidly multiply all sorts and conditions of saprophytic organisms, both pathogenic and non-pathogenic, may possibly create a source of greater danger than that which it is the object of the patentee [Adeney] to prevent."[64] Indeed, Wardle approved of recently developed processes which were expressly antiseptic, precipitation processes that promised sterile effluent. He noted the "mighty massacre" of microbes achieved by one, spoke sympathetically of the objective of another: "to kill these organisms, and to make the media in which they live and propagate impossible for them to feed upon; in fact, to extirpate and destroy them entirely."[65] According to Wardle, no method "could be commended which can permit the rapid growth of these organisms, the generators of putrefaction, and in many cases of disease."[66]

Others took the opposite view, arguing that indiscriminate destruction of microorganisms was likely to do more harm than good. In 1889 an "Amines" process of sewage treatment was brought onto the market. Its promoters claimed sewage was permanently disinfected by a substance they called aminol, obtained by mixing large amounts of lime with small quantities of trimethyl amine, obtained from herring brine. The chief selling point of the process was its sterile effluent, certified free of bacteria by no less an authority than Edward

---

[64] Ibid., p. 156.

[65] Ibid., pp. 67, 54.

[66] Ibid., p. 230.

E. Klein. According to one member of the "Amines Syndicate," no effluent was acceptable which did not destroy "life in its most dreaded form, viz., the various bacteria."[67]

Surely this should have been an effective sales pitch during years of germ hysteria, and the promoters were no doubt surprised to encounter widespread rejection of this goal among sanitarians. J.W. Tripe, medical officer of health for Hackney, observed that

> if they killed all the microbes, injurious and others -- for as yet they could not distinguish the injurious from the non-injurious -- what about the water? They knew that the oxidation of water went on to a great extent through these microbes; therefore, if they destroyed them all, they must have a water which would become bad.[68]

Alfred Carpenter, Croydon medical officer of health, agreed. Sterility "destroyed any value that might have been attached to the process as a means of purifying sewage." Carpenter feared sterile effluents would lead to "fermentation" rather than "digestion."[69]

Sterilization was also seen as an interference with the natural processes by which nuisances were removed, plants nourished, and

---

[67] R. Godfrey, "The Amines process of sewage treatment," TSIGB 10(1888-89): 205.

[68] Tripe in disc. of Godfrey, "Amines," p. 214.

[69] Ibid., pp. 214-15. In referring to microbe-free water going bad, Tripe was probably referring to a phenomenon known as secondary putrefaction in which effluents from chemical sewage treatment began to putrefy after discharge into a river as the bactericidal effects of chemicals were removed by dilution and the river deaerated by a large load of oxygen-demanding material. In referring to microbe-free sewage fermenting rather than digesting, Carpenter appears to have forgotten to consider that fermentation is also a microbial process.

elements recycled. Lt. Col. A.S. Jones, perhaps the most successful sewage farm manager in Britain, recognized that sewage purification and utilization involved a putrefactive stage and objected to attempts to prevent it.[70] Similarly, in a letter to Chemical News, "Cyclops" stressed the undesirability of a sterile sludge. The Amines inventors claimed their sterile sludge had fertilizer value; Cyclops argued that fertility and sterility were incompatible.

> Allowing that this [aminol] . . . can disinfect . . . sludge, so that . . . [it] will not decompose when applied to land, then, as it appears to me, he has over-reached his prime objective in putting it there; for he renders his sludge useless as a manure. The nitrogen of good sewage sludge is nearly all present as organic nitrogen, and therefore is not plant food until after decomposition or nitrification. "Permanent sludges," or undecomposable sludges, therefore, must remain on the land for ever as such.[71]

Syndicate representatives responded by suggesting biologically based objections were merely hypocritical whinings of sewage-treatment rivals grasping at any stick to beat a new and impressively successful rival.[72] The Schloesing-Muntz discovery and the subsequent (1887-1890) experiments by the Massachusetts State Board of Health on the purification of sewage by biological filtration had been taken all out of proportion to their real significance. In 1892 sanitary engineer Arthur Angell defended another new patent chemical sewage treatment process, the International System, in a similar

---

[70] Ibid., p. 216.

[71] Cyclops to CN, "The Amines Process," CN 60(1889): 245.

[72] Godfrey, "Amines," p. 218.

tone. According to Angell, "aerobian fermenting vats" were impractical. Nitrification was indeed a microbial process, but "that being so we must not get carried away into the biologists' dreamland."[73]

Discussions of the International System and the Amines process illustrate a theme considered before: the deftness with which sanitarians dodged substantive issues by assuming criticisms to represent "interests" and therefore to be illegitimate. Here biologically based objections to these processes were not rebutted on substantive grounds but through attacks on the motives and credibility of the critics. Yet one cannot blame Angell and the "amines" men for jumping to the conclusion that criticisms of their processes were commercial rather than scientific. As we have seen, Victorian water-and-sewage debates were saturated with ulterior motives: personal, political, religious, commercial, professional. Science was the idiom of these struggles. Data and scientific principles bore no inherent rightness or wrongness, but simply varying degrees of rhetorical utility. Moreover, as Angell suggested, "the friendly microbe" could prove very useful in commercial and political spheres.

The names William Dibdin and August Dupre figure prominently in the discussions about the propriety of allowing microorganisms to purify sewage. Both were chemists, both recognized the importance

---

[73] Arthur Angell, "The treatment and disposal of sewage and sewage sludge," <u>TSIGB</u> 13(1892): 212.

of biological processes in river self-purification, and both, but particularly Dibdin, reveal the difficulties of doing science in a context of distrust.

III. Dupre, Dibdin, and the Politics of Aquatic Saprophytism

During the 1880s and 1890s Dupre and Dibdin were the most important figures in Britain in the application of "the good microbe" to three connected problems: 1) the technological problem of treating sewage; 2) the scientific problem of understanding what happened to sewage in a river; 3) the political problem of making convincing excuses for the harmlessness of dumping raw sewage. Much of their work was joint; it will be well to sketch their activities before considering the significance of their work.[74]

August Dupre (1835-1907), a German-born-and-educated chemist (Ph.D. Heidelburg, 1855), emigrated to Britain in the late 1850s. In 1863 he was appointed lecturer in chemistry and toxicology at the Westminster Hospital and in 1873 became Westminster public analyst. From the mid-1870s he was a consultant chemist to the Local Government Board medical department.[75] In 1880-81 Dupre worked with R.D. Cory in the investigations which revealed the inability of existing processes of water analysis to distinguish water infected

---

[74] For a discussion of their work see Rideal, Sewage and the Bacterial Purification, pp. 43, 176-89.

[75] "August Dupre," DNB, 20th Century, pp. 535-36.

with excreta of typhoid sufferers. The crux of the problem, Dupre recognized, was that only living organic matters -- germs -- were harmful. Existing water analysis had no means for distinguishing living from dead organic matter. In 1884 Dupre suggested a means: deoxygenation of water was a sign of the presence of living organic matter.

Essentially, Dupre had developed a biological oxygen demand test, but for quite different purposes than such tests are now used. Dupre was not interested in measuring biologically oxidizable matter, but in determining the presence of the organisms which oxidized it. As a means of potable water analysis Dupre's method was a dead end, immediately obviated by Koch's plate culturing techniques, a much more direct means of determining the presence and even the type of organisms in water. More significant than the test itself was the assumption on which it rested and which Dupre misrepresented as a conclusion of his research: that **"the consumption of oxygen from . . . a natural water is, in the vast majority of cases at all events, due to the presence of growing organisms, and that in the complete absence of such organisms little or no oxygen would be thus consumed."**[76]

---

[76] August Dupre, "On changes in aeration of water as indicating the nature of the impurities present within it," in 14th Annual Report of the Local Government Board for 1884-85. Supplement containing the report of the Medical Officer for 1884, P.P., 1884-85, 33, [C.-4516.], Appendix B 11, p. 307. See also Rideal, Sewage and the Bacterial Purification, p. 43; and Earle B. Phelps, Stream Sanitation (New York: John Wiley, 1944), p. 66.

As the Frankland-Tidy debates indicate, in the early 1880s British sanitarians (Robert Warington being an exception) still conceived of self-purification as a process of spontaneous oxidation. The view of decay Pasteur had developed in the 1860s had not been widely applied. Dupre's assumption that deoxygenation in water was due to microorganisms is therefore significant as the first instance of a British sanitary scientist approaching the problem of self-purification from a recognizably Pasteurian perspective.

Dupre quickly recognized the implications of this perspective for sewage purification and river self-purification. In testimony to the Royal Commission on Metropolitan Sewage Discharge in 1882 he noted that most of the sewage in the lower Thames was not (spontaneously) oxidized, but "disposed of by animal and vegetable life."[77] In May 1886 he suggested that "a very good thing would be to cultivate the low organisms on a large scale, and to discharge them with the effluent into the river, as the power which these low organisms had [to oxidize sewage] was something remarkable."[78] He returned to the theme in an 1888 address to the Sanitary Institute: "Our [sewage] treatment should be such as to avoid the killing of these organisms or even hampering them in their actions, but rather to do everything to favour them in their beneficial work."[79]

---

[77] RCMSD, <u>1st Report</u>, <u>Evidence</u>, QQ 16100-2.

[78] Dupre in disc. of Tidy, "The treatment of sewage," <u>JRSA</u> 34(1885-86): 669.

[79] August Dupre, "Address to the section on chemistry, meteorology, geology," <u>TSIGB</u> 9(1888): 367.

In addition to his work for the Local Government Board, Dupre was a sewage-treatment consultant for the Metropolitan Board of Works periodically from 1878 until the late 1880s. His biological perspective regarding sewage treatment and self-purification developed during this period; significantly, it was a perspective that fit the political needs of the Board. Dupre offered alternative treatments, far cheaper than irrigation, which were justified by the principle that the action of microorganisms, both in the sewage works and the river, should be the principal means of sewage treatment.

While Dupre was chiefly responsible for the development of the principle of biological treatment, William Dibdin was mainly responsible for putting it into practice. William Dibdin (1850-1925), a native Londoner, studied chemistry at University College, London. In 1877 he became a gas examiner for the Board of Works, and in the next five years rose through the ranks of the chemical department to become chief assistant to T.W. Keates, department director. Keates died in early May 1882, just as the Board was organizing its defense for the anticipated hearings of the Sewage Discharge Commission. There was some feeling that Dupre, valued consultant, should be Keates' successor, but it was pointed out that Dupre was already being paid as a consultant for the upcoming hearings and that the present assistant, Dibdin, could supervise mundane business, mainly gas analyses, at less cost. In June 1882 Dibdin was appointed chief chemical assistant at £250/year.[80] He proved an aggressive administrator; during his 15 years as chief chemist the importance and salary of that position increased substantially.

At the end of 1883 the first report of the Royal Commission on Metropolitan Sewage Discharge appeared, carrying the conclusion that pollution of the lower Thames by the Board was indeed a nuisance. Refusing to accept the verdict, the Board declined to participate in the second phase of the hearings in which solutions for the problem would be considered. Nevertheless, it had already begun internal studies of solutions.[81] With a grant of £1000 (£200 more were provided in the beginning of 1884) Dibdin and Dupre built tanks and conducted experiments on sewage precipitation and sewage aeration in late 1883 and early 1884. Sewage treatment by forced aeration was Dupre's pet idea, a logical consequence of his work on biological water analysis for the Local Government Board. Dibdin was less enthusiastic: "Aerating the sewage in a similar manner to that in which aquarium tanks are aerated" was a terrible bother, gave negligible improvement.[82]

---

[80] GLCRO/Metropolitan Board of Works (MBW), Minutes, 1882, 1, 30 June 1882, 1057/6; GLCRO/MBW Works and General Purposes Committee, Minutes, sub-committee on officers, 26 June 1882, p. 282. See also C.J. Regan, "Early developments in London's sewage disposal with particular reference to the work of W.J. Dibdin," Journal and Proceedings, Institute of Sewage Purification, 1951, pp. 338-47; William Saunders, History of the first London County Council, 1889-1890-1891 (London: National Press Association, 1892), p. 621; SR n.s. 20 (1897): 575.

[81] GLCRO/MBW Annual Report, 1884, pp. 4-10.

[82] RCMSD, 2nd Report, Evidence, Q 19150; GLCRO/MBW 2400, Rpt. 1190, "Extracts from the report of the works and general purposes committee on the treatment of the Metropolitan sewage," 8 March 1886, p. 3; Rept. 1218A, "Letters from Dr. A. Dupre to officers of the

Dibdin was more interested in the sewage precipitation experiments. He agreed with most sewage treatment authorities that precipitation processes did not remove dissolved organic matter; that they clarified sewage rather than purifying it. Dibdin also found that dosing sewage with more chemicals did not produce corresponding increases in clarification; nearly all suspended matter could be innocuously removed by addition of 3.7 grains lime and 1 grain iron sulphate per gallon, a far smaller amount of chemicals than most authorities endorsed. Clarified sewage full of dissolved organic matter might safely be dumped into the river, Dibdin believed, except on excessively hot days when an artificial oxidant (sulfuric acid or manganate of soda) would be required to prevent the stink. In July 1885 Dibdin got more money for precipitation experiments by enticing the Board with the possibility that his treatment scheme would allow use of existing outfalls and permit dumping of sewage on any tide, a solution far cheaper than the alternative of vastly lengthened sewers leading to the mouth of the Thames.[83] In October Dibdin got permission to have the scheme considered by outside chemists. F.A. Abel, William Odling, A.W. Williamson, and Dupre reported that additional chemicals would not be beneficial, but had little else positive to offer. Dibdin attached to their report an

---

Board in the years 1883-4-5, on the subject of the sewage of the Metropolis," 29 June 1888, #1-5; GLCRO/MBW 684, Minutes of MBW sub-committee on Metropolitan sewage, 20 July 1886, pp. 235-40, 30 July 1886, p. 258.

[83] GLCRO/MBW 683, Minutes of the MBW sub-committee on Metropolitan sewage, 15 July 1885, pp. 678-79, 22 July 1885, pp. 688-92.

introduction which suggested a more enthusiastic opinion than was in fact the case, and by the spring of 1886, despite half-hearted endorsement by a friendly committee of chemists, the opposition of the government and the Board's own legal counsel, and ridicule by many other chemists, Dibdin got his scheme adopted -- but only temporarily.[84]

On the verge of success, Dibdin's progress was blocked. E. Rider Cook, a Board member who claimed knowledge of chemistry, objected that Dibdin's scheme was nonsense chemically and a waste of public money. Thus far Dibdin had lobbied without aid of Dupre's oxidizing organisms. His response to Cook was based on Dupre's philosophy that sewage treatment should encourage the aerobic organisms that ultimately destroyed sewage. The purpose of lime was to reduce the food supplied to these organisms to a level the river ecosystem could cope with, while the purpose of manganate addition was to prevent nuisance without sacrifice of the dissolved oxygen these organisms required. Thus, the traditional antiseptic or preservative treatments Cook advocated were the antithesis of what was

---

[84] Ibid., 7 October 1885, p. 718; MBW 684, Minutes of the MBW sub-committee on Metropolitan sewage, 14 October 1885, p. 3, 21 October 1885, pp. 19-20, 28 October 1885, pp. 30-35, 13 January 1886, pp. 83-86, 2 March 1886, p. 131, 10 March 1886, pp. 141-42, 8 June 1886, p. 192. Tidy argued that this group of chemists would not likely give an impartial answer since three of the four had testified for the MBW in the RCMSD hearing. This was like asking for a prescription from a doctor who claimed the patient was perfectly healthy, according to Tidy (in disc. "The treatment of sewage," JRSA 35(1886-87): 79-80). See also "Disposal of sewage," SR n.s. 7(1885-86): 489.

required.[85] About the same time as his contretemps with Cook, Dibdin made these views public in a January 1887 paper to the Institution of Civil Engineers, a paper which he later portrayed as the first statement of the principle of biological sewage treatment.[86]

Dibdin failed to convince his employers. He was relieved of sewage treatment responsibility and for the last two years (1887-89) of the Board's existence treatment operations were run by Sir Henry Roscoe, who despite his acknowledgment that sewage was purified by aquatic organisms, shared Cook's faith in antiseptic treatment and used chloride of lime at the outfalls. Roscoe was only marginally successful, and when the London County Council took over in 1889 Dibdin's authority was reinstated and his scheme for treatment put into effect.[87]

---

[85] GLCRO/MBW 2401, Rpt. 1225, "MBW. Report by the chemist upon the letter of Mr. E.R. Cook as to the treatment of the sewage in the sewers by manganate," 17 January 1887; Rept. 1226, "Statement made by Mr. E. Rider Cook to the Works and General Purposes Committee, 31 January 1887, on the Report by the chemist"; Report 1231, "MBW. Report by the chemist in reply to Mr. E.R. Cook's statement on the deodorization of the sewage in the sewers by manganate of soda," 28 February 1887; MBW 2403, Rpt. 1298, "MBW. Letters, reports, etc., by E.R. Cook, esq., and the chemist on the subject of the treatment of the sewage in the sewers by manganate of soda," 25 June 1888 (includes the above reports).

[86] William Dibdin, "Sewage sludge and its disposal," MPICE 88 (1886-87): 155-74.

[87] GLCRO/MBW 2403, Rpt. 1276, "MBW. Preliminary reports by Sir Henry Roscoe: 1) On the deodorisation of sewage in the metropolitan sewers, 2) On the deodorisation of sewage at the outfalls," 27 Feb. 1888; GLCRO/LCC/MD Ctte. Pres. Papers, Vol. 1, 18 April 1889, #16, H.E. Roscoe, "Report on the deodorisation of metropolitan sewage at the outfalls, May-December 1888";#25, W.J. Dibdin, "Deodorisation and disposal of the sewage of London"; Vol. 4, 31 January 1890, #30, Roscoe to MD Ctte., 24 January 1890.

Thereafter Dibdin's work on sewage treatment and river purification became increasingly biological. In 1889 he did a microscopical examination of foreshore muds along the lower Thames, studying both estuarine organisms and non-living sewage components in the mud, much as Sorby had done seven years earlier.[88] In 1891 the L.C.C. resolved to experiment on the filtration of chemically-treated effluents. Similar studies done by the Massachusetts State Board of Health in 1888-90 had proved sewage filtration was a biological process and had shown in detail how this biological activity could be controlled. It is probable that Dibdin and his L.C.C. colleagues based their experiments not on the Massachusetts work, however, but were inspired by the common practice in British sewage plants of submitting effluent from chemical treatment to a final filtration as well as to the availability of new patent effluent filters, the International Company's "polarite," for example.[89] Between 1892 and 1895, Dibdin, his assistant George Thudichum, and L.C.C. engineer Alexander Binnie shared supervision of the experiments which began in 1892 with four 22 sq. yard filters, and were scaled up in a one-acre filter bed in September 1893. Unlike the

---

[88] GLCRO/LCC/MD Cttee. Pres. Papers, Vol. 1, 9 April 1889, #22, H.E. Roscoe, "Report on the chemical examination of the foreshores and mud deposits of the river Thames and its estuary . . . together with microscopical examinations . . . by the chemist to the Board."

[89] GLCRO/LCC/MD Cttee. Pres. Papers, Vol. 4, 19 February 1890, #12, "Experiments on filtration"; and L.C.C., "Main Drainage of London. Joint Report of Sir Benjamin Baker and the Chief Engineer of the Council," p. 15.

Massachusetts experiments in which parameters were systematically varied to determine optimal filtration conditions, the London experiments were unsystematic comparisons of arbitrarily selected filtering media fed at different rates. Dibdin made mistakes and learnt from them. By late 1893, the Binnie-controlled one-acre filter had become a putrid mess, and Dibdin, pleading to his superiors for control of the project, argued that successful filtration required "proper ecological conditions."[90] The feed rate of sewage had to be maintained in an equilibrium with the strength of the sewage, the size of the microbial population, and the rates of digestion and re-oxygenation.

From 1892 to 1895 Dibdin was also conducting thorough chemical and biological studies of the lower Thames. Each week data was collected on the amount of upstream water in the lower river, on the distribution of oxygen and fish, and on the general condition of the river. Such data from 6400 samples were the basis of an 1894 report on the "Character of the River Thames from Teddington to the Nore." Here Dibdin considered in broad scope the self-purifying capacity of the Thames estuary. By studying the residence time of upstream water in the estuary, the input of pollution, and the amount of organic matter in the water, Dibdin was able to discuss the oxygen balance of the estuary, to form rough estimates of the oxygen coming from upstream and diffusing from the air, and the

---

[90] GLCRO/LCC/MD Cttee. Pres. Papers, Vol. 18, 7 December 1893, #9.

amount of oxidation that was occurring.[91]

As Dibdin gained experience in running a biological filter and in observing the response of the lower river to changing conditions, his biological perspective became increasingly broad. In August 1892, for example, he explained L.C.C. sewage treatment strategy to a committee from the London Chamber of Commerce. Biological purification, Dibdin pointed out, was not simply a process of feeding sewage slops to aquatic pigs, but in fact required a systematic understanding of the kinetics of several physical and biochemical processes: feed rate, metabolic rate, and the rate of reaeration. Dibdin understood that in running a filter he was dealing with an ecosystem: "By reason of the multitude of organisms to be dealt with, a filter may be fittingly compared to a great animal. If you overfeed it, the result will be disastrous."[92] The same applied to dumping effluent into the river:

> If you overfeed the river with effluent, then you put more organic matter into the river than the aquatic life is capable of dealing with. . . . The point at which the aquatic life becomes unable to deal with the food supply is that at which the oxygen dissolved in the water, and which is necessary to support that aquatic life, is absorbed at a greater rate than [that at which] the water will renew its supply of oxygen from the air.[93]

---

[91]GLCRO/LCC/MD Cttee. Pres. Papers, Vol. 22, 27 September 1894, #50, "Results of the examination of the character of the River Thames from Teddington to the Nore." See also Dibdin, <u>Purification of Sewage and Water</u>, pp. 285-88.

[92] London Chamber of Commerce, Thames Sewage Committee, "Report of the special commission on the visit of inspection to the Thames sewage precipitation works at Barking and Crossness on 2 June 1892," Appendix ("Questions put to, and answered by LCC officials, August,

Thus, more organisms did not necessarily mean better and faster purification. The river did not stink in cool weather because the metabolization of sewage (equally the removal of dissolved oxygen) occurred sufficiently slowly that the very low oxygen levels which led to putrefaction did not occur. On the other hand, "during warm weather the multiplication of the organisms proceeds at such a rate that the supply of oxygen is insufficient and consequently they become suffocated. The whole then becomes corrupt, that is, both the food and the fed begin to putrefy. The remedy is therefore obviously to prevent the too rapid multiplication of organisms, by adjusting their food supply."[94] As this statement, together with the Teddington-Nore report and comments in Dibdin's filtration reports illustrate, Dibdin had thoroughly incorporated Dupre's perspective into his day-to-day work for the L.C.C. by the mid-1890s.

In 1897 Dibdin resigned from the L.C.C. to become a sewage-treatment consultant. The biological processes involved in sewage treatment and river purification had been widely recognized. In that year appeared the first edition of Dibdin's <u>Purification of Sewage and Water</u>, a combination of detailed technical advice, philosophical reflections, and historical observations. Dibdin saw himself as founder and elder statesman of biological sewage purification. Look-

---

1892"), Q 11, in GLCRO/LCC/MD Cttee. Pres. Papers, Vol. 17, March 16, 1893, #56.

[93] Ibid.

[94] Ibid.

ing back on the previous fifteen years he recognized what seems to us a familiar pattern of the Whig history of science -- one man's diligent kindling of truth in the face of profound ignorance and widespread prejudice. His "teachings of modern science" had been "thrown aside because they interfere[d] with the views of interested persons," they had been denounced by "the lecture-table chemist," and he had been "laughed at and thought to be a madman" for proposing to treat London's sewage biologically and in the river.[95] According to Dibdin, he had been the prophet rejected in his own land (Sorby and Dupre also got honorable mentions).

Dibdin's description of the response to his policies and ideas was accurate; however, to a large degree it was justified. Before becoming "modern science" biological purification had been a political stratagem. In the mid-1880s Dibdin's motives were probably far less noble than he later imagined them to have been; certainly they were perceived that way. Dibdin introduced the biological purification argument in his January 1887 paper as a justification for the scheme of chemical treatment he had proposed to the MBW. The chemicals Dibdin proposed to use on the sewage, lime and iron sulphate, were not new as sewage precipitants. A patent for lime sewage treatment had been granted as early as 1802, and because it was cheap, lime was a widely used, though only marginally successful

---

[95] William Dibdin, "The purification of sewage and water," SR n.s. 19(1897): 213, 215; Arthur Angell, "The treatment and disposal of sewage and sewage sludge," TSIGB 13(1892): 212-13; C.H. Cooper, "Notes on sewage treatment," TSIGB 13(1892): 207.

sewage precipitant.[96]

Instead, what was novel about Dibdin's plan was the "homeopathic" doses of chemicals advocated.[97] Since lime treatment was barely successful when 30 grains were used, it was hard to accept Dibdin's claim that lime in excess of 3.7 grains did no good. It appeared to most sanitarians that MBW hirelings Dibdin and Dupre were simply advocating a sham treatment whose only advantage was cheapness. Charles Meymott Tidy noted that Dupre, in giving advice on sewage treatments at other inquiries, favored much larger amounts of the same chemicals. It didn't help that Dibdin and Dupre boasted repeatedly of the economy of their scheme. Given the immense quantity of metropolitan sewage even a one grain per gallon increase in treatment chemicals represented an enormous sum of money.[98] Thus to most sanitarians, long outraged that the MBW had apparent legal warrant to pollute while all other towns were required to purify sewage, Dibdin's scheme seemed only an excuse for avoiding the irrigation solution recommended by the Royal Commission of 1882-84.[99]

---

[96] Tidy, "The treatment of sewage," JRSA 34(1885-86): 1155-58.

[97] R.W. Peregrine Birch, "Mr. Dibdin's experiments," SR n.s. 9 (1887-88): 53; A.S. Jones, "Pollution of the Thames," Fortnightly Review o.s. 46(1886): 87.

[98] Tidy, "The treatment of sewage," JRSA 35(1886-87): 79; Dupre in Tidy, "The treatment of sewage," JRSA 34(1885-86): 669; Tidy in disc. of Dibdin, "Sewage sludge and its disposal," MPICE 88(1886-87): 211-12.

[99] D.J. Ebbets, "Metropolitan sewage discharge: A costly failure!" SR n.s. 8(1886-87): 437-39; John Bailey Denton and A.S. Jones, "The purification of the Thames," SR n.s. 8(1886-87): 28; Henry Robinson, "Sewage disposal," TSIGB 10(1888-89): 195-96.

Worse, Dibdin advocated ineffective methods because they were ineffective. Stronger treatments would interfere with biological processes in the river; biology therefore justified inadequacy.

The context in which Dibdin's biological suggestions were made is clear from an examination of his January 1887 paper to the Institution of Civil Engineers. Although his scheme had been discussed at earlier meetings of sanitarians, this paper represents the first comprehensive presentation of the work Dupre and Dibdin had been doing. Only two pages of the 17 page paper dealt with the harmfulness of antiseptic forms of sewage treatment, such as large amounts of lime. Dibdin called these the "reverse of nature's method" and prophesied that in future strains of specially cultivated organisms would be added to sewage to destroy it.[100] Most of those who took part in the discussion commented on the viability of the chemical treatment Dibdin proposed. Among those who considered the paper's biological aspects, Dupre expounded at length on the concept of biological purification and a few others accepted it as a possibility. A.S. Jones cynically noted how well the biological doctrine complemented MBW politics. He praised Dibdin's ingenuity in adapting theories to meet the exigencies: "He [Jones] did not say that theory was invented to meet the state of things, but that appeared to be the dominant idea in the mind of Mr. Dibdin."[101] Jones thought Dibdin

---

[100] Dibdin, "Sewage sludge and its disposal," pp. 160-62.

[101] Jones in disc. of Dibdin, "Sewage sludge and its disposal," pp. 195-97.

took biology "a little too far" and was misusing Warington's discoveries. He raised an objection considered earlier: what would become of the excreta and dead bodies of the sewage-consuming organisms? Another critic, James Edmunds, medical officer of health of St. James, Westminster, interpreted Dibdin's biological suggestions in terms of pathogen destruction. This was like a "ferret sucking eggs," and Edmunds thought there must surely be a better way to destroy pathogens.[102]

It is this 1887 paper that Dibdin later cited as revolutionary in the history of sewage treatment. Here, in the face of public ridicule, he had made the suggestion that had inspired the Massachusetts State Board of Health experimenters and thus inaugurated the age of bacterial sewage treatment. Whatever one may say about the integrity of Dibdin's self-image at this stage in his career, it is clear that his observations about biological purification, like Edward Frankland's germs, fell victim to the political context in which British water science was done. What happened at the 1887 meeting was what usually happened at meetings where papers on sewage treatment were read. The audience assumed the speaker's purpose was to advertise a particular treatment strategy or to advocate a particular position on water issues, and responded accordingly. Speaking to the Society of Arts in 1886, Charles Meymott Tidy tried, probably in vain, to persuade his audience that this time he came

[102] Edmunds in ibid., p. 239.

as judge, not advocate. In effect, Tidy was admitting that usually when he spoke about sewage treatment his remarks were not to be regarded as impartial; doubtless the first time in this paper he offered praise or criticism some concluded that he had returned to advocacy.[103]

Political motives underlay Dibdin's position but they also underlay the positions taken by many of his critics. For example, A.S. Jones and sanitary engineer John Bailey Denton were trying to win the approval of parliament and the Metropolitan Board for a plan to treat metropolitan sewage on a huge sewage farm on Canvey Island near the mouth of the Thames. Their plan required that the MBW abandon the existing outfalls and at its expense extend the main sewers far down river to Hole Haven, at which point they would take responsibility for the sewage and pipe it the short distance to Canvey Island. Thus Jones' criticism was motivated not only by his long-standing commitment to sewage farming, but also by his opposition to the major obstacle that stood in the way of his enterprise: use of the existing outfalls. Other critics had connections with patent sewage precipitation processes which Dibdin had decided not to use: J.C. Mellis represented the Rivers Purification Association and W.C. Sillar the Native Guano Company.

Revolutions in sewage treatment technology, in attitudes towards the phenomena of decomposition, and in the statement of the questions

---

[103] Tidy, "The treatment of sewage," JRSA 34(1885-86): 612.

of sanitary science did occur during the 1890s. As with his own enlightenment, Dibdin invested these revolutions too with a rationality of development they did not possess. What Dibdin saw as the triumph of Pasteur's conception of decomposition in sanitary affairs had several components.

First, "nature's methods" of sewage treatment generated excitement because they were believed to be cheap. As one authority put it, the "heart of every town councillor leapt with joy" at the realization that sewage could be treated without expensive chemicals or large amounts of land.[104] Universal adoption of biological sewage treatment in Britain would save £60,000,000 per year, Dibdin boasted.[105] Developers of other biological methods of treatment, such as septic tank developers Donald Cameron and W.D. Scott Moncrieff, also stressed the economical operation of their processes.[106] By contrast with biological methods, precipitation works and sewage farms represented "a grievous burden on the rate-payer."[107] These claims for cheapness, though much overblown, brought bacterial sewage treatment a political constituency. A 1902 London County Council

---

[104] W. Dunbar, Principles of Sewage Treatment, translated with the author's sanction by H.T. Calvert (London: Charles Griffin, 1908), pp. 157-59.

[105] Ibid. Also see S. Rideal in disc. of Dibdin, "The purification of the Thames," MPICE 129(1896-97): 152.

[106] Donald Cameron, "Some recent experiments in sewage treatment at Exeter," Engineering 62(1892): 257; "Report on the Scott-Moncrieff system for the bacteriological purification of sewage," Builder 68 (1895): 217.

[107] Dibdin, "Purification of the Thames," p. 168.

report listed 30 British cities using bacterial cities, at least on an experimental basis, with success.[108]

Luckily, what was cheap was also, in the broadest sense of the word, good. An issue of natural theology -- the relationship between God's creation and man's urbanization -- had always been a fundamental concern of British sanitarians. As Stevenson has pointed out, the germ theory shook the faith of those who believed nature and urban man were compatible.[109] As J.B.C. Kershaw put it,

> The life of many of the human race has become a burden to them since they heard of the close connection that exists between bacteria and disease. They realise in a most vivid manner that the food, the water they drink, and even the very air they breathe are rarely free from these "pestilential microbes." Life as these nervous ones see it is a long-drawn-out conflict between the human unit and countless myriads of invisible but deadly enemies. The attack is delivered at every moment and at every step of the man's path through life; and the end is forgone -- the microbes invariably win though their triumph may be long delayed.[110]

The discoveries of nitrification, of septic sewage treatment, of the ecological importance of microbes generally, were basis for a counter germ theory. Literary periodicals informed the public that bacteriology showed the world was good, and, by implication, the creation wise. Kershaw followed the above-quoted passage with a description of microbes as the makers of alcohol and vinegar and the cleaners-up

---

108 Dibdin, Purification of Sewage and Water, pp. 69-74.

109 Lloyd G. Stevenson, "Science down the drain -- on the hostility of certain sanitarians to animal experimentation, bacteriology, and immunology," Bull. Hist. Med. 29(1955): 2, 25-26.

110 J.B.C. Kershaw, "Bacteria in harness," Chambers' Journal 76(1899): 419.

of the world.[111] A wide variety of celebratory metaphors appeared. One writer noted that like savages bacteria could be trained to do good, another that they worked for free, took no breaks, and never struck.[112] Discussion of Pasteur's perspective towards decomposition in the sanitary and popular literature of the 1890s represents not so much the publicizing of scientific results as the public enjoyment of their implications. By confirming the faith that the creation was the best of all possible worlds, biological purification acquired a religious constituency.

One scientific discovery of the mid-1890s seemed marvelously to complement this optimism, to suggest that in nature were resolutions to all the anxieties bacteriology had raised, that an appreciation of man-nature harmony required only filling in a few more pieces of the picture. This was the observation that pathogenic bacteria could not usually withstand septic conditions. British sanitarians had recognized that putrefaction was a process in which organic matter became inorganic, and one of them, Robert Angus Smith, had believed that putrefaction had the particular effect of destroying morbid poisons.[113] The fate of pathogens under putrefactive conditions had been one of the issues of bacterial survival considered

---

[111] Ibid., pp. 420-21.

[112] E.A. de Schweinitz, "The mission of the microbe," SR n.s. 19 (1897): 449; W. Noble Twelvetrees, "Disinfection: physical, chemical, mechanical," SR n.s. 19(1897): 483.

[113] Robert Angus Smith, "Presidential address to the section on chemistry, meteorology, and geology," TSIGB 5(1883-84): 274-77.

by the 1892-94 Royal Commission on Metropolitan Water Supply. So suggestive was the evidence that putrefaction destroyed pathogens that in 1894 Edward E. Klein, alarmed that on this basis sewage-polluted water might be regarded as safe, carried out thorough investigations of the limits within which this phenomenon occurred.[114] Cameron and Scott-Moncrieff, developers (separately) of the septic tank, gave pathogen destruction equal billing with economical operation, as reasons for adopting the septic route of sewage treatment. As Cameron explained it,

> Micro-organisms in general, and those of disease in particular, are peculiarly sensitive beings, requiring certain well-defined conditions as to food, temperature, and so on. During the passage of the sewage through the tank and filters, any organisms in it are subjected to complete changes of environment. . . . During each of these stages any organisms originally present in the sewage are liable to be preyed on by others better adapted to the conditions in which they are placed.[115]

Though nuisance prevention had replaced pathogen destruction as the main priority of sewage treatment, it was doubtless reassuring to know that the new sewage treatments destroyed pathogens incidentally while preventing nuisance.

Finally, the development of biological processes of sewage treatment was not essentially a technological application of Pasteur-

---

[114] Edward E. Klein, "On the behaviour of the typhoid bacillus and of Koch's vibrio in sewage," in 29th Annual Report of the Local Government Board. Supplement containing the report of the Medical Officer for 1894-95, P.P., 1895, 51, [C.-7906.], Appendix B 2, p. 407.

[115] Cameron, "Some recent experiments," p. 257. See also W.D. Scott Moncrieff in The Builder 68(1895): 245.

ian science as Dibdin suggested. In his books on sewage treatment, Dibdin began with a consideration of the biochemical job to be done and went on to examine how this job was done naturally and artificially. The implication was that technologists had applied a scientific understanding of decomposition in their designs of biological filters and tanks. In fact such job descriptions as Dibdin offered were usually applied only after filters were built and working. In 1897 German Sims Woodhead lamented that sewage treatment engineers were ignoring a large body of chemical and bacteriological research on the stages of decomposition.[116] W. Dunbar, the leading German sewage treatment expert of the early twentieth century, gave a far more accurate description of the development of biological sewage treatment than Dibdin when he suggested that the biological processes of the 1890s were variations on the theme of intermittent filtration, a process which Edward Frankland had developed in 1869-70, but whose biological mechanisms had not been recognized until recently.[117] The notion that the age of biological sewage treatment had begun in 1887 when Dibdin announced the proper scientific principles of sewage treatment, and the claim that the later experiments of the Massachusetts State Board of Health were inspired by this announcement was, to Dunbar, absurd.[118]

---

[116] Woodhead in disc. of Dibdin, "Purification of the Thames," pp. 142-45.

[117] Dunbar, Principles of Sewage Treatment, p. 120.

[118] Ibid., p. 121.

Historiographical failings aside, Dibdin is not lightly dismissed. His career, like those of Edward and Percy Frankland, illustrates how scientific inquiry could work in a social environment which in some ways seemed so strongly to discourage it. As revealed by the Teddington-Nore study, and a similarly impressive set of reports on the chemistry and bacteriology of sewage by Dibdin, F.H. Andrewes and J. Parry Laws, Dibdin was successful working within, and even manipulating the science-in-politics context, if not in transcending it.[119] A comparison with Percy Frankland's filtration studies is particularly useful. For Frankland political implications came to dominate science. In Dibdin's case, the political context set the agenda for science and paid the bill for a kind of scientific research whose massive data requirements made it rarely practicable for Victorian scientists. Again, those studies of the Thames which culminated with the Teddington-Nore report exemplify this. Their original purpose was to provide a data base in the event the London sewage question should again become an issue. Before beginning precipitation treatment at the outfalls, for example, an internal report was produced for the MBW by Dibdin and Roscoe, showing how dreadful was the condition of the lower river. The apparent purpose of this report was to serve as a baseline against which the

---

119 GLCRO/LCC/MD Cttee. Pres. Papers, Vol. 21, 7 December 1893, #6, J. Parry Laws, "Report to the Main Drainage Committee of the L.C.C. on sewer air investigations"; Vol. 23, 13 December 1894, #48, "Investigations on the micro-organisms in sewage"; Vol. 22, 18 January 1894, #24, "Investigations of sewer and sewer air."

improved conditions which would presumably result from sewage treatment could be compared.[120] But while filling his employers' order for what was essentially political information, Dibdin also used his data to develop a thorough understanding of the relations between biological and chemical phenomena in the Thames estuary, an understanding which was both politically useful and scientifically important.

[120] GLCRO/LCC/MD Cttee. Pres. Papers, Vol. 1, 9 April 1889, #22, H.E. Roscoe, "Report on the chemical examination of the foreshores and mud deposits of the River Thames and its estuary . . . together with microscopical examinations . . . by the chemist to the Board."

CONCLUSION: SOCIETY, SCIENCE, AND SELF-PURIFICATION

This has been a story mainly of failure, of a massive dose of science failing to provide the technical guidance needed to solve social problems of water supply, wastes disposal, and river use as well as failing to achieve a scientific understanding of the factors that affect the quality of water in rivers. Some of that failure is clearly a result of the limited ability of science to produce the appropriate facts. Key pieces of knowledge, such as the identity of the cause of cholera, were unavailable to water-supply advisors for much of the period under study. As George Rosen has noted, sanitary reform was pursued despite the "little real knowledge" in regard to some medical matters of central importance in public health.[1] Nevertheless, the failure of science to provide guidance in these matters is in sharp contrast with the rapid progress of biology, chemistry, and medical science that characterizes the second half of the nineteenth century. During these years the foundations of epidemiology were laid, the germ theory was proving itself as the basis both for curative and preventive medicine, the role of microorganisms as decomposers was recognized, and the chemistry of putrefaction was being worked out. These discoveries and others have been celebrated for so long, and in retrospect appear so

---

[1] George Rosen, "Economic and social policy in public health," in George Rosen, From Medical Police to Social Medicine: Essays in the History of Health Care (New York: Science History Publications, 1974), p. 197.

significant in their impacts on society, that it is hard to recognize that they did not automatically bring with them the software necessary for successful application to social problems. Again Rosen is helpful:

> the mere accretion of medical ideas and knowledge cannot of itself assure application. Social environment and intellectual milieu must provide favorable conditions and patterns of behavior in terms of which knowledge can be put to use.[2]

Some responsibility for this failure, therefore, must lie with the institutions through which basic science was applied to public health problems.

I have suggested two factors as particularly important in accounting for this failure:

1) the essentially political (or trans-scientific) nature of the public health problems scientists were asked to solve, and the failure of political institutions to solve them even when scientific information was available.

2) the subversion of the "republic of science" by trans-scientific issues that damaged the workings of the scientific community by undermining an assumption vital for cooperative endeavor: that the work of one's colleagues was founded on objectivity, honesty, and a "philosophic spirit," rather than on a political basis.

That there are important links between science and its social

---

[2] Ibid., p. 187.

environment has become a platitude of modern history of science, and is even more fundamental in the history of medical science. Rosen describes the task of the historian of public health as "to investigate and demonstrate how economic, social, medical, and scientific events intertwine and interact to create specific public health developments."[3] Yet problems arise when one begins to probe the nature of these intertwinings. I have been continually vexed by two issues which I suspect are hazards that characteristically complicate the study of science-society interactions.

1) The Cause-and-Effect Problem: Do scientists take positions on social issues which are compatible with their views of the nature of nature or do they adopt views of nature which justify their views on social policy? For example, did Chadwick advocate a particular set of housing, burial, sewage, and water reforms because he believed in the association of disease with the products of putrefaction, or was he drawn to that theory of disease because it justified the reforms he thought necessary (Chapters 3 and 4)? Similarly, was it simply coincidence that Dibdin's views of biological self-purification upheld the interests of his employer, the Metropolitan Board of Works, or did he advocate that concept in order to defend his employer (Chapter 10)?

Wherever possible, I have avoided these kinds of questions because they can only be answered by detailed and intimate evidence of

---

[3] Ibid., p. 200.

a kind that is rarely available. Moreover, I suspect that this binary perspective misrepresents the way most people's minds reconcile facts, scientific principles, and beliefs. It seems more likely that Chadwick and Dibdin pursued a peace of mind in which views of nature, human nature, and duty were mangled into compatibility. According to Walter Houghton this struggle to achieve unity, to see, for example, all forms of progress, whether religious, scientific, social, economic, or moral as compatible and lying in the same direction, is characteristic of Victorian thought.[4] One can detect a consistency and compatibility among Charles Meymott Tidy's beliefs: that nature was benign, that the status quo was just, that the common people could be trusted to come to the best decision about water supplies when given the facts, and that polluted river water was safe after a certain distance of flow. Likewise there is a consistency in Edward Frankland's thought: in his faith in technocracy, in his acceptance of the manipulation of public opinion as a proper means of achieving sanitary reform, in his fear that nature was at best neutral toward humanity, and in the assertion that no distance of flow could be trusted to make polluted water safe to drink. I suspect both men were unable to keep scientific facts and social welfare distinct, especially in issues like river self-purification where scientific and social questions were so closely linked. The resulting amalgam of opinion each man pro-

---

[4] Walter Houghton, <u>The Victorian Frame of Mind</u> (New Haven: Yale University Press, 1957).

duced reflects a compromise for the sake of unity -- a melting together of science and morality, yet not to the degree that either element was totally obscured. Though the conclusions of the self-purification studies undertaken by Frankland in 1868-69 and Tidy in 1880-81 are consistent with the alternative concepts of the social duty of the expert advisor that each man represented, both men recognized that as scientists it was necessary to equip morality with facts on the changing composition of natural and artificial waters (Chapter 8).

2) The On-Off Problem: Do social factors affect:

-- only the rate of scientific activity?

-- the rate of scientific activity and indirectly the content of science through encouragement of certain disciplines or the presentation of certain problems to be solved?

-- the rate of scientific activity, and, directly, the content of science?

George Rosen writes of "the great, and in many cases, over-riding importance of non-medical factors in creating the structure and channels within which . . . [medical science] may operate."[5] His implication is that basic science develops relatively independently along intrinsically rational lines (though perhaps at a rate determined by society's support for science) and will be applied to

[5] Rosen, "Economic and social policy," p. 200. For similar statements see pp. 176, 187, and 197.

social problems when conditions are favorable. Besides the portrayal of science as an efficient problem-solving machine that can be turned on or off, this perspective suggests that the impact of society on science will either be favorable or neutral, that the rational world of facts can be recognized or not recognized but that it cannot be obscured by societal inputs. Both axioms are violated in the self-purification controversy.

First, no independent discipline of water-and-sewage science, complete with facts, methods, and principles, existed for sanitary scientists to apply to local water-and-sewage problems. Facts were acquired within a political context; methods and principles adopted on the basis of their political implications. Between 1850 and 1900 most of the investigations of self-purification were undertaken with clear political purposes. Edward Frankland's 1868-69 studies of self-purification, for example, were the first experimental studies of that phenomenon, yet Frankland's conclusion (in defiance of his data) was political: that self-purification virtually did not exist (Chapter 8). Likewise, Tidy's 1880-81 papers on "River Water," while they were empirical advances, were essentially expressions in a scientific idiom of the view that river water was acceptable for public water supplies (Chapter 8). The profound difference between investigations done within and outside a political framework can be seen by comparing Percy Frankland's report and testimony on water bacteriology to the Royal Commission on Metropolitan Water Supply with the report he and Marshall Ward prepared for the Royal

Society (Chapter 9).

Second, the social environment of self-purification science had a prejudicial effect on the ability of the scientific community to obtain knowledge. An adversary context permeated water-and-sewage science. It is evident not only where scientists served as witnesses in pollution litigation, but also in conflicts over the merits of rival processes of water and sewage purification, such as that which developed between Frankland and Crookes, or over rival processes of water analysis, such as that between Frankland and Wanklyn (Chapter 8). That context led to a condition in which science was unable to function effectively as a cooperative institution because water policy disputes and associated professional or personal rivalries took precedence over the pursuit of knowledge.

The social context of science also perpetuated vagueness. The imprecise concept of filth-disease equivalency was useful as a justification for sanitary reform. At the Medical Department of the Local Government Act Office and later the Local Government Board, John Simon's "scientific inquirers," particularly pathologist John Burdon Sanderson, attempted to refine the filth-disease association, while in the field his sanitary inspectors exploited it to support their calls for local sanitary reform (Chapter 3). New concepts or techniques were obscured or compromised in this social environment of science. Robert Warington's 1881 observation that Pasteur's understanding of decomposition could resolve the dispute between Edward Frankland and Tidy on the oxidation of organic matter in water remained unappreciated because it did not fit easily into the political nature of the dispute (Chapter 8). Koch's procedures for water bacteriology acquired political tarnish when used in connection with the London water controversy (Chapter 9). Indeed, the effect of the institutions in which science existed on the content of science are so important here that it is better to talk of the application of social problems to science rather than the application of science to social problems as is usually done.

Originally this study was to compare the self-purification issue in Britain with its handling in Massachusetts. That state was one of the first places in America where intensive studies of polluted rivers were done (by the Board of Health) and the comparison also seemed worthwhile because the State Board of Health of Massachusetts, along with William Dibdin of the Metropolitan Board of Works in London, were rival claimants in an 1890s dispute over priority for the discovery of biological sewage treatment, a technology whose discovery in Britain was closely linked to the self-purification controversy (Chapter 10). A detailed comparison was impractical, but how different things were in Massachusetts!

There, the discovery of the concept of biological sewage treatment was a model of interdisciplinary cooperation, rather than an excuse for the unwillingness of the Metropolitan Board of Works to adopt more expensive sewage treatment processes.[6] Adversary pro-

---

[6] Massachusetts State Board of Health, Report on Water Supplies and Sewerage, pt. ii, Experimental Investigations by the State Board of Health of Massachusetts upon the Purification of Sewage by

ceedings did not dominate scientific inquiry on water-and-sewage matters. For the chemist William Ripley Nichols, who surveyed the pollution of Massachusetts rivers for the Board of Health in the mid 1870s, self-purification was an observation, not a dogmatic assertion presented as a defense of pollution or of inadequate water supply.[7] An 1870 Massachusetts observer concluded a description of the vitriolic sewage debates going on in England with the words: "strange as this may seem to some people in Massachusetts."[8]

There are several reasons for ending this study in 1900. First, by 1900 an international outlook is far more apparent in British sanitary science than it was in the mid 1870s or even the mid 1880s. German sanitary science had a major influence on the recognition of the importance of stream biology by the scientists of the West Riding of Yorkshire Rivers Board about 1910.[9]

Filtration and Chemical Precipitation, and upon the Intermittent Filtration of Water. Made at Lawrence, Massachusetts, 1888-90 (Boston: Wright & Potter, 1890); and H.W. Clark and Stephen DeM. Gage, "A review of twenty-one years' experiments upon the purification of sewage at the Lawrence Experiment Station," Fortieth Annual Report of the State Board of Health of Massachusetts (Boston: Wright & Potter, 1909), pp. 253-538.

7 William Ripley Nichols, "On the present condition of certain rivers in Massachusetts," Fifth Annual Report of the State Board of Health of Massachusetts, January 1874 (Boston: Wright & Potter, 1874), pp. 76-82.

8 H. Bowditch, "Letter on sewage," Second Annual Report of the State Board of Health of Massachusetts, January 1871 (Boston: Wright & Potter, 1871), pp. 233-34.

9 Yorkshire Water Authority Library, Leeds, Reports of the West Riding of Yorkshire Rivers Board, #98, H.M. Wilson and H.T. Calvert, "Report on a visit to Germany," (Wakefield, 1906); #101, "Report upon a visit to Germany in connection with the 14th International Congress of Hygiene, September 1907," (Wakefield, 1907).

Second, the technical problems of water-and-sewage science were being solved. The significance of bacteriological evidence had been an open question during the 1880s and even the early 1890s. Trust in bacteriological tests, -- the acceptance of coliform counts as the measure of water quality, for example -- developed only after procedures had been refined and standardized, and sanitarians had acquired experience running such tests and correlating their results with health statistics. Biological sewage treatment and water chlorination (introduced about 1912, though not quickly adopted in Britain) were effective technical solutions for problems that had been the source of intense political debate during the nineteenth century.

Finally, there were significant changes in the institutions of self-purification debate around 1900. The controversy over the self-purification of the Thames and Lea that was linked to attempts to obtain public control of the London water supply ended in 1901, when the London water companies were bought out under an act of parliament (the sources of supply remained unchanged). A Royal Commission on Sewage Disposal was established in 1898 and sat until 1911. That commission appears to have avoided the partisanship which characterized many earlier commission reports on water and sewage issues, notably the reports of the second Rivers Pollution Commission (Chapter 8) and those of Chadwick's General Board of Health (Chapter 5). The 1898 commission is noteworthy for the breadth of expertise its membership represented, and for the thoroughness of its work. Controversial aspects of water-and-sewage policy, such as the effect of marine sewage discharge on edible shellfish, were investigated thoroughly by scientists hired by the commission, a contrast to the practice of earlier commissions (the Royal Commission on Metropolitan Water Supply, for example) of determining facts by allowing scientists representing opposing parties to present contradictory scientific studies.

APPENDIX A: SAMPLES OF WATER ANALYSES, 1828-72

ANALYST, DATE, CITATION, SUBSTANCES TESTED FOR

1. Letheby, Henry. 1868. RPPC/1868/1st Report, P.P., 1870, 40, p. 203
   - carbonate of lime and magnesia
   - sulphate of lime
   - chloride of sodium
   - nitrate of soda
   - silica and alumina
   - organic matter
   - ammonia
   - hardness

2. Crace-Calvert, Frederick. 1868. RPPC/1868/1st Report, p. 215 (item 1)*
   - mineral matter:
     - chloride of calcium, of sodium, & sulphate of magnesium, nitrates
     - sulphate of lime
     - carbonate of lime, carbonate of magnesia, carbonate of iron protoxide
   - hardness
   - organic matter and volatile salts

3. Crace-Calvert, Frederick. 1867. RPPC/1868/5th Report, P.P., 1874, 33, Evidence, pp. xxxvii-iii
   - total hardness
   - permanent hardness
   - organic matter
   - mineral matter
     - chloride of magnesium, chloride of sodium, chloride of calcium, (sulphate of magnesium, sulphate of soda, sulphate of potash)
     - sulphate of lime (qual)
     - weight of sulphate of lime
     - carbonate of lime, carbonate of magnesium, carbonate of iron

4. Miller, William. 1867. RPPC/1868/5th Report, Evidence, p. xlii (item 3)
   - carbonate of lime
   - carbonate of magnesia

---

* "Item" refers to the sample number under which a complete citation for the source of the analysis can be found.

sulphate of magnesia
chloride of magnesium
chloride of sodium
combustible and organic matter
nitric acid
ammonia
hardness before boiling
hardness after boiling

5. Herapeth, William. 1868. RPPC/1868/5th Report, Evidence, p. xliv (item 3)
   chlorate of magnesium
   nitrate of magnesia
   chloride of sodium
   sulphate of magnesia
   carbonate of magnesia and lime
   sulphate of lime
   oxide of iron
   humus
   hardness before boiling
   hardness after boiling

6. Collins, J.H. 1872. RPPC/1868/5th Report, Evidence, p. xliv (item 3)
   turbidity, taste, smell (qual)
   hardness
   organic
   inorganic
     chloride of sodium, carbonate of lime, and sulphate of magnesia with traces of iron
     nitric acid

7. Letheby, Henry. 1869. RPPC/1868/3rd Report, P.P., 1871, 25, Evidence, p. 62
   total solid matter
   chloride of sodium
   earthy sulphates
   hardness
   hardness after boiling
   oxygen required by organic and other oxydisable matter
   ammonia
   nitrogen as nitrates

8. Manning, F.A. & J.T. Way. n.d. RPPC/1868/3rd Report, Evidence, p. 67 (item 7)
   solid residue
     mineral
     organic

hardness
  permanent
  temporary

9. Letheby, Henry. n.d. RPPC/1868/3rd Report, Evidence, p. 74 (item 7)
   carbonate of lime and magnesia
   sulphate of lime
   chloride of sodium
   silica, alumina, and oxide of iron
   organic matter
   hardness
   hardness after boiling

10. Voeckler, Augustus. n.d. RPPC/1868/3rd Report, Evidence, p. 77 (item 7)
   solid matter, dried at 300°F
   organic matter
   carbonate of lime
   sulphate of lime
   magnesia
   oxides and traces of phosphates
   alkaline salts
   hardness before boiling
   hardness after boiling

11. Richardson & Marreco. n.d. RPPC/1868/3rd Report, Evidence, p. 273 (item 7)
   solid matters
   earthy salts
   hardness
   color (qual)

12. Thomson, Robert Dundas. c. 1867. RCWS, P.P., 1868-69, 33, Appendix E, p. 31
   total impurity
   organic impurity
   hardness

13. Macadam, Stevenson. n.d. RPPC/1868/4th Report, Evidence, p. 57 (item 8)
   A. [for more or less bad waters]
      saline matter, principally carbonate of lime
      sulphate of lime and chloride of sodium
      organic matter and nitrates
      total dissolved matter
      hardness

   B. [for good water]

chloride of sodium
chloride of magnesium
sulphate of lime
chloride of potassium
carbonate of lime
carbonate of magnesia
phosphate and oxide of iron
soluble silica
organic
  organic matter of vegetable origin
total matter dissolved
hardness

14. Gamgee, Arthur. 1865. RPPC/1868/4th Report, Evidence, p. 323 (item 8)
  hardness
  total solids
  volatile and combustible matters and fixed salts
  oxygen (yielded by permanganate of potash) required to oxidize the organic matter
  chlorine
  sulphuric acid
  odor, taste, color (qual)
  nitric acid, ammonia, and nitrites (qual)

15. Way, J.T. 1866. RCWS, Appendix F, p. 36 (item 12)
  Pt. A
    lime
    magnesia
    soda
    chlorides of sodium and potassium
    oxide of iron, silica, etc.
    sulphuric acid
    carbonic acid
    organic matter
    total impurity
    hardness before boiling
    hardness after boiling

  Pt. B ("these substances [part A] are probably combined as follows:")
    carbonate of lime
    carbonate of magnesia
    carbonate of soda
    sulphate of soda
    chlorides of sodium and potassium
    oxide of iron, silica, etc.
    organic matter

16. Hofmann, A.W., Thomas Graham, & W.A. Miller. 1851. J. Chem. Soc. 4(1851): 378-79.
    Pt. A [acids and bases separately]
      lime
      magnesia
      potassium
      sodium
      iron, alumina, and phosphates
      sulphuric acid (SO_3)
      chlorine
      carbonic acid
      silica
      nitric acid
      ammonia

    Pt. B [acids and bases combined]
      carbonate of lime
      sulphate of lime
      nitrate of lime
      carbonate of magnesia
      chloride of sodium
      sulphate of soda
      chloride of potassium
      sulphate of potassa
      carbonate of potassa
      silica
      iron, alumina, and phosphates
      ammonia
      organic matter
      total
      solid residue on evaporation
      free carbonic acid (in inches and grains)
      suspended matter
      hardness

17. Brande, W.T. 1849. GBH Water Supply Report, P.P., 1850, 22, Appendix I, p. 32
      Clark's hardness
      total solids

18. Phillips, Richard. n.d. GBH Water Supply Report, Appendix III, p. 189 (item 17)
      carbonate of lime
      sulphate of lime and common salt

19. Letheby, Henry, F.A. Abel, & William Odling. 1867. RCWS, App. AG & AH, pp. 72-74 (item 12)

General Properties
- hardness
- hardness after boiling
- dissolved gases
  - % carbonic acid
  - % oxygen
  - % nitrogen
- oxygen required to oxidise nitrite and protosalt iron
- oxygen required to oxidise organic matter
- total oxygen required
- oxygen dissolved (grains)
- ammonia

Analysis
- lime
- magnesia
- potassa
- soda
- ammonia
- iron and alumina
- carbonic acid
- sulphuric acid
- nitric acid
- chlorine
- silica

Total
- total minus oxygen equivalent of chlorine = total saline matter
- organic and other volatile matter by incineration
- total solid matter by analysis
- total solid matter by evaporation
- suspended matter organic
- suspended matter mineral
- total per gallon

20. Miller, W.A. 1865. Medical Officer of the Privy Council, <u>8th Annual Report</u>, <u>P.P.</u>, 1866, 33, Appendix 6, p. 244.
- color, odor, taste
- hardness
- hardness after boiling 1 hour
- specific gravity
- total solids
  - fixed salts
  - volatile and combustible matter
- nitric acid
- ammonia
- total gas/gallon
  - oxygen
  - nitrogen
  - carbonic acid

ratio of oxygen to nitrogen

21. Parkes, E.A. 1865. M.O. Privy Council, <u>8th Report</u>, Appendix 14, p. 398 (item 20)
    total solids
    organic matter
    chloride of sodium
    hardness
    grains of oxygen required to oxidize organic matter @ 140°F

22. Pearson & Gardner. 1828. GBH Water Supply Report, Appendix I, p. 32 (item 17)
    specific gravity
    muriate of magnesia
    muriate of soda
    sulphate of lime
    carbonate of lime
    silica and vegetable or carbonaceous matter

23. Campbell, Dugald. n.d. RCWS, Evidence, p. 208 (item 12)
    solids
    mineral
    volatilized matter by carbonization
    oxidizable organic matter
    hardness

24. Parkes, E.A. 1865. M.O. Privy Council, <u>8th Report</u>, Appendix 14, pp. 399, 404 (item 20)
    mineral solids
    volatile solids
    total solids
    chloride of sodium
    hardness
    grains of oxygen required to oxidize organic matter @ 140°F

25. de Chaumont, F.S.B. 1865. M.O. Privy Council, <u>8th Report</u>, Appendix 14, pp. 410-11 (item 20)
    physical
        clarity, taste, sediment, color
    qualitative
        lime
        magnesia
        sulphuric acid
        ammonia
        iron
        organic matter (by chloride of gold)
        hardness
            fixed
            removable

quantitative
- fixed solids
  - carbonate of lime
  - sulphate of lime
  - carbonate of magnesia
  - carbonate of soda
  - chloride of sodium
- volatile solids
  - organic matter by permanganate of potash
  - other volatile matter
- total solids; by evaporation, analysis
- iron
- silica
- free carbonic acid

APPENDIX B: BIOGRAPHICAL DIGEST OF SOME IMPORTANT BUT NOT WELL-KNOWN FIGURES PROMINENT IN THE HISTORY OF RIVER SELF-PURIFICATION

Bidder, George Parker, jr., 1836-96. Barrister active in sanitary litigation, known for ability to perform rapid mental calculations. B.A. Cambridge, 1858, 7th wrangler. (Engineering 61 (1896): 202)

Brande, W.T., 1788-1866. Chemist active as a consultant on water matters during the 1850s. Professor of chemistry at the Royal Institution. (DNB vol. 2, pp. 1124-26)

Brodie, Sir Benjamin Collins (the younger), 1817-80. Theoretical chemist. Educated Harrow and Balliol College, Oxford. Professor of chemistry at Oxford, 1865-72. (DNB vol. 2, pp. 1288-89)

Crookes, William, 1832-1919. Consulting chemist, water analyst, sewage treatment entrepreneur, science editor and journalist. Student and assistant, Royal College of Chemistry, 1848-54. (Proc. Royal Society of London, series A, 96(1920): l-ix)

Dibdin, William, 1850-1925. Chemist active in water-and-sewage matters. Educated University of London [?]. Chemist for the Metropolitan Board of Works, later the London County Council, 1875-97. (C.J. Regan, J. and Proc., Inst. Sewage Purification, 1951, pp. 338-47)

Dupre, August, 1835-1907. Chemist active in water-and-sewage matters. b. Mainz, educated at Giessen (Liebig, Will), and Heidelberg (Bunsen, Kirchoff). Ph.D., 1855. Lecturer in chemistry and toxicology at the Westminster Hospital, 1863-1907; public analyst for Westminster, 1873-1901. (DNB, 20th Century, pp. 535-36)

Frankland, Edward, 1825-99. Chemist active in water-and-sewage matters. b. Lancaster, student of Playfair (1845), at Marburg (with Bunsen, 1847-49, Ph.D.), and Giessen (with Liebig). Professor of chemistry at the Royal College of Chemistry (later joined with the Royal School of Mines), 1865-85. (W.H. Brock, DSB, Vol. 5, pp. 124-27)

Frankland, Percy, 1858-1948. Chemist and bacteriologist. Pioneer in water bacteriology. Second son of E. Frankland, educated Würzburg (with Wislicenus). Professor of chemistry, Birmingham, 1894 on. (W.H. Garner, J. Chemical Society, 1948, iii, pp. 1996-2005)

Hassall, Arthur Hill, 1817-94. Microscopist and physician active in water politics and water analysis during the 1850s. Studied medicine in Dublin; M.R.C.S., 1839, M.R.C.P., 1851, M.B., M.D., (London), 1848, 1851. (E.G. Clayton, Arthur Hill Hassall, Physician and Sanitary Reformer [London: Balliere, Tindall, and Cox, 1908])

Hawksley, Thomas, 1807-93. Sanitary engineer, one of the main proponents of self-purification. Trained as an architect and surveyor. (Proc. Royal Society of London 55(1894): xvi-xvii).

Hofmann, August Wilhelm, 1818-92. Chemist active in water-and-sewage matters. Ph.D., Giessen (with Liebig), 1841. Professor of chemistry, Royal College of Chemistry, 1845-65. (W.H. Brock, DSB, Vol. 6, pp. 461-64)

Klein, Edward Emanuel, 1844-1925. Histologist and bacteriologist active in water-and-sewage matters. b. Essek, Slovonika; M.D. Vienna. Director, Brown Institution; professor of histology, physiology, and advanced bacteriology, St. Bartholomew's Hospital. (Proc. Royal Society of London, series B, 98(1925): xxv-xxix)

Lankester, Edwin, 1814-74. Physician and microscopist active in sanitary matters. Educated University of London and Heidelberg (M.D.). Editor of the Quarterly J. of Microscopical Science (1853-71), medical officer of health, St. James Westminster, 1856-74. (DNB vol. 11, pp. 578-80)

Letheby, Henry, 1816-76. Chemist and medical man active in sanitary matters. Educated London University (M.B. and Ph.D.). Lecturer in chemistry at the London Hospital, medical officer of health for the City of London, 1856-76. (DNB vol. 11, p. 1010)

Miller, William Allen, 1817-70. Chemist active in water-and-sewage matters. Educated Kings College, London, and Giessen (with Liebig). Professor of chemistry, Kings College, London. (DNB vol. 13, pp. 429-30)

Odling, William, 1829-1901. Chemist active in water-and-sewage matters. Educated at London University (M.D.) and at Guy's Hospital (with A.S. Taylor). Professor of chemistry at the Royal Institution, 1867; at Oxford, 1872 on. (W.H. Brock, DSB, Vol. 10, pp. 177-79)

Sanderson, John Burdon, 1829-1905. Pathologist and microscopist. Educated in Edinburgh (with J.H. Bennett) and Paris (with A. Wurtz and Claude Bernard). Professor of human physiology and physiology at University College, London, 1870-82, later professor at Oxford. Supervisor of the Brown Institution; one of John

Simon's main "scientific inquirers" at the Local Government Board Medical Department. (DNB, 20th Century, pp. 1489-90)

Smith, Robert Angus, 1817-74. Chemist active in sanitary matters. Educated Giessen (with Liebig, Ph.D.). Inspector under Rivers Pollution Prevention Act (1876). (DNB vol. 18, pp. 520-22)

Rawlinson, Robert, 1810-98. Sanitary engineer. Trained under Robert Stephenson. Chief engineering inspector, Local Government Board, and predecessor organizations until 1888. (MPICE 134(1897-98): 348)

Sorby, Henry Clifton, 1826-1908. Microscopist active in metropolitan sewage controversies, 1882-84. Privately educated. One of the most successful amateur scientists. (Proc. Royal Society of London, series B, 80(1908): lvi-lxvi)

Tidy, Charles Meymott, 1843-92. Chemist and medical man active in water-and-sewage matters. Educated at London Hospital (with Letheby) and at Aberdeen (M.B.). Professor of chemistry, public health and medical jurisprudence at the London Hospital, 1876-92. (DNB vol. 19, pp. 864-65)

Wanklyn, James Alfred, 1834-1906. Chemist active in water-and-sewage matters. Educated at Owens College, Manchester (with Frankland) and Heidelberg (with Bunsen). Professor of chemistry at the London Institution, 1863-70. (W.H. Brock, DSB, Vol. 14, pp. 168-70)

Warington, Robert (the elder), 1807-67. Chemist, inventor of the aquarium. Educated by J.T. Cooper and at University College, London (with E. Turner). Chemical operator for the Society of Apothecaries. (Proc. Royal Society of London 16(1868): xlix)

Warington, Robert (the younger), 1838-1907. Chemist active in the study of nitrification and agriculture generally. Educated at Apothecaries Hall (with father). Employed by J.B. Lawes at Rothamstead; Professor of Rural Economy at Oxford, 1894-97. (Proc. Royal Society of London, series B, 80(1908): xv-xxiv)

BIBLIOGRAPHY

I. MANUSCRIPT SOURCES

Greater London County Council Record Office:

1. Metropolitan Board of Works Papers, 1865-88: Main Drainage Committee, Minutes; Sub-committee on the sewage of the Metropolis, Minutes; Minutes of the Board; Works and General Purposes Committee, Minutes; Reports to the Board; Annual Reports; Materials relating to the Royal Commission on Metropolitan Sewage Discharge.

2. London County Council Papers, 1889-94: Main Drainage Committee, Agenda Papers, Minutes, Presented Papers.

3. Thames Navigation Act, 1870. In Arbitration. The Thames Conservators and the Metropolitan Board of Works. Minutes of Proceedings, Report, and Determination of Arbitrators.

Public Record Office

HLG 50 2078, 2079: River Lee Pollution, 1874-86
MH 19 67: Correspondence, Home Office to Local Government Board
HLG 1 40-44: West Riding of Yorkshire Rivers Board Materials
HO 74 3: Home Office, Correspondence with commissions of inquiry
MH 29 1-20: Metropolitan Water Works Papers, 1875-94

Liverpool University Library

T. Mellard Reade Correspondence, 14 letters Percy and Edward Frankland to T. Mellard Reade, 1876-85.

Royal Institution

Tyndall Correspondence, 5 letters, Tyndall to Edward Frankland and Edward Frankland to Tyndall, 1871-80.

General Papers, Box 17, file 211, Miscellaneous materials relating to Edward Frankland.

Royal Society of London

Miscellaneous Manuscripts vol. 10 90-102, 13 letters, Edward Frankland to H.E. Armstrong, 1867-96.

Imperial College

Armstrong Papers, series 2, C241-46, 6 letters, J.J. Day to H.E. Armstrong, 1867-69.

Yorkshire Water Authority Library

Reports and Annual Reports, 1895-1911.

House of Lords Record Office

Minutes of Evidence, House of Commons, 1872, vol. 5, (Birmingham Sewerage).

Minutes of Evidence, House of Commons, 1878, vol. 5, (Cheltenham Corporation Water Bill).

Select Committee on the Pollution of Rivers Bill, Report with evidence, Lords Papers, vol 9, 1873, (132.).

Note on Parliamentary Papers

Parliamentary Papers have been the single most valuable source in this project. Two publications by P. and G. Ford are especially valuable aids in the use of these papers. Their Guide to Parliamentary Papers (3rd ed. [Totowa, N.J., 1972]) describes the various kinds of materials published as Parliamentary Papers and explains their organization. Their Select List of British Parliamentary Papers, 1833-99 (Oxford, 1953) provides an excellent subject index of these papers. Official subject indices are also available (annual, decennial, and fifty-year)but are difficult to use. Parliamentary Papers are cited by the session, the volume number within that session (1868, 40 is the fortieth volume of the Parliamentary Papers for session 1868), and by the session or command number, those listed in parentheses or brackets (usually) and followed by a period. Since title citations often vary, these numbers identify the paper. Pagination here is by paper, rather than by volume.

II. PARLIAMENTARY PAPERS

1828 Royal Commission on Water Supply, Report and Minutes of Evidence, 9, (267.).

1846 Select Committee on Metropolitan Sewage Manure, Minutes of Evidence, Appendix, and Index, 10, (474.).

1850 General Board of Health
Report on the Epidemic Cholera of 1848 and 1849, 21, [1273.].
Report on the Supply of Water to the Metropolis, 22, [1218.].
Report on the Supply of Water to the Metropolis. Appendix 1, Returns of the Water Companies, 22, [1281.].
Report on the Supply of Water to the Metropolis. Appendix 2, Analyses of several waters put in by F. Braithwaite, 22, [1282.].
Report on the Supply of Water to the Metropolis. Appendix 3, Reports of Evidence, Medical, Chemical, Geological, and Miscellaneous, 22, [1283.].

1851 Select Committee on Metropolis Water Supply Bill, Report and Minutes of Evidence, 15, (643.).

1852 Select Committee on Metropolis Water Bills
Report, 12, (395.).
Minutes of Evidence, 12, (395-I.).

1854- Select Committee on the Public Health Bill and the Nuisances
1855 Removal Bill, Report and Minutes of Evidence, 13, (244.).

General Board of Health. Medical Council.
Report of the Committee for Scientific Inquiries in Relation to the Cholera Epidemic of 1854, 21, [1980.].
Committee for Scientific Inquiries in Relation to the Cholera Epidemic of 1854, Appendices, 21, [1996.].

T.E. Blackwell, Report of T.E. Blackwell, C.E., to the President of the General Board of Health, on the Drainage and Water Supply of Sandgate in connexion with the outbreak of cholera in that town, 45, (82.).

1856 Reports to the Rt. Hon. William Cowper, M.P., president of the General Board of Health on the Metropolis Water Supply, under the Provisions of the Metropolis Water Act, 52, [2137.].
A.W. Hofmann and Lyndsay Blyth, "Report on the Chemical Quality of the Water supplied to the Metropolis,"
William Ranger, Henry Austin, and Alfred Dickens, "Examination of the Thames."

1857 session 1 A.H. Hassall, Report to the Rt. Hon. William Cowper, president of the General Board of Health on the Microscopical Examination of the Metropolitan Water Supply; under the Provisions of the Metropolis Water Act, 8, [2203.].

1857 session 2 Henry Austin, Report on the Means of Deodorizing and Utilizing the Sewage of Towns, addressed to the Rt. Hon. President of the General Board of Health by Henry Austin, C.E., Chief Superintendant Inspector, 20, [2262].

Douglas Galton, James Simpson, and Thomas E. Blackwell, Report on the Metropolitan Main Drainage, 36, (233.).

1857-58 Thomas Hawksley, G.P. Bidder, and Joseph Bazalgette, Report to the Metropolitan Board of Works by Messrs. Hawksley, Bidder, and Bazalgette, 48, (419.).

Copy of a letter to the Rt. Hon. Lord John Manners, M.P., 1st Commissioner of H.M. Works, from the Government Referees for the Main Drainage of the Metropolis, in answer to the report made by Messrs. Bidder, Hawksley, and Bazalgette to the Metropolitan Board of Works, upon the Report of the Referees, 48, (403.).

Brief Observations of Messrs. Bidder, Hawksley, and Bazalgette on the answer of the government referees to their report to the Metropolitan Board of Works, 48, (471.).

Select Committee on the River Thames, Report with proceedings of the committee, minutes of evidence, appendix, and index, 11, (442.).

Royal Commission on the Sewage of Towns, Preliminary Report with appendices, 32, [2372.].

1860 Select Committee on the Serpentine, Report with Minutes of Evidence, 20, (192.).

Medical Officer of the Privy Council, Second Annual Report for 1859, 29, [2376.].

1861 Royal Commission on the Sewage of Towns, Second Report with appendices, 33, [2882.].

1862 Select Committee on the Sewage of Towns
First report with evidence, 14, (160.).
Second report with evidence, 14, (469.).

1864 Select Committee on Sewage (Metropolis), Report, together with proceedings of the committee, Minutes of evidence, appendix, and Index, 14, (487.).

Letter . . . Relating to the Pollution of streams, by Lords Ebury and Shaftesbury, representing the Sanitary Association of Great Britain, and Lords Saltoun and Llanover, representing the Fisheries Preservation Association to the Rt. Hon. Viscount Palmerston, 1st Lord of the Treasury, 50, (224.).

1865 Royal Commission on the Cattle Plague, Third Report, with appendices, 22, [3656.].

Royal Commission on the Sewage of Towns, Third Report with appendices, 27, [3472.].

1866 Select Committee on the Thames Navigation Bill, Report with evidence, 12, (391.).

Royal Commission on Rivers Pollution, 1865
First Report (River Thames), with appendix, and plans, 33, [3634.].
Minutes of Evidence, 33, [3634-I.].

1867 Select Committee on East London Water Bills, Report, together with proceedings of the committee, minutes of Evidence, appendix, and index, 9, (399.).

Medical Officer of the Privy Council, Ninth Annual Report, for 1866, 37, [3949.].

Royal Commission on Rivers Pollution, 1865
Second Report (The River Lee) with appendices, 33, [3835.].
Minutes of Evidence, 33, [3835-I.].

Third Report (Rivers Aire and Calder) with appendices, 33, [3850.].
Minutes of evidence and index, 33, [3850-I.].

A Return of Royal Commissions issued from the year 1861 to the present time . . . with total expenses, 40, (261.).

Correspondence between the Board of Trade and the East London Water Works Company with reference to Captain Tyler's report on the water supplied by the company, 58, (574.).

Report by Captain Tyler to the Board of Trade on the Quantity and Quality of the water supplied by the East London Water Works Company, 58, (339.).

1867-68 Select Committee on the Lee River Conservancy Bill, Report with evidence, 11, (306.).

Medical Officer of the Privy Council, Tenth Annual Report, with appendices, for 1867, 36, [4004.].

William Farr, Report on the Cholera Epidemic of 1866 in England. Supplement of the 29th Annual Report of the Registrar General of Births, Deaths, and Marriages in England, 37, [4072.].

1868-69 Royal Commission on Water Supply
Report, 33, [4169.].
Minutes of Evidence, 33, [4169-I.].
Appendices, 33, [4169-II.].

1870 Medical Officer of the Privy Council, Twelfth Annual Report for 1869, 30, [C.-208.].

Royal Commission on Rivers Pollution, 1868
First Report (Mersey and Ribble Basins)
Report and Plans, 40, [C.-37.].
Minutes of Evidence, 40, [C.-109.].

Second Report (The A.B.C. Process of Treating Sewage), 40, [C.-180.].

1871 Third Report (Pollution from the Woollen Industry), 25, [C.-347.].
Minutes of Evidence, 25, [C.-347-I.].

Medical Officer of the Privy Council, Thirteenth Annual Report for 1870, 31, [C.-349.].

1872 Royal Commission on Rivers Pollution, 1868
Fourth Report (Rivers of Scotland), Report and Plans, 34, [C.-603.].
Minutes of Evidence, 34, [C.-603.I.].

Letter from the Royal Commission on Rivers Pollution to the Board of Trade of 10th April 1872; and, a Report thereon by the Water Examiner, 49, (186.).

Francis Bolton, Copy of any Reports to the Board of Trade made by the Water Examiner appointed under the Metropolis Water Act, 1871, 49, (82.).

Edward Frankland, Report on the Analysis of the Waters supplied by the Metropolitan Water Companies during the several months of the years 1869, 1870, and 1871, by Professor Frankland; copy of his letter to the Registrar General, dated 10th July 1869, and Analyses of the Metropolitan Water Supply for October 1871 and January 1872, 49, (99.).

1873 Return of the Names of Boroughs, Local Boards, Parishes, and Special Drainage Districts which have through loans provided Sewage farms or other means for the disposal of sewage by filtration and precipitation, 56, (134.).

1874 Medical Officer of the Privy Council and the Local Government Board, n.s. 1, Annual Report to the Local Government Board, 31, [C.-1066.].

Royal Commission on Rivers Pollution, 1868
Fifth Report (Pollution arising from Mining Operations and Metal Manufactures)
Report and Plans, 33, [C.-951.].
Minutes of Evidence, 33, [C.-951-I.].

Sixth Report (The Domestic Water Supply of Great Britain), 33, [C.-1112.].

1875 Medical Officer of the Privy Council and the Local Government Board, n.s. 4, Annual Report for 1874, 40, [C.-1318.].

n.s. 5, Papers concerning the European Relations of the Asiatic Cholera, 40, [C.-1370.].

n.s. 6, Report on Scientific investigations in pathology and medicine, 40, [C.-1371.].

1876 Report of a Committee appointed by the President of the Local Government Board to Inquire into the several modes of Treating Town Sewage, 38, [C.-1410.].

Medical Officer of the Privy Council and the Local Government Board, n.s. 8, Scientific Investigations in Pathology and Medicine, 38, [C.-1608.].

1877 Local Government Board

Sixth Annual Report, "The Rivers Pollution Prevention Act, 1876," 37, [C.-1865.].

1878 Seventh Annual Report, "The Rivers Pollution Prevention Act, 1876," 37 [C.-2130.].

Seventh Annual Report, Supplement containing the Report of the Medical Officer, 37 pt. 2, [C.-2130.-I.].

1878-79 Eighth Annual Report, Supplement containing the Report of the Medical Officer, 29, [C.-2452.].

Eighth Annual Report, "The Rivers Pollution Prevention Act, 1876," 28, [C.-2372.].

1880 Ninth Annual Report, "The Rivers Pollution Prevention Act, 1876," 26, [C.-2681.].

Ninth Annual Report, Supplement containing the Report of the Medical Officer for 1879, 27, [C.-2681-I.].

1881 Tenth Annual Report, "Rivers Pollution Prevention Act, 1876," 46, [C.-2982.].

Dr. Robert Angus Smith, Rivers Pollution Prevention Act, 1876, Report to the Local Government Board, 23, [C.-3080.].

1882 Local Government Board

Eleventh Annual Report, "Rivers Pollution Prevention Act, 1876," 30 pt. 1, [C.-3337.].

Eleventh Annual Report, Supplement containing the Report of the Medical Officer for 1881, 30 pt. 2, [C.-3337-I.].

1884 Thirteenth Annual Report, "Rivers Pollution Prevention Act, 1876," 37, [C.-4166.].

Dr. Robert Angus Smith, Rivers Pollution Prevention Act, 1876, Second Annual Report to the Local Government Board by Dr. R. Angus Smith, one of the inspectors under the act, on the Examination of Wastes, 19, [C.-4085.].

Royal Commission on Metropolitan Sewage Discharge
First Report, 41, [C.-3842.].
Minutes of Evidence, from July 1882 to July 1883, together with a selection from the appendices, and a digest of the evidence, 41, [C.-3842.-I.].

1884-85 Second and final report, 31, [C.-4253.].
Minutes of Evidence, May-October 1884, 31, [C.-4253.-I.].

Local Government Board
Fourteenth Annual Report, "Rivers Pollution Prevention Act, 1876," 32, [C.-4515.].

1885-86 Fifteenth Annual Report, "Report of the Water Examiner," 31, [C.-4844.].

Fifteenth Annual Report, "Rivers Pollution Prevention Act, 1876," 31, [C.-4844.].

1886 sess. 1 Select Committee on Rivers Pollution (River Lee), Report together with Proceedings of the Committee, Minutes of Evidence, appendix, and index, 11, (207.).

1887 Local Government Board
Sixteenth Annual Report, "Rivers Pollution Prevention Act, 1876," 36, [C.-5131.].

1888 Alfred E. Fletcher, Report to H.M. Secretary for Scotland on the Rivers Pollution Prevention Act, [Scotland], 1876, 59, [C.-5346.].

Return of all Royal Commissions issued from the year 1866 to the year 1874 . . . , 81, (426.).

1889 Local Government Board
Eighteenth Annual Report, "Report of the Water Examiner," 35, [C.-5813.].

1892 Twenty-First Annual Report, "Report of the Water Examiner," 38, [C.-6745.].

1893-94 Royal Commission on Metropolitan Water Supply
Report, 40 pt. 1, [C.-7172.].
Minutes of Evidence, 40 pt. 1, [C.-7172.-I.].
Appendices, 40 pt. 2, [C.-7172.-II.].

1894 <u>Return showing the Urban Sanitary Districts and Contributary Places in Rural Sanitary Districts in England and Wales in which systems for treating sewage by precipitation are in operation, and whether in such cases the precipitate is of a saleable value</u>, 70, (298.).

1895 Local Government Board

<u>Twenty-Fourth Annual Report, Supplement containing the Report of the Medical Officer</u>, 51, [C.-7906.].

III. PERIODICALS

Major articles from these periodicals are cited separately. A great deal of valuable material came from book reviews, editorials, letters to the editor, filler, abstracts of papers given at professional meetings and of the discussions that followed them, abstracts of papers in foreign journals, reports on the meetings of official bodies, and reports on the activities of municipal governments which were inapplicable for separate citation in the bibliography. Full references are given in footnotes.

British Association for the Advancement of Science, *Reports*, 1856-76.

British Medical Journal [*BMJ*], 1858-78, 1892.

The Builder, 1850-56, 1861, 1865-69, 1872, 1875, 1877, 1881.

Chemical News [*CN*], 1860-93.

The Field, 1867-75, 1894.

Food, Water, and Air, 1871-74.

Institution of Civil Engineers, *Minutes of Proceedings* [*MPICE*], 1848-1900.

Journal of Public Health and Sanitary Review, 1855-58.

The Lancet, 1848-51, 1857, 1864, 1866-67, 1878, 1884, 1893.

Association of Municipal Engineers and Surveyors (later Association of Municipal and Sanitary Engineers), *Proceedings*, 1873-75, 1889-90.

Punch, 1858-59.

Royal Society of Arts, *Journal* [*JRSA*], 1853-87.

Sanitary Institute of Great Britain, *Transactions* [*TSIGB*], 1880-94.

Sanitary Record [*SR*], 1874-97.

The Times of London, 1849-51, 1867, 1870-71.

IV. PRIMARY SOURCES

Acland, H.W. Notes on Drainage, with special reference to the Sewers and Swamps of the upper Thames. London: John Murray and James Parker, 1857.

Adeney, W.E. "The bacterio-chemical examination of polluted waters." Engineering 61(1896): 728-30, 762-64.

——— "The chemical bacteriology of sewage: its hygienic aspect." J. State Medicine 1(1892-93): 78-89.

——— "The course and nature of fermentative changes in natural and polluted waters, and in artificial solutions, as indicated by the composition of dissolved gases." Scientific Transactions, Royal Dublin Society 5(Sept. 1895): 539-620 (parts 1-3).

——— "On the chemical examination of organic matters in river waters." Scientific Proceedings, Royal Dublin Society, n.s. 8(1893-98): 337-45.

——— "On the reduction of manganese peroxide [MnO_2] in sewage." Scientific Proceedings, Royal Dublin Society, n.s. 8 (1893-98): 247-51.

Aikin, Arthur, and Alfred Swaine Taylor. "Appendix to the report on the analysis of the Leicester waters — remarks on the purity of water." In Thomas Wicksteed, Preliminary Report upon the Sewerage, Drainage, and Supply of Water from the borough of Leicester and a Report upon the Analysis of Sewage Water, and the Water of the Streams in the neighbourhood of Leicester by Arthur Aikin and Dr. A.S. Taylor. London, 1850.

Allen, A.H. "On some points in the analysis of water, and the interpretation of the results." The Analyst 2(1877): 61-65.

Anderson, William. "The Antwerp Water Works." MPICE 72(1882-83): 22-44; discussion, 45-83.

Angell, Lewis. Sanitary Science and the Sewage Question. A Lecture addressed to the Department of Applied Sciences in Kings College, London. London: Spon, 1871.

Anonymous. The A.B.C. Sewage Process, being a report of the experiments hitherto made at Leicester, Tottenham, and Leamington, on the Purification and Utilization of Sewage. 2nd ed. London: E. Stock, 1868.

——— "Bacteriology." Nature 31(1884): 49-50.

——— "Domestic Chemistry III: Domestic waters." Knight's Penny Magazine 7(1838): 54-56.

——— "The flora of bank-notes." Nature 32(1885): 8-9.

——— "The government and the pollution of rivers." Nature 13 (1875): 81-82.

——— "The pollution of rivers." Manchester Guardian, 26 May 1875, p. 661.

——— "The pollution of rivers." Nature 9(1874): 197-98.

——— "The pollution of rivers." Van Nostrand's Eclectic Engineering Magazine 5(1871): 131-33.

——— "The pollution of rivers." Van Nostrand's Eclectic Engineering Magazine 7(1872): 506-8.

——— "Pollution of rivers, its prevention." Van Nostrand's Eclectic Engineering Magazine 10(1874): 171-76.

——— "The pollution of rivers by chemical manufacturers." Van Nostrand's Eclectic Engineering Magazine 7(1872): 172-74.

——— "Reports on the examination of Thames water." JRSA 31 (1882-83): 74-76, 87-90.

——— "The river Thames." Van Nostrand's Eclectic Engineering Magazine 19(1878): 342-45.

——— "The ruin of rivers." Van Nostrand's Eclectic Engineering Magazine 9(1873): 550-53.

——— "The self-purification of rivers." Van Nostrand's Eclectic Engineering Magazine 2(1870): 638-40.

——— "Sewage precipitation and the A.B.C. process vindicated." Quart. J. of Science, 3rd series, 7(1885): 473-78.

——— "The silver Thames." Littell's Living Age 58(1858): 376-78.

——— "The Southampton Congress of the Sanitary Institute." Engineering News 42(1899): 170-71.

——— "The Thames and its difficulties." Fraser's Magazine 58 (1858): 167-72.

——— "The Thames in his glory." Littell's Living Age 58(1858): 375-76.

——— "The Thames: its uses and abuses." Fraser's Magazine 38(1848): 685-88.

——— "Thames water." Knight's Penny Magazine 7(1838): 139.

——— "The Thames water question." Westminster Review 12 (1830): 31-42.

——— The Utilisation of the Metropolitan Sewage and Reduction of Local Taxation, together with a brief review of the evidence taken before the select committee of the House of Commons on the Sewage of Towns and their final report. A letter to Frederick Douton, esq., M.P., Chairman of the Main Drainage Committee of the Metropolitan Board of Works from a Ratepayer. London: Kent, [c. 1865].

——— "Waste." Quart. J. of Science, 3rd series, 5(1883): 377-83.

——— "The whole duty of a chemist." Nature 33(1885-86): 73-74.

Armstrong, H.E. "The alkaloids, the present state of knowledge concerning them and the methods employed in their investigation." J. Society for Chemical Industry 6(1887): 482-91.

Ashby, Alfred. "Water analysis." The Analyst 6(1880): 108-9.

Austin, C.E. On the Cleansing of Rivers. London: R.J. Mitchell, 1872.

Ayres, P.B. "On the disposal of the fecal and other refuse of London on rational principles, in contrast with the schemes of sewerage now before the public." JPH & SR 3(1857): 17-28.

Bazalgette, C.N. "The sewage question." MPICE 48(1876-77): 105-59; discussion, 160-250.

Bazalgette, J.W. "On the main drainage of London and the interception of the sewage from the R. Thames." MPICE 24(1865): 280-94; discussion, 295-358.

Beale, L.S. "The constituents of the sewage mud of the Thames." J. Royal Microscopical Society, 2nd series, 4(1884): 1-19.

[Beardmore, Nathaniel]. "Water supply." Westminster Review 54 (1851): 185-96.

Beloe, Charles. "Sewage purification." Prof. Papers, Corps of Royal Engineers 21, no. 5(1895): 1-22.

Bennett, Alfred W. "Vegetable growths as purifiers of sewage." St. Thomas's Hospital Reports, n.s. 13(1884): 39-42.

Berkeley, M.J. "Observations on the recent investigations into the supposed cholera fungus." Monthly Microscopical J. 2(1869): 12-15.

Berzelius, J.J. "Some ideas on a new force which acts in organic compounds." In H.M. Leicester and H.S. Klickstein, A Source Book in Chemistry, 1400-1900. Cambridge: Harvard University Press, 1952.

Bevan, G. Phillips, The London Water Supply, its Past, Present, and Future. London: Edward Stanford, 1884.

Birch, R.W. Peregrine. The Disposal of Town Sewage. London: Spon, 1870.

Bischof, Gustaf. "Extension of time of culture in Dr. Koch's bacteria water test by partial sterilisation with special reference to the metropolitan water supply." CN 57(1888): 15.

------- "Notes on Dr. Koch's water test." J. Soc. for Chemical Industry 5(1886): 114-21.

------- "The purification of water." JRSA 26(1877-78): 486-96.

Blyth, Alexander Wynter. A Dictionary of Hygiene and Public Health, comprising Sanitary Chemistry, Engineering, and Legislation, the Dietetic Value of Foods, and the Detection of Adulterations, on the plan of the "Dictionaire d'Hygiene Publique" of Prof. Ambroise Tardieu. London: Griffin, 1876.

Bolton, Francis. London Water Supply, including a History and Description of the London Water Works, Statistical Tables, and Maps, new edition, entirely revised and enlarged with a short exposition of the law relating to water companies generally, an alphabetical digest of the leading decisions of the courts, the statutes, and a copious index by Philip A. Scratchely. London: Clowes, 1888.

------- and Percy Frankland. Lectures on the Collection, Storage, Purification, and Examination of Water, delivered to the School of Military Engineering at the Royal Engineers Institute,

Chatham, on the 24th and 25th of March, 1886. London: Harrison, 1886.

Booth, Abraham. A Treatise on the Natural and Chymical Properties of Water and on various British Mineral Waters. London: Wightman, 1830.

Bowditch, Henry. "Letter from the chairman of the State Board of Health concerning houses for the people; convalescent homes, and the sewage question." In 2nd Annual Report of the Massachusetts State Board of Health, January 1871. Boston: Wright and Potter, 1871.

Brande, W.T. Manual of Chemistry. 6th ed., Vol. II, Pt. ii, Organic Chemistry. London: Parker, 1848.

------- and Alfred Swaine Taylor. Chemistry. London: John Davies, 1863.

Bree, C.R. Popular Illustrations of the lower forms of life. London: H. Cox, 1868.

[Broderip, William]. "The aquarium." Fraser's Magazine 50(1854): 190-203.

Brown, S. "Review of The Human Body and its Connexion with Man, illustrated by the Principal Organs, by James John Garth Wilkinson, (London, 1851)." North British Review 17(1852): 131-44.

Buck, Alfred, ed. A Treatise on Public Health and Hygiene. 2 vols. London: Sampson, Low, Marston, Searle, and Rivington, 1879. Vol. 1.

Buckland, Frank. The Natural History of British Fishes; their Structure, Economic Uses, and Capture by Net and Rod. London: S.P.C.K., 1880.

------- Log-Book of a Fisherman-Zoologist. London: Chapman & Hall, 1876.

------- The Pollution of Rivers and its effects upon the Fisheries and Supply of Water to Towns and Villages, an address to the Sanitary Institute of Great Britain, 3 July 1878. London: C.L. Marsh, 1878.

Bunce, John T. History of the Corporation of Birmingham with a sketch of the earlier Government of the Town. 2 vols. Birmingham: Cornish, 1878, 1885.

Burchell, William. River Sanitation. A Short Essay to induce a public movement for the Prevention of River Pollution. London: Krough, [1884]

Burke, U.R. A Handbook of Sewage Utilization. 2nd ed. London: Spon, 1873.

Burnell, George. "On the present condition of the water supply of London." JRSA 9(1860-61): 169-77.

Butler, W.F. "River pollution — the reports of the commission." Van Nostrand's Eclectic Engineering Magazine 10(1874): 538-41.

Byrne, Edward. "Experiments on the removal of organic and inorganic substances in water." MPICE 27(1867): 1-5; discussion, 6-54.

Cameron, Charles A. A Manual of Hygiene, Public and Private, and Compendium of Sanitary Laws. Dublin: Hodges, Foster, 1874.

Cameron, Donald. "Some recent experiments in sewage treatment at Exeter." Engineering 62(1896): 206-7.

Cargill, Thomas. Sewage and its general application to Grass, Cereal, and Roots Crops, showing the Results obtained by Actual Experience down to the present time. London: Roberton, Brooman, & Co., 1869.

Carpenter, Alfred. "The power of soil, air, and vegetation to purify sewage." SR 2(1875): 285-89, 304-6.

Carpenter, W.B. "The germ theory of zymotic disease considered from the natural history point of view." Nineteenth Century 15 (1884): 317-36.

——— The Microscope and its Revelations. 1st ed. London: Churchill, 1856.

——— The Microscope and its Revelations. 6th ed. London: Churchill, 1881.

——— Principles of Human Physiology. 4th ed. London: Churchill, 1853.

Carter, W. Allen. Sanitary Papers read to the Edinburgh and Leith Society and the Royal Scottish Society of Arts, Edinburgh. Edinburgh: Bell & Bradfute, 1877.

Cassall, Charles E., and B.H. Whitelegge. "Remarks on the examination of water for sanitary purposes." SR n.s. 5(1883-84): 427-29, 479-82.

Chadwick, Edwin. "Progress in sanitation: in preventative as opposed to curative science." Transactions, National Association for the Promotion of Social Science, Dublin 1881. London: Longmans, Green, & Co., 1882. Pp. 625-49.

Chauveau, A. "Nature du virus vaccin. Nouvelle demonstration de l'inactivite du plasma de la sérosité vaccinale virulente." Compts Rendus de l'Académie des Sciences 66(1868): 317-21.

------- "Nature du virus vaccin. Détermination expérimentale des éléments qui constituent le principe actif de la sérosité vaccinale virulente." Compts Rendus de l'Académie des Sciences 66(1868): 289-93.

Child, Gilbert W. The Present State of the Town Sewage Question. Oxford and London: John Henry and James Parker, 1865.

Church, Arthur H. Plain Words about Water. London: Chapman & Hall, 1877.

/A Civil Engineer 7. The London Water Supply, being an Examination of the alleged Advantages of the Schemes of the Metropolitan Board of Works and of the inevitable Increase of Rates which would be required thereby. London: Spon, 1878.

Clowes, Frederick. "General conclusions observed from the experimental bacterial treatment of raw sewage at the outfalls of the London sewage into the Thames." J. Sanitary Institute 21 (1900): 308-10; discussion, 310-20.

/Conder, Francis R. 7. "On the water circulation of great cities." Scottish Review 7(1886): 264-86.

------- "The question of the Thames." Fraser's Magazine, n.s. 18 (1878): 726-36.

------- "The sewage problem." Manchester Guardian, 1 Oct. 1885, 5 Oct. 1885.

------- "The water supply of London." Fraser's Magazine, n.s. 22(1880): 185-98.

Cooke, W. Fothergill. "On the utilisation of the sewage of towns by the deodorisation process established at Leicester and on its economical application to the metropolis." JRSA 5(1856-57): 49-62.

Corfield, William H. A Digest of Facts relating to the Treatment and Utilisation of Sewage. 2nd ed., revised and corrected. London: MacMillan, 1871.

------- Water and Water Supply. New York: D. van Nostrand, 1875.

Cornish, C.J. The Naturalist on the Thames. London: Seeley & Co., 1902.

Crace-Calvert, Frederick. "Carbolic or phenic acid and its properties." JRSA 15(1867): 729-33.

------- "On the purification of polluted streams." JRSA 4(1856): 505-7.

Cresswell, Charles N. "River conservancy." JRSA 29(1881): 274-83.

[Crookes, William]. "The economy of nitrogen." Quart. J. of Science n.s. 8(1878): 145-66.

Crookes, William. "Presidential address." 68th Annual Report of the B.A.A.S. (Bristol, 1898). London: John Murray, 1899.

[Crookes, William]. "A solution to the sewage problem." Quart. J. of Science n.s. 3(1873): 55-73.

Crookes, William. Twelve Months Experience with the A.B.C. Process of Purifying Sewage. A Letter addressed to a Shareholder in the Native Guano Co., Ltd. London: Chemical News Office, 1872.

Crookshank, Edgar. Manual of Bacteriology. New York: J.H. Vail, 1887.

Danchell, Frederick H. Concerning Sewage and its Economical Disposal. London: Simpkin, Marshall, & Co., 1872.

Daubeny, Charles. "On ozone." J. Chemical Society 20(1867): 1-28.

de Chaumont, F.S.B.F. "On certain points with reference to drinking water." TSIGB 1(1880): 64-70.

Deering, W.H. "On some points in the examination of waters by the ammonia method." J. Chemical Society 28(1875): 679-83.

Dibdin, William J. The Purification of Sewage and Water. London: Sanitary Pub. Co., 1903.

------- "The purification of the Thames." MPICE 129(1896-97): 80-111; discussion, 112-87.

------- "Sewage sludge and its disposal." MPICE 88(1886-87): 155-74; discussion 194-298.

Dickens, Charles. Bleak House. 1853; New York: Laurel Classics, 1965.

Dumas, J.A., and Auguste Cahours. "Sur les matières azotées neutres de l'organisation." Annales de la Chemie et de Physique, 3rd series 6(1842): 385-448.

Dunbar, William. Principles of Sewage Treatment. Trans. with the author's sanction by H.T. Calvert. London: Griffin, 1908.

Dupre, August. "Advances in water analysis and sewage treatment." TSIGB 9(1888): 352-69.

------- "Remarks on some points in water analysis." The Analyst 5(1880): 215-18.

Durand-Claye, A. "Assainissement de la Seine. Rapport fait au nom de la Commission chargeé de proposer les measures a prendre pour remedier a l'infection de la Seine aux abords de Paris." Annales d'Hygiène Publique et de Médecine Légale, 2nd series 44(1875): 241-92.

Duthie, James. A Treatise on the Utilisation of Towns' Sewage, particularly with reference to Preston. Preston: Dobson, 1870.

F.L. "A human skull." Cornhill Magazine 2(1860): 718.

Fenwick, Andrew. The Truth about Sewage in a few words. London: Ibister, 1874.

Fergus, Andrew. Excremental Pollution, a Cause of Disease; with hints as to remedial measures. London, 1872.

------- The Sanitary Aspect of the Sewage Question, with remarks on a little suspected, and not easily detected source of Typhoid and other Zymotics. Glasgow: Porteous, 1872.

Fisheries Preservation Association. On the Pollution of the Rivers of the Kingdom; the enormous magnitude of the evil, and the urgent necessity in the interest of the Public Health and the Fisheries for its Suppression by immediate Legislative Enactment. London: The Association, 1868.

Flower, Lamorock. "Sewage treatment: more especially as affecting the pollution of the River Lea." Public Health: A Journal of Sanitary Science and Progress 2(1876): 396-99.

------- "Pure air and pure water." Public Health: A Journal of Sanitary Science and Progress 3(1877): 400-3.

Folkard, Charles W. "The analysis of potable water, with special reference to previous sewage contamination." MPICE 68(1881-82): 57-69; discussion, 72-115.

Folsom, Charles. "The disposal of sewage." Seventh Annual Report of the Massachusetts State Board of Health, January 1876. Boston: Wright and Potter, 1876. Pp. 276-401.

Forbes, Urquhart. "River pollution." Westminster Review 118 (1882): 153-63.

/Foster, Michael/. "Animals and plants." Quarterly Review (American ed.) 126(1869): 130-42.

Fowler, Alfred M. Utilisation of Sewage Water and Water from Polluted Streams. Manchester, /1874/.

Fox, Cornelius B. Sanitary Examinations of Water, Air, and Food. A Handbook for the Medical Officer of Health. Philadelphia: Lea & Blakiston, 1878.

Frankland, Edward. Experimental Researches in Pure, Applied, and Physical Chemistry. London: van Voorst, 1877.

------- "On a simple apparatus for determining the gases incident to water analysis." J. Chemical Society 21(1868): 109-20.

------- "On river pollution." Van Nostrand's Eclectic Engineering Magazine 12(1875): 534-39.

------- "On some points in the analysis of potable waters." J. Chemical Society, 3rd series 1(1876): 825-51.

------- "On the development of fungi in potable water." J. Chemical Society 24(1871): 66-76.

------- "On the proposed water supply of the metropolis." Proc. Royal Institution 5(1866-69): 346-70.

------- "On the spontaneous oxidation of organic matter in water." J. Chemical Society 37(1880): 517-46.

------- "On the water supply of the metropolis." Proc. Royal Institution 5(1866-69): 109-26.

------- Water Analysis for Sanitary Purposes. With Hints for the Interpretation of the Results. Philadelphia: Presley Blakiston, 1880.

------- "The water supply of London, and the cholera." Quart. J. of Science 4(1867): 313-29.

------- "The water supply of the metropolis during the year 1865-1866." J. Chemical Society 19(1866): 239-48.

------- and H.E. Armstrong. "On the analysis of potable waters." J. Chemical Society 20(1868): 77-108.

Frankland, Grace. Bacteria in Daily Life. London: Longmans, Green, & Co., 1903.

Frankland, Percy. "The bacterial examination of drinking water." Seventh International Congress of Hygiene and Demography, Reports of Meetings and Discussions held in London, August 10-17, 1891. London: The Society of Medical Officers of Health/E.W. Allen, [1891]. P. 168.

------- "The bacterial purification of water." J. Sanitary Institute 16(1895): 383-92; discussion, 392-97.

------- "The bacterial purification of water." MPICE 127(1896-97): 83-111; discussion, 112-59.

------- "The cholera and our water supply." Nineteenth Century 14 (1883): 346-55.

------- "The filtration of water for town supply." TSIGB 8(1886-87): 276-84.

------- "The micro-organisms of air and water." Nature 38(1888): 232-36.

------- "New aspects of filtration and other methods of water treatment; the gelatine process of water examination." J. Society of Chemical Industry 4(1885): 698-907.

------- "On the application of bacteriology to questions relating to water supply." TSIGB 9(1888): 369-77.

------- Our Secret Friends and Foes. 2nd ed. London: S.P.C.K., 1894.

------- "The present state of our knowledge concerning the self-purification of rivers." Seventh International Congress of Hygiene and Demography, Reports of Meetings and Discussions held in London, August 10-17, 1891. London: The Society of Medical Officers of Health/E.W. Allen, [1891]. P. 147.

------- "Recent bacteriological research in connection with water supply." J. Soc. for Chemical Industry 6(1887): 316-26.

------- "The removal of micro-organisms from water." CN 52(1885): 27-29.

------- "The selection of domestic water supplies." SR n.s. 6 (1884-85): 547-51.

------- "The upper Thames as a source of water supply." JRSA 32 (1883-84): 428-53.

------- "Water purification; its biological and chemical basis." MPICE 85(1885-86): 197-220; discussion, 221-63.

------- and Grace Frankland. Micro-organisms in Water: Their Significance, Identification, and Removal. London: Longmans, Green, and Co., 1894.

------- and H. Marshall Ward. "First report to the Water Research Committee of the Royal Society, on the present state of our knowledge concerning the bacteriology of water, with especial reference to the vitality of pathogenic schizomycetes in water." Proc., Royal Society of London 51(1892): 183-279.

Fresenius, K.R. Instruction in Chemical Analysis. Quantitative. Edited by J. Lloyd Bullock. London: Churchill, 1846.

------- Quantitative Chemical Analysis. 6th ed. (rpt. from 4th ed.). Translated from the 5th German edition by A. Vacher. London: Churchill, 1873.

------- A System of Instruction in Quantitative Chemical Analysis. Edited by O.D. Allen and S.W. Johnson. New York: John Wiley, 1896. (Eighth thousand of 1881 edition)

Friswell, R.J. "Review of Clement Higgins, A Treatise on the Law Relating to Obstruction and Pollution of Water Courses, together with a brief summary of the various sources of Rivers Pollution." Nature 16(1877): 225-26.

Galton, Douglas. "The lessons to be learnt from the experimental investigations of the State Board of Health of Massachusetts upon the purification of sewage." J. Sanitary Institute 17(1896): 1-22.

[Galton, Douglas]. "The purification of sewage and water." Edinburgh Review 188(1898): 151-74.

Gerardin, Alphonse. "Altération, corruption, et assainissement des rivières." Annales d'Hygiène Publique et de Médecine Légale, 2nd series 43(1875): 5-41.

Godfrey, R. "The amines process." TSIGB 10(1888-89): 203-18.

Graham, Thomas, W.A. Miller, and A.W. Hofmann. "Chemical report on the supply of water to the metropolis." J. Chemical Society 4 (1851): 375-413.

Gregory, William. Elementary Treatise on Chemistry. Edinburgh: Adam & Charles Black, 1855.

[Gregory, William]. "Review of Liebig's Animal Chemistry." Quarterly Review 70(1842): 98-128.

------- "Review of Liebig, Organic Chemistry in its application to Agriculture and Physiology." Quarterly Review 69(1842): 329-45.

Groome, J.H. The Defecation and Utilization of the Sewage Stream. A Lecture to the Framingham Farmers' Club. London: William Hunt, 1866.

Halcrow, Lucy, and Edward Frankland. "On the action of air upon peaty water." J. Chemical Society 37(1880): 506-17.

Hart, Ernest. "Cholera and our protection against it." Nineteenth Century 32(1892): 632-51.

Hartley, W.N. "The self-purification of peaty rivers." JRSA 31 (1882-83): 469-84.

[Hassall, A.H.], (as Lancet Analytical Sanitary Commission). "Records of the results of microscopical and chemical analyses of the solids and fluids consumed by all classes of the public." Lancet, 1851, i, pp. 187-93, 216-25, 253-61, 279-84.

Hassall, A.H. A Microscopic Examination of the Water Supplied to London and Suburban Districts. London: S. Highley, 1850.

------- The Narrative of a Busy Life. An Autobiography. London: Longmans, Green, and Co., 1893.

Hatton, Frank. "On the action of bacteria on gases." J. Chemical Society 39(1881): 247-58.

------- "On the oxidation of organic matter in water by filtration through various media; on the reduction of nitrates by sewage, spongy iron, and other agents." J. Chemical Society 39(1881): 258-76.

Hawksley, Thomas, M.D. Matter -- its ministry to Life in Health and Disease; and Earth, -- as the Natural Link between Organic and Inorganic Matter. London: Churchill, 1866.

------- "The power for good or evil of refuse organic matter." In John Hitchman, ed., The Sewage of Towns, Papers by Various Authors read at a Congress on the Sewage of Towns, held at Leamington Spa, Warwickshire, October 25th and 26th, 1866. London: Simpkin, Marshall, and Co., 1867.

------- "A proposal for the drainage of London." JPH & SR 3(1857): 28-36.

Haviland, Alfred. "The rivers pollution bill." Medical Examiner 2 (1877): 406-8.

Heaton, C.W. "The future water supply of London." Quart. J. of Science 6(1869): 225-45.

Hehner, Otto. "Notes on water analysis." The Analyst 2(1877): 177-80.

Heisch, Charles. "On organic matter in water." J. Chemical Society 23(1870): 371-75.

Hering, Rudolph. "Notes on the pollution of streams." Public Health: Reports and Papers of the American Public Health Association 13(1887): 272-79.

------- "Report of the results of an examination made in 1880 of several sewerage works in Europe." In Annual Report of the National Board of Health for 1881. Washington, D.C.: G.P.O., 1882. Pp. 99-223.

Higgins, Clement. A Treatise on the Law relating to the Pollution and Obstruction of Watercourses; together with a brief summary of the Various Sources of Rivers Pollution. London: Stevens & Haynes, 1877.

Hitchman, John, ed. The Sewage of Towns, Papers by various Authors read at a Congress on the Sewage of Towns held at Leamington Spa, Warwickshire, October 25th and 26th, 1866. London: Simpkins, Marshall, & Co., 1867.

Hoffert, H. A Guide to the Sewage Question for 1876, treated from a Sanitary, Economical, and Agricultural Point of View. Weymouth: Sherren & Son, 1876.

Hofmann, A.W. The Life-Work of Liebig. The Faraday Lecture for 1875. London: MacMillan & Co., 1876.

------- and Edward Frankland. "The deodorisation of sewage." JRSA 7(1859): 661-64.

Hogg, Jabez. "River pollution and the government bill." Public Health: A Journal of Sanitary Science and Progress 3(1877): 387-88.

------- "River pollution with special reference to the impure water supply of towns." JRSA 23(1875): 579-92.

Homersham, Samuel, ed. A Microscopical Examination of Certain Waters submitted to Jabez Hogg and a Chemical Analysis by Dugald Campbell with introductory notes by S.C. Homersham. London: Trounce, 1874.

------- Review of the Report by the General Board of Health on the Supply of Water to the Metropolis; contained in a Report to the Directors of the London (Watford) Spring Water Company. London: John Weale, 1850.

Hope, William. Food Manufacture vs. River Pollution. A letter addressed to the Newspaper Press of England. London: Edward Stanford, 1875.

------- "On the use and abuse of town sewage." JRSA 18(1869-70); 298-308.

------- Sewage Irrigation. A Lecture delivered to the Rate Payers of West Derby. London: Edward Stanford, 1871.

------- Three Letters on the Sewage of the Metropolis. London: Edward Stanford, 1865.

Houston, A.C. Report on the Scott-Moncrieff System for the Bacteriological Purification of Sewage. London: Waterlow, [1893].

Hull, E. "The future water supply of London." Quart. J. of Science 4(1867): 51-58.

Huxley, T.H. "On the border territory between the animal and vegetable kingdoms." MacMillan's Magazine 33(1876): 373-84.

Jacob, A. "The treatment of town sewage." MPICE 32(1870-71): 371-420.

Jesse, Edward. "The Thames." Once a Week 3(1860): 108-10.

Johnston, James F.W. The Chemistry of Common Life. 2 vols. New York: D. Appleton & Co., 1855.

------- "The circulation of matter." Blackwood's Magazine 73 (1853): 550-60.

Jones, Alfred. "Pollution of the Thames." Fortnightly Review 46 (1886): 79-90.

[Kemp, T.]. "Liebig's philosophy, a review of Liebig's Researches on the Chemistry of Food." Dublin Review 25(1848): 179-204.

[Kennedy, T.F.]. Papers relating to the Disposal of Sewage from Houses in the Country, the Prevention of the Pollution of Rivers, and the unsatisfactory Action of the Local Authority. Edinburgh: Douglas, 1878.

Kershaw, G.B. Modern Methods of Sewage Purification. London: Griffin, 1911.

Kershaw, J.B.C. "Bacteria in harness." Chambers' Journal 76(1899): 419-22.

Kimmins, C.W. The Chemistry of Life and Health. London: Methuen, 1892.

Kingsley, Charles. Two Years Ago. New York: MacMillan, 1882.

------- Water Babies. New York: Dutton, 1949.

[Kingsley, Charles]. "The water supply of London." North British Review 15(1851): 228-53.

Kingsley, Charles. Yeast -- A Problem. New York: J.F. Taylor, 1903.

Kingzett, C.T. "The chemistry of infection, or the germ theory of disease from a chemical point of view." JRSA 26(1877-78): 311-20.

------- "Contributions to the history of putrefaction, I." J. Chemical Society 37(1880): 15-22.

Kirkwood, James P. "A special report on the pollution of rivers; an examination of the water-basins of the Blackstone, Charles, Taunton, Neponset, and Chicopee Rivers; with general observations on water supplies and sewerage, with an appendix giving chemical analyses by William Ripley Nichols." Seventh Annual Report of the State Board of Health of Massachusetts, January 1876. Boston: Wright and Potter, 1876. Pp. 21-174.

Klein, Edward. "Bacteria, their nature and function." Nature 48 (1893): 82-87.

------- "Bacteriological research from a biologist's point of view." J. Chemical Society 49(1886): 197-205.

------- Microorganisms and Disease. An Introduction to the Study of Specific Microorganisms. 3rd ed. London: MacMillan, 1886.

Konthack, A.A. "The bacteria which we breathe, eat, and drink." Nature 55(1896): 209-14.

Krepp, Frederick C. The Sewage Question: Being a general review of all systems and methods hitherto employed in various cities for draining cities and utilising sewage: treated with reference to Public Health, Agriculture, and Nation Economy generally. London: Longmans, Green, & Co., 1867.

Lankester, Edwin. The Aquavivarium, fresh and marine; being an account of the Principles and Objects involved in the domestic culture of Water Plants and Animals. London: Robert Hardwicke, 1856.

------- Cholera: What is it? and how to prevent it. London: George Routledge, 1866.

------- Half-Hours with the Microscope. 3rd ed., illustrated by Tuffen West. London: Hardwicke, [1863].

------- "On the drinking waters of the metropolis." Proc. Royal Institution 2(1854-58): 466-70.

------- "The public health: the east end of London." Quart. J. Science 4(1867): 65-75.

------- A School Manual of Health. London: Groombridge, 1868.

------- and Peter Redfern. Reports made to the Directors of the London (Watford) Spring Water Company on the Results of Microscopical Examinations of the Organic Matters and Other Solid Contents of Waters Supplied from the Thames and Other Sources. [London], 1852.

Latham, Baldwin. A Chapter in the Local History of Croydon. Presidential Address to the Croydon Natural History Society. London: West, Newman, 1909.

[Letheby, Henry]. "Processes of analysis of potable water." BMJ, 1869, i, pp. 379-80, 432.

Letheby, Henry. "Report to the City of London Commissioners of Sewers, 14 September 1858. Sewage and sewer gases." JPH & SR 4(1858): 275-96.

[Letheby, Henry]. The Sewage Question: Comprising a Series of Reports: Being Investigations into the Condition of the Principal Sewage Farms and Sewage Works of the Kingdom. From Dr. Letheby's Notes and Chemical Analyses. London: Balliere, Tindall, and Cox, 1872.

Letheby, Henry, William Odling, and F.A. Abel. Metropolis Drinking Water Inquiry, 1867. Tables of the General Properties of the Water supplied by the Metropolitan Water Companies before and after Filtration. London: Bircham, Dalrymple, Drake, Bircham, & Burt, 1867.

Liebig, Justus. Animal Chemistry or Organic Chemistry in its Application to Physiology and Pathology. Edited from the author's Mss by William Gregory with notes, additions and corrections by Dr. Gregory and John W. Webster, M.D. (Cambridge, Mass.: John Owen, 1842), with a new introduction by Fredric L. Holmes. New York: Johnson Reprint, 1964.

------- Animal Chemistry, or Chemistry in its Application to Physiology and Pathology. Edited from the author's Ms by William Gregory, from the 3rd London edition, revised and greatly enlarged. New York: John Wiley, 1852.

------- Chemistry in its Application to Agriculture and Physiology. Edited from the Ms of the Author by Lyon Playfair. 2nd ed. London: Taylor & Walton, 1842.

------- Familiar Letters on Chemistry and its Relation to Commerce, Physiology, and Agriculture. Edited by John Gardner. Philadelphia: Campbell, 1843.

------- Familiar Letters on Chemistry in its Relations to Physiology, Dietetics, Agriculture, Commerce, and Political Economy. 3rd ed., revised and much enlarged. London: Taylor, Walton, and Maberly, 1851.

------- Lettres sur la Chemie considérée dans ses applications à l'Industrie, à la Physiologie et l'Agriculture. Nouvelle edition Française, pub. par M. Charles Gerhardt. Paris: Masson, 1847.

------- The Natural Laws of Husbandry. 1863; rpt., New York: Arno, 1972.

Lubbock, John. "The London water supply." Nineteenth Century 31 (1892): 224-32.

Macadam, Stevenson. The Chemistry of Common Things. London: T. Nelson, 1866.

MacDonald, J.D. A Guide to the Microscopical Examination of Drinking Water. 2nd ed. London: Churchill, 1883.

McWeeney, E.J. Bacteria; or the Study of Germs. Dublin, 1893.

Mallet, J.W. "The determination of the organic matter in potable water." CN 46(1882): 63-66, 72-75, 90-92, 101-2, 108-12.

Manning, J.A. The Utilization of Sewage; being a reply to Baron Liebig's Letter to Lord Robert Montagu. London: Hatchard, 1865.

Mansergh, James. The Thirlmere Water Scheme of the Manchester Corporation with a few remarks on the Longendale Works and Water Supply generally. Lecture delivered at Queenswood College to the Queenswood Mutual Improvement Society, 10 April 1878. London: Spon, 1878.

[Mansfield, Charles], as Ithi Kefalende. "Hints from Hygea." Fraser's Magazine 41(1850): 295-306.

Mantell, Gideon. Thoughts on Animalcules; or a Glimpse of the Invisible World revealed by the Microscope. London: John Murray, 1846.

Massachusetts State Board of Health. Report on Water Supplies and Sewerage, pt. 1, Examinations by the State Board of Health on the Water Supplies and Inland Waters of Massachusetts, 1887-90. Boston: Wright and Potter, 1890.

------- Report on Water Supplies and Sewerage, pt. 2, Experimental Investigations by the State Board of Health of Massachusetts upon the Purification of Sewage by Filtration and Chemical Precipitation, and upon the Intermittent Filtration of Water. Made at Lawrence, Massachusetts, 1888-90. Boston: Wright and Potter, 1890.

A Medical Practitioner. A New Plan yet an Old Plan for relieving London of its Sewage in a letter to Viscount Palmerston. London: James Ridgway, 1858.

Miller, W.A. Elements of Chemistry, Theoretical and Practical, ii Inorganic Chemistry. 2nd ed. London: J. Parker, 1860.

------- Elements of Chemistry, Theoretical and Practical, iii, Organic Chemistry. 2nd ed. London: Parker, Son, and Bowen, 1862.

------- Elements of Chemistry, Theoretical and Practical, ii, Inorganic Chemistry. 4th ed. London: Longmans, Green, Reader, and Dyer, 1868.

------- Elements of Chemistry, Theoretical and Practical, iii, Organic Chemistry. 4th ed. London: Longmans, Green, Reader, & Dyer, 1869.

------- "Observations on some points in the analysis of potable waters." J. Chemical Society 18(1865): 117-32.

Mills, E.J. "On potable waters." J. Chemical Society 33(1878): 57-70.

[Mills, E.J.]. "The use of refuse." Quarterly Review (American edition) 124(1868):173-85.

Monson, Edward. Metropolitan Sewage and What to do with it. London: Batford, 1883.

Mooney, John A. "Our little enemies." Catholic World 48(1888): 227-37.

Moule, Henry. Town Refuse the Remedy for Local Taxation. 2nd ed. London: William Ridgway, 1872.

Mulder, G.J. The Chemistry of Vegetable and Animal Physiology. Trans. by P.F.H. Fromberg; introduction and notes by J.F.W. Johnston. Edinburgh: William Blackwood, 1849.

Murchison, Charles. A Treatise on the continued Fevers of Great Britain. London: Parker, Son, and Burn, 1862.

------- A Treatise on the continued Fevers of Great Britain. 2nd ed. London: Longmans, Green, & Co., 1873.

Napier, James. "The rivers pollution scheme." The Sanitary Journal n.s. 3 (1879-90): 53-56.

Naylor, W. Ribble Joint Committee. Report on the Nature and Treatment of Manufacturers' Waste Effluents. Preston: Snipe, 1893.

Nelson, T.J. An Incredible Story, told in a letter to the Rt. Hon. Earl of Beaconsfield, Prime Minister. London: Truscott, 1879.

Newton, William J. Bourough of Accrington. Report on the Survey of the River Hyndburn and its tributaries, with particulars of the various polluting agencies by the borough surveyor. Accrington: Bowker, 1891.

Nichols, W.R. "On the present condition of certain rivers of Massachusetts together with considerations touching on the Water Supply of Towns. A Report to the State Board of Health of Massachusetts." Fifth Annual Report of the State Board of Health of Massachusetts, January 1874. Boston: Wright and Potter, 1874. Pp. 61-152.

------- and George Derby. "Sewerage; sewage; the pollution of streams; the water supply of towns. A report to the State Board of Health of Massachusetts." Fourth Annual Report of the State Board of Health of Massachusetts, January 1873. Boston: Wright and Potter, 1873. Pp. 19-132.

[O'Brien, William]. "The supply of water to the metropolis." Edinburgh Review 91(1849-50): 377-408.

Odling, Elizabeth M. Memoir of the late Alfred Smee, F.R.S., by his Daughter. London: Bell, 1878.

Odling, William. "Micro-organisms in drinking water." J. Soc. for Chemical Industry 5(1886): 544.

------- Report on the Effects of Sewage Contamination upon the River Thames to the Vestry of St. Mary, Lambeth. Lambeth: The Vestry/ G. Hill, 1858.

------- "On the chemistry of potable waters." JRSA 32(1883-84): 930-38, 976-86.

Parkes, Edmund A. A Manual of Practical Hygiene. 6th ed. Edited by F.S.B. Francois de Chaumont, with an appendix giving the American practice in matters relating to hygiene prepared by and under the supervision of Frederick M. Owen. 2 vols. New York: William Wood, 1883.

Parkes, Louis. "The possibility of the spread of disease through river waters supplied to London: a review of the evidence given before the Royal Commission on Water Supply, 1893." TSIGB 15 (1894): 243-56.

------- "Water analysis." TSIGB 9(1887-88): 377-94.

Parry, W. Kaye. The Application of recent Advances in the Study and Treatment of Sewage. Dublin, 1895.

------- "A new method of sewage purification." J. State Medicine 1(1892): 90-97.

Pasteur, Louis. Correspondence de Pasteur, 1840-95, réunie et annotée par Pasteur Vallery-Radot. Paris: Flammarion, 1951.

Paterson, E.J. The Hermite System of Sanitation of Towns by Electricity. London, [1895].

Paterson, M.M. The Pollution of the Aire and Calder: How to deal with it. London: Spon, 1893.

Paul, B.H. Manual of Technical Analysis: A Guide for Testing and Valuation founded upon the Handbuch der Technisch-Chemischen Untersuchungen of Dr. P.A. Bolley. London: H.G. Bohn, 1857.

[Paul, B.H.]. "Water analysis for sanitary purposes." BMJ, 1869, i, pp. 427-28, 495-97, 543-44; 1869, ii, pp. 32-33.

Pettenkofer, Max. The Value of Health to a City. Two Lectures delivered in 1873, trans. with an introduction by H.E. Sigerist. Baltimore: Johns Hopkins University Press, 1941.

[Pole, William]. "The water supply of London." Quarterly Review (American edition) 127(1869): 234-51.

------- "The water supply of London." Quarterly Review 174(1892): 63-94.

Potter, E.C. The Pollution of Rivers: by a Polluter. A letter to the Rt. Hon. G. Sclater-Booth, president of the Local Government Board. Manchester: Johnson & Rawson, 1875.

Preece, William. "Modern methods of sewage disposal." Engineering News 42(1899): 171-72.

Preston-Thomas, Herbert. The Work and Play of a Government Inspector. Edinburgh: Blackwood, 1909.

Prestwich, Joseph. On the Geological Conditions affecting the Water Supply to Houses and Towns with special reference to the modes of supplying Oxford. London: Parker, 1876.

Priestley, Eliza. "The bacteria beds of modern sanitation." Nineteenth Century 49(1901): 624-29.

------- "The realm of the microbe." Nineteenth Century 29(1891): 811-31.

Pritchard, Andrew. A History of Infusoria, Living and Fossil, arranged according to Die Infusionstierchen of C.G. Ehrenberg. London: Whittaker & Co., 1842.

Procter, William. The Hygiene of Air and Water: being a popular Account of the Effects of the Impurities of Air and Water, their detection and the modes of remedying them. London: Hardwicke, 1872.

Prout, William. Chemistry Meteorology and the Function of Digestion considered with reference to Natural Theology. London: William Pickering, 1834.

Rafter, George. "On some recent advances in water analysis and the use of the microscope for the detection of sewage contamination." American Monthly Microscopical J., May 1893, pp. 127-39.

------- "Sewage irrigation." U.S.G.S. Water Supply and Irrigation Papers #3. Washington, D.C.: G.P.O., 1897.

------- "Sewage irrigation, pt. 2." U.S.G.S. Water Supply and Irrigation Papers #22. Washington, D.C.: G.P.O., 1899.

Redfern, Peter. Report to the Directors of the London (Watford) Spring Water Company, on the Organic and other solid matters found by Microscopical Examination of the Water supplied from the Thames and other sources. London, 1852.

Richards, H.C. and W.H.C. Payne. London Water Supply, being a compendium of the History, Law, and Transactions relating to the Metropolitan Water Companies from earliest times to the Present Day. 2nd ed., edited by P.J.H. Soper. London: P.S. King, 1899.

Richardson, B.W. The Cause of the Coagulation of the Blood: being the Astley Cooper Prize Essay for 1856. London: Churchill, 1858.

------- Clinical Essays. Vol. 1. London: Churchill, 1862.

------- The Field of Disease. A Book of Preventative Medicine. London: MacMillan, 1883.

------- Hygeia -- A City of Health. London: MacMillan, 1876.

------- "On the poisons of the spreading diseases, their nature and their mode of distribution." In John Hitchman, ed., The Sewage of Towns, Papers by various Authors, read at a Congress on the Sewage of Towns, held at Leamington Spa, Warwickshire, October 25th and 26th, 1866. London: Simpkins, Marshall & Co., 1867.

------- "The Thames." Fortnightly Review 48(1887): 800-17.

------- "Water supply in relation to health and disease." JPH & SR 1(1855): 130-40.

Rideal, Samuel. Sewage and the Bacterial Purification of Sewage. 2nd ed. London: Sanitary Pub. Co., 1901.

Robinson, Henry. Sewage Disposal with reference to Rivers Pollution and Water Supply. London, 1891.

------- Some recent phases of the sewage question with remarks on ensilage, a Lecture to the Surveyors Institute, 9 February 1885. London: Spon, [1885].

Roechling, H.A. "The present state of sewage irrigation in Europe and America." J. Sanitary Institute 17(1896): 483-95.

------- Rivers Pollution and Rivers Purification, read to the Annual Meeting of the Municipal and County Engineers, Bury, 22 July 1892. London [?], 1892.

------- "The sewage farms of Berlin." MPICE 109(1892): 179-228.

Roscoe, H.E. The Life and Experiences of Sir Henry Enfield Roscoe . . . written by himself. London: MacMillan, 1906.

------- Mersey and Irwell Joint Committee, Preliminary Report, 26 March 1892.

------- and Joseph Lunt. "Contributions to the chemical bacteriology of sewage." Abstracts of Proceedings, Royal Society of London 49 (1890-91): 455-57.

Salamon, A. Gordon, and W. DeVere Mathew. "The purification of water." J. Society for Chemical Industry 5(1886): 261-73.

Sanderson, J. Burdon. "Cholera: its cause and prevention." Contemporary Review 48(1883): 171-87.

Sandwith, Humphrey. "Public health." Fortnightly Review 23(1875): 254-70.

Saunders, C.E. "Legislation for the purification of rivers and its failures." SR n.s. 8(1886-87): 343-47, 363.

Saunders, William. History of the First London County Council, 1889-1890-1891. London: National Press Agency, 1892.

Sayer, A. Metropolitan and Town Sewage, Their Nature, Value, and Disposal. London: F.W. Calder, 1857.

------- London Main Drainage. The Nature, Value, and Disposal of Sewage, with considerations on Drainage, Sewers, and Sewerage; Sketches of Water Supply, and on Legislation on Sewers, Ancient and Modern. 2nd ed. London: F.W. Calder, 1858.

Schloesing, Theophile. "Assainissement de la Seine. Épuration et utilisation des eaux d'egout. Rapport fait au nom d'une commission." Annales d'Hygiène Publique et de Médecine Légale, 2nd series 47(1877): 193-273.

------- and Achille Muntz. "Sur la nitrification par les ferments organises." Compts Rendus de l'Académie de Sciences 84(1877): 301-3.

Scott, H.Y.D. The Sewage Question, and the Lime and Cement Process of Major General H.Y.D. Scott, C.B., with the opinion of many chemists, engineers, and agriculturalists, pt. 1. London: The Company, 1873.

Shelford, William. "The treatment of sewage by precipitation." MPICE 45(1875-76): 144-66; discussion, 167-206.

Shenstone, W.A. Justus von Liebig -- his Life and Work. New York: MacMillan, 1895.

Shepard, George. London Sewage and its Application to Agriculture. The London Drainage and how to apply the sewerage to the land with advantage to the farmer and the capitalist. London: Effingham, Wilson, 1857.

Simon, John. General Pathology. A Course of Lectures delivered at St. Thomas's Hospital. London: Henry Renshaw, 1850.

------- English Sanitary Institutions, reviewed in their Course of Development, and in some of their Political and Social Relations. 1890; rpt., New York: Johnson Reprint, 1970.

------- Personal Reflections of Sir John Simon, K.C.B. London: The Wiltons, 1894.

Slater, J.W. "The composition of sewage." Quart. J. of Science, 3rd series 6(1884): 385-90.

------- "Observations on polluted waters." Quart. J. Science, 3rd series 5(1883): 386-91.

------- Sewage Treatment, Purification, and Utilization. A Practical Manual for the Use of Corporations, Local Boards, Medical Officers of Health, Inspectors of Nuisances, Chemists, Manufacturers, Riparian Owners, Engineers, and Ratepayers. London: Whittaker & Co., 1888.

Smith, Robert Angus. Air and Rain. The Beginnings of a Chemical Climatology. London: Longmans, Green, and Co., 1872.

------- Disinfectants and Disinfection. Edinburgh: Edmonston and Douglas, 1869.

------- "On the air and water of towns." Eighteenth Report of the B.A.A.S. (Swansea, 1848). London: John Murray, 1849. Pp. 16-31.

------- "On the air and water of towns -- action of porous strata, water, and organic matter." Twenty-first Annual Meeting of the B.A.A.S. (Ipswich, 1851). London: John Murray, 1852. Pp. 66-77.

------- "On the detection of living germs in water." SR n.s. 4 (1882-83): 344-47.

------- "On the examination of water for organic matter." Proc. Manchester Literary and Philosophical Society, 3rd series 4 (1871): 37-88; amended version, CN 19(1869): 278-82, 304-6; 20 (1869): 26-30, 112-15.

------- "Presidential address to section III (Chemistry, Meteorology, Geology) of the Sanitary Institute of Great Britain." TSIGB 5(1883-84): 266-96.

------- "Science in our courts of law." JRSA 8(1859-60): 135-49, 185-86.

Smith, Thomas Southwood. A Treatise on Fever. Philadelphia: Carey and Lea, 1830.

Snow, John. "On continuous molecular changes, more particularly in their relation to epidemic diseases." London: Churchill, 1853. In Snow on Cholera -- a Reprint of Two Papers by John Snow, M.D., together with a biographical memoir by B.W. Richardson, M.D., and an introduction by Wade Hampton Frost. New York: Commonwealth Fund, 1936.

------- "On the mode of communication of cholera." 2nd ed. London: Churchill, 1855. In Snow on Cholera -- a Reprint of Two Papers by John Snow, M.D., together with a biographical memoir by B.W. Richardson, M.D., and an introduction by Wade Hampton Frost. New York: Commonwealth Fund, 1936.

Sorby, Henry C. "Detection of sewage contamination by the use of the microscope and on the purifying action of minute animals and plants." J. Royal Microscopical Society, 2nd series 4(1884): 988-91.

South, J.L. "Zoology." Encyclopedia Metropolitania, or Universal Dictionary of Knowledge. London: Fellowes . . . , 1845.

Spence, Frank. "How to stop river pollution." Contemporary Review 64(1893): 427-33.

[Spencer, Thomas]. A Brief Description of Spencer's Patent Magnetic System of Purifying Water. London: Magnetic Filter Co., 1869.

Spencer, Thomas. Liverpool Corporation Water Works. Report on the Quality of the Rivington Water, the Reservoirs, and the Physical Characteristics of the District. London: Nichols, 1857.

Stallard, J.H. On the Sanitary Requirements of Liverpool. Liverpool: Holden, 1871.

Symons, G.J. "On the floods of England and Wales during 1875, and on water economy." MPICE 45(1875-76): 1-18; discussion, 49-100.

Taylor, Francis. Human Manure -- its Collection and Conversion to Guano. London: Churchill, 1861.

Taylor, J.E. The Aquarium; its Inhabitants, Structures, and Management. 2nd ed. London: David Bogue, 1881.

Thomson, Thomas. Chemistry of Organic Bodies. Vegetables. London: J.B. Balliere, 1838.

Thorp, William. "River pollution with special reference to the late commission." JRSA 23(1874-75): 382-91.

Thresh, John C. "The interpretation of the results obtained upon the chemical and bacteriological examination of potable waters." The Analyst 20(1895): 80-91, 97-111.

Thudichum, J.F.W. The Discoveries and Philosophy of Liebig. The Cantor Lectures. London: Trounce, 1876.

------- "The relation of microscopic fungi to great pathological processes, particularly the process of cholera." Monthly Microscopical J. 1(1869): 14-27.

Tidy, Charles Meymott. Handbook of Modern Chemistry, Inorganic and Organic for the Use of Students. London: Churchill, 1878.

------- The London Water Supply, being a report submitted to the Society of Medical Officers of Health on the Quality and Quantity of the Water supplied to the Metropolis during the past ten years. London: Churchill, 1878.

------- "The processes for determining the organic purity of potable waters." J. Chemical Society 35(1879): 46-106.

------- "River water." J. Chemical Society 37(1880): 267-327.

------- "River water." [second paper] CN 43(1881): 113-4.

------- "On the treatment of sewage." JRSA 34(1885-86): 612-25, 1127-89; discussion, 664-73; 35(1886-87): 41-50.

Tyndall, John. Essays on the Floating Matter of the Air in Relation to Putrefaction and Infection. New York: D. Appleton & Co., 1888.

Vallery-Radot, Pasteur, ed. Oeuvres de Pasteur. 7 vols. Paris: Masson, 1922-38.

Voelcker, Augustus. "On the composition of farmyard manure and the changes which it undergoes on keeping under different circumstances." J. Royal Agricultural Society 17(1856): 191-260.

Wanklyn, J.A. "Verification of Wanklyn, Chapman, and Smith's water analysis on a series of artificial waters." J. Chemical Society 20(1867): 591-95.

------- Water Analysis: A Practical Treatise on the Examination of Potable Water. Revised by W.J. Cooper with a Memoir and Portrait of the Author. London: Kegan, Paul, Trench, Trubner & Co., 1907.

------- and E. Therophon Chapman. "On the action of oxidizing agents on organic compounds in the presence of excess alkali." J. Chemical Society 21(1868): 161-72.

------- Water Analysis: A practical Treatise on the Examination of Potable Water. 6th ed. London: Trubner, 1884.

------- and Miles H. Smith. "Note on Frankland and Armstrong's Memoir on the analysis of potable waters." J. Chemical Society 21(1868): 152-60.

------- "Water analysis; determination of the nitrogenous organic matter." J. Chemical Society 20(1867): 445-54.

[Ward, F.O.]. "Metropolitan water supply." Quarterly Review 87 (1850): 468-502.

Warden, C.J.H. "The biological examination of water." CN 52 (1885): 52-54.

Wardle, Thomas. On Sewage Treatment and Disposal: for Cities, Towns, Villages, Private Dwellings, and Public Institutions. Manchester: John Heywood, [c. 1893].

Warington, Robert [the elder]. "Maintenance of the balance between the animal and vegetable kingdoms." The Builder 11(1853): 231-32.

------- "Memoranda of observations made in small aquaria, in which the balance between animal and vegetable organisms was permanently maintained." Annals and Magazine of Natural History, 2nd series 14(1854): 366-73.

------- "Notice of observations on the adjustment of the relations between the animal and vegetable kingdoms, by which the vital functions of both are permanently maintained." J. Chemical Society 3(1850): 52-54.

------- "Observations on the natural history of the water-snail and fish kept in a confined and limited portion of water." Annals and Magazine of Natural History, 2nd series 10(1852): 273-80.

------- "On preserving the balance between animal and vegetable organisms in sea water." Annals and Magazine of Natural History 2nd series 12(1853): 319-24.

------- "On the aquarium." Proc. Royal Institution 2(1854-58): 403-8.

Warington, Robert [the younger]. "Observations upon Dr. Tidy's paper on river water." CN 41(1880): 265.

------- "On nitrification." J. Chemical Society 33(1878): 44-50; pt. ii, 35(1879): 429-56.

Watson, Thomas. "The abolition of zymotic disease." Nineteenth Century 1(1877): 380-96.

Way, John T. "On the composition of waters of land-drainage and rain." J. Royal Agricultural Society 17(1856): 123-62.

------- "On the use of town sewage as manure." J. Royal Agricultural Society 15(1854): 135-67.

Whitehead, H. "The influence of impure water on the spread of cholera." MacMillan's Magazine 13(1866): 182-90.

------- "The Broad Street Pump: an episode in the cholera epidemic of 1854." MacMillan's Magazine 13(1865-66): 113-22.

[Wigner, G.W.] "Analysis of public water supplies in England. Instructions for analysis prepared by a committee appointed by the Society of Public Analysts." The Analyst 6(1880): 127-39.

Wigner, G.W. "On the mode of statement of the results of water analysis and the formation of a numerical scale for the valuation of the impurities in drinking waters." The Analyst 2(1878): 208-20.

------- "On the valuation of the relative impurities of potable waters." The Analyst 6(1880): 111-25.

------- "Thames water." The Analyst 3(1878): 364-65.

Williams, M. Whitley. "On the estimation of organic carbon and nitrogen in water analysis, simultaneously with the estimation of nitrous acid." J. Chemical Society 39(1881): 144-49.

Wilson, George. A Handbook of Hygiene. 2nd ed. London: Churchill, 1873.

------- A Handbook of Hygiene and Sanitary Science. 4th ed. London: Churchill, 1879.

Winsor, Frederick. "A special report on the water-supply, drainage, and sewerage of the state, from the sanitary point of view." Seventh Annual Report of the State Board of Health of Massachusetts, January 1876. Boston: Wright & Potter, 1876. Pp. 175-275.

Witt, H.M. "On a peculiar power possessed by porous media (sand and charcoal) of removing matter from solution in water." Philosophical Magazine, 4th series 12(1856): 23-34.

Wood, J.G. The Fresh and Salt Water Aquarium. London: George Routledge, 1868.

Woodhead, German Sims. Bacteria and their Products. London: Walter Scott, 1891.

Young, W.C. "A comparison of the organic carbon and nitrogen results obtained by Dr. Frankland and the companies' analysts from the waters supplied by the metropolitan water companies." The Analyst 20(1895): 159-64.

V. SECONDARY SOURCES

Ackerknecht, Erwin. "Anticontagionism, 1821-1867." Bull. Hist. Med. 22(1948): 562-94.

Armstrong, Ellis, ed. History of Public Works in the United States, 1776-1976. Chicago: American Public Works Association, 1976.

Aulie, Richard P. "Boussingualt and the Nitrogen Cycle." Diss., Yale University, 1968.

Baker, M.N. British Sewage Works. New York: Engineering News Publishing Co., 1904.

------- The Quest for Pure Water. The History of Water Purification from the Earliest Times to the 20th Century. New York: American Water Works Association, 1948.

Balls, A.K. "Liebig and the chemistry of enzymes and fermentation." In F. Moulton, ed., Liebig and after Liebig -- A Century of Progress in Agricultural Chemistry. Washington, D.C.: A.A.A.S., 1942.

Barton, Ruth. "The X-Club: Science, Religion, and Social Change in Victorian England." Diss., University of Pennsylvania, 1976.

Bell, Samuel P. A Biographical Index of British Engineers in the Nineteenth Century. New York and London: Garland, 1975.

Bereano, Philip L. "Courts as institutions for assessing technology." In William A. Thomas, ed., Scientists in the Legal System, Tolerated Meddlers or Essential Contributors. Ann Arbor: Ann Arbor Science Publications, 1974.

Blake, Nelson M. Water for the Cities -- A History of the Urban Water Supply Problem in the United States. Syracuse: Syracuse University Press, 1956.

Bond, M. Guide to the Records of Parliament. London: H.M.S.O., 1971.

Brand, Jeanne L. Doctors and the State: The British Medical Profession and Government Action in Public Health, 1870-1912. Baltimore: Johns Hopkins University Press, 1965.

Briggs, Asa. History of Birmingham, vol. 2, Borough and City, 1865-1938. London: Oxford University Press, 1952.

------- Victorian Cities. Harmondsworth, U.K.: Pelican, 1968.

Brock, W.H. "The spectrum of scientific patronage." In G.L'e. Turner, ed., The Patronage of Science in the Nineteenth Century. Leyden: Noordhoff International, 1976.

Brown, P.E. "John Snow -- the autumn loiterer." Bull. Hist. Med. 35(1961): 519-28.

Browne, C.A. "Justus von Liebig -- man and teacher." In F. Moulton, ed., Liebig and after Liebig -- A Century of Agricultural Chemistry. Washington, D.C.: A.A.A.S., 1942. Pp. 1-9.

Bulloch, William. The History of Bacteriology. London: Oxford University Press, 1938

Chirnside, R.C., and J.H. Hamence. The Practising Chemists. A History of the Society for Analytical Chemistry, 1874-1974. London: The Society, 1974.

Clark, H.W., and Stephen deM. Gage. "A review of twenty-one years' experiments upon the purification of sewage at the Lawrence Experiment Station." Fortieth Annual Report of the State Board of Health of Massachusetts. Boston: Wright and Potter, 1909. Pp. 253-538.

Clark, Ian D. "Expert advice in the controversy about supersonic transport in the United States." Minerva 12(1974): 416-32.

Clayton, E.G. Arthur Hill Hassall, Physician and Sanitary Reformer. London: Balliere, Tindall, and Cox, 1908.

Crellin, J.K. "Airborne particles and the germ theory: 1860-1880." Annals of Science 22(1966): 49-60.

------- "The dawn of the germ theory: particles, infection, and biology." In F.N.L. Poynter, ed., Medicine and Science in the 1860s, Proceedings of the Sixth British Congress on the History of Medicine, University of Sussex, 6-9 Sept. 1967. London: Wellcome Trust, 1968. Pp. 57-76.

------- "The disinfectant studies by F. Crace-Calvert and the introduction of phenol as a germicide." Die Vortrage der Hauptversammulung der Internationalen Gesellschaft fur Geschichte der Pharmazie, Bd. 28. Stuttgart: Wissenschaftliche Verlagsgesellschaft, 1966. Pp. 61-67.

Cullen, M.J. The Statistical Movement in Early Victorian Britain. The Foundations of Empirical Social Research. New York: Harvester/ Barnes and Noble, 1975.

Dibdin, W.J. The Rise and Progress of Aerobic Methods of Sewage Disposal. London: Sanitary Publishing Co., 1911.

Doetsch, Raymond, ed. Microbiology -- Historical Contributions from 1776 to 1908. New Brunswick, N.J.: Rutgers University Press, 1960.

Duclaux, Emile. Pasteur: The History of a Mind. Translated by Edwin F. Smith and Florence Hodges. Philadelphia: W.B. Saunders, 1920. Reprint with a forward by Rene Dubos. Metuchen, N.J.: Scarecrow Reprints, 1973.

Dyer, Bernard. "Some reminiscences of its first fifty years." In Bernard Dyer and C. Ainsworth Mitchell, The Society of Public Analysts and Other Analytical Chemists, some Reminiscences of its first fifty years and a Review of its Activities. Cambridge: The Society/W. Heffer, 1932. Pp. 1-71.

Eyler, John M. "The Conversion of Angus Smith: the changing role of chemistry and biology in sanitary science, 1850-1880." Bull. Hist. Med. 54(1980): 216-24.

------- Victorian Social Medicine. The Ideas and Methods of William Farr. Baltimore: Johns Hopkins University Press, 1979.

Ferguson, Thomas. Scottish Social Welfare, 1864-1914. Edinburgh: Livingstone, 1958.

Finer, S.E. The Life and Times of Sir Edwin Chadwick. London: Methuen, 1952.

Fisher, Richard B. Joseph Lister, 1827-1912. New York: Stein & Day, 1977.

Fitter, R.S.R. London's Natural History. London: Collins, 1945.

Ford, G. and P. Ford. Select List of British Parliamentary Papers, 1833-1899. Oxford: Basil Blackwell, 1953.

------- A Guide to Parliamentary Papers, What They Are, How to Find Them, How to Use Them. 3rd ed. Totowa, N.J.: Rowman & Littlefield, 1972.

Fournier D'Albe, E.E. The Life of Sir William Crookes. London: T. Fisher Unwin, 1923.

Frankland, Edward. Sketches from the Life of Edward Frankland. London: Spottiswoode, 1901.

------- Sketches from the Life of Edward Frankland, born January 18, 1825, died August 9, 1899, edited and concluded by his two daughters, M.N.W. and S.J.C. London: Spottiswoode, 1902.

Fraser, Derek. Urban Politics in Victorian England. Leicester: Leicester University Press, 1976.

Frazer, W.M. A History of English Public Health, 1834-1939. London: Balliere, Tindall, and Cox, 1950.

French, R.D. Antivivisection and Medical Science in Victorian Society. Princeton: Princeton University Press, 1975.

Fruton, Joseph S. Molecules and Life -- Historical Essays on the Interplay of Chemistry and Biology. New York: Wiley-Interscience, 1972.

Gibbon, Gwilym, and Reginald W. Bell. History of the London County Council, 1889-1939. London: MacMillan, 1939.

Gibson, A., and W.V. Farrar. "Robert Angus Smith, F.R.S., and sanitary science." Notes and Records of the Royal Society of London 28(1974): 241-62.

Gill, Conrad. History of Birmingham, vol. 1, Manor and Borough to 1865. London: Oxford University Press, 1952.

Glick, Thomas F. "Science, technology, and the urban environment: the Great Stink of 1858." In Lester J. Bilsky, ed., Historical Ecology, Essays on Environment and Social Change. Port Washington, N.Y.: National University Publishers, 1980.

Goodfield-Toulmin, J. "Some aspects of English physiology, 1780-1840." J. History of Biology 2(1969): 283-320.

Goodman, D.C. "Chemistry and the two organic kingdoms of nature in the 19th century." Medical History 16(1972): 113-30.

Green, J. Reynolds. A History of Botany, 1860-1900. Oxford: Clarendon Press, 1909.

Gutchen, Robert M. "Local improvements and centralization in 19th century England." Historical J. 4(1961): 85-96.

Hamlin, Christopher. "Edward Frankland's early career as London's water analyst, 1865-76: the public relations of 'previous sewage contamination.'" Bull. Hist. Med. 56(1982): 56-76.

------- "Recycling as a goal of sewage treatment in mid-Victorian Britain." Proceedings of the Annual Conference of the Society for the History of Technology, Milwaukee, October 1981. In press.

Hardin, Garrett. Exploring the New Ethics for Survival. The Voyage of the Spaceship 'Beagle.' Baltimore: Pelican, 1973.

Hartley, Dorothy. Water in England. London: George Allen & Unwin, 1964.

Hawkes, H.A. The Ecology of Waste-Water Treatment. New York: MacMillan, 1963.

Houghton, Walter. The Victorian Frame of Mind. New Haven: Yale University Press, 1957.

Hynes, H.B.N. The Biology of Polluted Waters. Liverpool: Liverpool University Press, 1960.

Jacobson, Daniel. "The pollution problem of the Passaic River." Proc. New Jersey Historical Society 76(1958): 186-98.

Keith-Lucas, Bryan. "Some influences affecting the development of sanitary legislation in England." Econ. Hist. Review, 2nd series 6(1954): 290-96.

Knight, David. "Agriculture and chemistry in Britain around 1800." Annals of Science 33(1976): 187-96.

Lambert, Royston. "Central and local relations in mid-Victorian England: the Local Government Act Office, 1858-71." Victorian Studies 6(1962): 121-50.

------- Sir John Simon, 1816-1904, and English Social Administration. London: McGibbon & Kee, 1963.

Lewis, R.A. Edwin Chadwick and the Public Health Movement. London: Longmans, Green, & Co., 1952.

Lilienfeld, David. "A corner of history: contagium vivum and the development of water filtration: the beginning of the sanitary movement." Preventive Medicine 6(1977): 361-75.

Lipman, Timothy O. "Vitalism and reductionism in Liebig's physiological thought." Isis 58(1967): 167-88.

Lipschutz, D.E. "The water question in London, 1827-31." Bull. Hist. Med. 42(1968): 510-26.

Lockett, W.T. "Louis Pasteur: his life and work." Proc. Institute of Sewage Purification, 1932, pp. 191-200.

Longmate, Norman. King Cholera -- the Biography of a Disease. London: Hamish Hamilton, 1966.

Luckin, William. "The final catastrophe: cholera in London, 1866." Medical History 21(1977): 32-42.

Mclaughlin, Terence. Dirt. A Social History as seen through the Uses and Abuses of Dirt. New York: Stein and Day, 1971.

McLeod, Roy M. "Government and resource conservation: the Salmon Acts administration, 1860-1886." J. British Studies 7(1968): 114-50.

Mazur, Allan. "Disputes between experts." Minerva 11(1973): 243-62.

Mitchell, C. Ainsworth. "A review of its activities." In Bernard Dyer and C. Ainsworth Mitchell, The Society of Public Analysts and Other Analytical Chemists, some Reminiscences of its first fifty years and a Review of its Activities. Cambridge: The Society/Heffer, 1932. Pp. 75-236.

Palmer, Roy. The Water Closet: A New History. Newton Abbot: David & Charles, 1973.

Partington, J.R. A History of Chemistry, vol. 4. London: MacMillan, 1964.

Peacock, Thomas Love. Gryll Grange. New York: A.M.S. Press, 1967.

Pelling, Margaret. Cholera, Fever, and English Medicine, 1825-1865. Oxford: Oxford University Press, 1978.

Pentelow, F.T.K. River Purification. A Legal and Scientific Review of the Last 100 Years, being the Buckland Lectures for 1952. London: Edward Arnold, 1953.

Phelps, Earle B. Stream Sanitation. New York: John Wiley, 1944.

Polanyi, Michael. "The republic of science, its political and economic theory." Minerva 1(1962): 54-73.

Read, Brian. Healthy Cities -- A Study in Urban Hygiene. Glasgow: Blackie, 1970.

Redlich, Josef, and Francis W. Hirst. The History of Local Government in England, being a reissue of vol. 1 of Local Government in England, edited, with an introduction and epilogue by Bryan Keith-Lucas. London: MacMillan, 1958.

Regan, C.J. "Early developments in London's sewage disposal with particular reference to the work of W.J. Dibdin." J. and Proc. Institute of Sewage Purification, 1951, pp. 338-47.

Rehbock, Philip F. "The Victorian aquarium in ecological and social perspective." In M. Sears and D. Merriman, eds., Oceanography in the Past, Proceedings of the 3rd International Conference on the History of Oceanography, held 22-26 September 1980 at Woods Hole. New York: Springer, 1980. Pp. 522-39.

Roach, John. "Liberalism and the Victorian intelligentsia." Cambridge Historical J. 13(1957): 58-81.

Robson, William. The Government and Misgovernment of London. 2nd ed. London: George Allen & Unwin, 1948.

Rose, R.N. The Field, 1853-1953. London: Michael Joseph, 1953.

Rosen, George. "The concept of medical police, 1780-1890." Centaurus 5(1957): 97-113.

------- "Economic and social policy in public health." In George Rosen, From Medical Police to Social Medicine: Essays in the History of Health Care. New York: Science History Publications, 1974. Pp. 176-200.

------- A History of Public Health. New York: MD Publications, 1958.

Rosenberg, Charles. The Cholera Years, the United States in 1832, 1849, and 1866. Chicago: University of Chicago Press, 1962.

Rosencrantz, Barbara G. "Cart before horse -- theory, practice, and professional image in American public health." J. Hist. Med. 29(1974): 55-73.

Russell, Colin A. "Percy Frankland: the iron gate of examination." Chemistry in Britain 13(1977): 425-27.

-------, N.G. Coley, and G.K. Roberts. Chemists by Profession -- the Origins and Rise of the Royal Institute of Chemistry. Milton Keynes, U.K.: Open University/Royal Institute of Chemistry, 1977.

Russell, Sir E. John. A History of Agricultural Science in Great Britain, 1620-1954. London: George Allen and Unwin, 1966.

Salomon, Hermann. Die Städtische Abwässerbesserung in Deutschland, Wörterbuch angeordnete Nachrichten und Beschreibungen städtischer Kanalisations-und Kläranlagen in deutschen Wohnplätzen. 2 vols.

Jena: Gustav Fischer, 1906-9.

Sharratt, A., and K.R. Farrar. "Sanitation and public health in 19th century Manchester." Memoirs and Proceedings of the Manchester Literary and Philosophical Society 114(1971): 50-69.

Shryock, R.H. The Development of Modern Medicine. rept., Madison: University of Wisconsin Press, 1979.

Slater, Jay. "The self-purification of rivers and streams, a study of the transformation of water safety standards in the late 19th century." Synthesis 2 no. 3 (1974): 41-55.

Simson, John von. "Die Flussverunreinigungsfrage im 19th Jahrhundert." Vierteljahrschrift für Sozial-und Wirtschaftgeschichte 65(1978): 370-90.

Small, F.L. The Influent and the Effluent, the History of Urban Water Supply and Sanitation. Winnipeg: Underwood McClellan, 1974.

Smith, Paul. Disraelian Conservatism and Social Reform. London: Routledge and Kegan Paul, 1967.

Sonntag, Otto. "Religion and science in the thought of Liebig." Ambix 24(1977): 159-69.

Stevenson, Lloyd. "Science down the drain -- on the hostility of certain sanitarians to animal experimentation, bacteriology, and immunology." Bull. Hist. Med. 29(1953): 1-26.

Stieb, Ernst W. (with the collaboration of Glenn Sonnedecker). Drug Adulteration in 19th Century Britain. Madison: University of Wisconsin Press, 1959.

Strell, Martin. Die Abwasserfrage in ihrer Geschichtlichen Entwicklung von den ältesten Zeiten bis zur Gegenwart. Leipzig: F. Leineweber, 1913.

Tarr, Joel A., and Francis McMichael. "Historic turning points in municipal water supply and wastewater disposal, 1850-1932." Civil Engineering Special Historical Issue, 1978, pp. 25-29.

Temkin, Owsei. "An historical analysis of the concept of infection." In Johns Hopkins University History of Ideas Club, Studies in Intellectual History. Baltimore: Johns Hopkins University Press, 1953.

Turing, H.D. River Pollution, being the Buckland Lectures for 1950. London: Edward Arnold, 1952.

Vallery-Radot, Rene. The Life of Pasteur. Translated by Mrs. R.L. Devonshire. 2 vols. New York: McClure, Phillips, & Co., 1902.

Volhard, Jakob. Justus von Liebig. 2 vols. Liepzig: Barth, 1909.

Waksman, Selman. Principles of Soil Microbiology. 2nd ed. Baltimore: Williams and Wilkins, 1932.

Warren, Charles E. Biology and Water Pollution Control. Philadelphia: W.B. Saunders, 1971.

Webb, Sidney, and Beatrice Webb. Statutory Authorities for Special Purposes with a Summary of the Development of Local Government Structure, vol. 4 of English Local Government. London, 1922. Reprint with a new introduction by B. Keith-Lucas. Hampden, Ct.: Archeon, 1963.

Weinberg, Alvin M. "Science and trans-science." Minerva 10(1972): 209-22.

Winslow, C.-E. A. The Conquest of Epidemic Disease. A Chapter in the History of Ideas. 1943; Madison: University of Wisconsin Press, 1980.

------- Man and Epidemics. Princeton: Princeton University Press, 1952.

------- and Earle B. Phelps. Investigations on the Purification of Boston Sewage with a History of the Sewage-Disposal Problem, U.S.G.S. Water Supply Paper #185. Washington, D.C.: G.P.O., 1906.

Wisdom, A.S. The Law on the Pollution of Waters. 2nd ed. London: Shaw & Sons, 1966.

Wylie, J.C. The Wastes of Civilization. London: Faber, 1959.

Young, R.M. "Malthus and the evolutionists: the common context of biological and social theory." Past and Present, #43 (May 1969): 109-45.

INDEX